AF508608

Veterinary Microbiology & Parasitology

Veterinary Microbiology & Parasitology

Editor

Robert W. Li

MDPI • Basel • Beijing • Wuhan • Barcelona • Belgrade • Manchester • Tokyo • Cluj • Tianjin

Editor
Robert W. Li
USDA
ARS
Beltsville
United States

Editorial Office
MDPI
St. Alban-Anlage 66
4052 Basel, Switzerland

This is a reprint of articles from the Special Issue published online in the open access journal *Animals* (ISSN 2076-2615) (available at: www.mdpi.com/journal/animals/special_issues/Veterinary_Microbiology_Parasitology).

For citation purposes, cite each article independently as indicated on the article page online and as indicated below:

LastName, A.A.; LastName, B.B.; LastName, C.C. Article Title. *Journal Name* **Year**, *Volume Number*, Page Range.

ISBN 978-3-0365-1520-5 (Hbk)
ISBN 978-3-0365-1519-9 (PDF)

Contents

Preface to "Veterinary Microbiology & Parasitology"

Numerous pathogens, both microscopic and macroscopic, affect animal species of veterinary relevance, including companion animals, livestock, and exotic animals. The term pathogen was originally referred to infectious microorganisms, such as viruses and viroids, bacteria, fungi, and protozoans. While prions do not contain genetic materials, they are transmissible and are often classified as pathogens. The modern concept of pathogens includes both infectious microbes and macroscopic parasites. The scientific disciplines that study microscopic organisms of veterinary significance are generally classified as veterinary microbiology, while parasitology refers to the scientific study of parasites and their hosts. However, the boundaries between these two scientific disciplines are not always rigidly defined. For example, some single-cell microscopic protozoan pathogens behave much like infectious microorganisms, but they belong to eukaryotic phyla and are often the subject of parasitology studies.

Animal pathogens affect animal health and wellbeing and production efficiency. These pathogens also have a considerate impact on social economics, food safety and security, and human health. Numerous historical events have demonstrated the extreme negative potential of infectious pathogens on animal health and wellbeing. For example, the recent high pathogenic avian influenza outbreak in the U.S. resulted in the depopulation of 7.5 million turkeys and 42 million chickens. The birds infected by high-risk virus pathogens had extremely high morbidity and mortality; and associated economic losses reached billions of dollars. In addition to infectious viral agents, many parasite species are deleterious to human and animal health and agricultural efficiency. Approximately 70% of production animals in developing countries are estimated to experience severe parasite transmission. In ruminants, parasitic infections result in a liveweight reduction up to 14% and milk yield reduction up to 2.2 kg milk/cow per day. Parasitism is not just a disease affecting productivity, but also a leading cause of mortality in young animals. The annual losses of crop yield due to parasitic nematodes are approximated 12% of the total, over 100 billion USD. As a result, parasitism represents one of the key factors threatening global food availability.

Studies of pathogens of veterinary significance do not just result in a better understanding of animal diseases and promote animal welfare, but also inspire the discovery of new human pathogens and improve human health. It is the pioneer work in dissecting etiology of avian sarcoma that led to the discovery of multiple important human oncogenes, such as MYC and JUN. The causal association between rotaviruses and diarrhea was first discovered in calves, long before they were identified as main causal agents for diarrhea in infants and children around the globe. Unfortunately, the significance and importance of many studies on animal pathogens and their profound impact on medicine and public health are not widely appreciated outside the veterinary community.

Infectious diseases that originate from both domestical animals and wildlife represent one of the greatest threats to human health, as evidenced by the devastating consequence of the COVID-19 pandemic. Each year, over 1 billion cases of human illness are attributable to zoonotic pathogens. As a result, identifying wild reservoirs of zoonotic pathogens has been an urgent public health priority. Recent studies show that domesticated species harbor approximately 84 times more zoonotic viruses than wild species. Eight of the top 10 mammalian species with the highest number of zoonotic viruses are domestical, such as pigs, cattle, and horses, each with 31 zoonotic viruses, followed by sheep, dogs, goats, cats, and camels. Moreover, many of animal parasites are also zoonotic, constituting an additional burden on human health.

Furthermore, the rapid emergence and spread of drug resistant pathogen strains are alarming. Climate changes will undoubtedly alter the interactions between animals and between animals and humans, which will have a considerate impact on the transmission rate of existing pathogens and the emergence of new pathogens or the reemergence of old pathogens.

This special collection covers a broad range of research topics, which likely represent frontiers in the contemporary veterinary microbiology and parasitology. The host species under investigation included wild animals (rodents), companion animals, such as dogs and horses, and livestock species, such as camels, poultry, ruminants, and swine. Almost all major pathogen types, including viruses, bacteria, mites and flies, protozoans, and helminths, have been discussed. The tools and approaches used range from epidemiological survey, immunology, to cutting-edge microbiome studies and metabolomics. This collection provides a broad knowledge base that will encourage dialogue and enhance interactions across the wide distribution of the research community in microbiology and parasitology.

The highest proportion of zoonotic viruses have been identified among species in the order Rodentia, followed by Chiropetra, and Primates. Many rodent species are hyper-reservoirs, carrying between 2 and 11 zoonotic pathogens, including viral pathogens and zoonotic helminths. Kalogianni and colleagues present a review article on Maedi-visna (MV) in sheep, a disease caused by maedi-visna virus, a small ruminant lentivirus (SRLV). The predominant clinical manifestations of MV are pneumonia and mastitis. MV has a worldwide distribution with distinct viral transmission patterns. Nevertheless, the prevalence rate of MV has been increasing globally. Currently, gaps in understanding the epizootiology of MV, the continuous mutation of existing and the emergence of new SRLV strains, lack of an effective detection protocol, and the inefficiency of preventive measures render the elimination of MV an unrealistic target. Therefore, modifications on the existing MV surveillance and control schemes on an evidentiary basis are necessary. Updated control schemes require the development of diagnostic protocols for the early and definitive diagnosis of SRLV infections. These authors summarize the current knowledge in the epizootiology and control of MV in dairy sheep and describe the research framework to close existing knowledge gaps. In another report, Piegari and colleagues conduct a pathological and microbiological evaluation of "sudden and unexpected death (SUD)" cases in young dogs. They conclude that viral infection with Canine parvovirus type 2 is the most common cause of SUD.

When pathogenic bacteria, such as staphylococci, streptococci, Gram-negative bacilli, and anaerobes, enter the animal's body and multiply, they directly disrupt the normal function of the cells, tissues, or organs involved. The toxins released by these infectious agents can also result in harmful effects, affecting animal health and wellbeing. Furthermore, numerous bacteria are zoonotic. In this collection, at least two bacterial-related diseases are discussed. Anaplasma phagocytophilum, a Gram-negative bacterium transmitted by ticks, is the causal agent of pasture fever, a disease affecting domestic ruminants, such as cattle, sheep, and goats. This pathogen can also infect humans in temperate and tropical regions. Atkas and colleagues present a PCR-based diagnostic technique that can be used to distinguish two variants of this important pathogen. The phylogeny and prevalence of A. phagocytophilum and related variants were then investigated in the Mediterranean region of Turkey. Their findings highlight the importance of detection of variants and provide a basic took kit for studying molecular epidemiology of caprine and ovine anaplasmosis. Bovine digital dermatitis (BDD), first reported in 1974, is arguably the most important foot infection causing lameness in cattle, which has a serious impact on animal welfare and productivity. Treponema species have been considered as the primary causal agent. To understand BDD pathogenesis, Espiritu and colleagues compared the microbial composition and diversity between BDD lesions and normal cattle skin using 16S rRNA gene-based sequencing. These authors conclude that rich microbial diversity and the overabundance of opportunistic bacterial pathogens are likely associated with BDD.

Three articles in this collection discuss the biology of mites and flies. Oestrus ovis, the sheep bot fly, is an obligate parasite having a worldwide distribution with the high prevalence rates in Mediterranean and West African countries. While generally not considered a life-threatening parasite, this species can cause significant losses in productivity, particularly in tropical and Mediterranean regions. Metwally and colleagues investigated the prevalence rate of this species in small ruminants in Saudi Arabia and found that the infestation rate is two times higher in goats than in sheep. Furthermore, male hosts tend to have a significantly higher infestation rate than females. Dr. Arun De's group presents a report to examine host-parasite interactions and genetic characterization of Sarcoptes scabiei, a burrowing mite affecting humans and many other mammals. The disease scabies is one of the earliest diseases with a known cause recorded in the Bible. This group evaluated the effect of S. scabiei infestation on host physiology with special emphasis on serum biochemical parameters, lipid profile, oxidant/antioxidant balance, stress parameters, and immune responses using in a porcine model. They found that S. scabiei triggered stress response, elevated levels of serum cortisol and heat shock proteins, increased the serum concentration of immunoglobulins as well as IL-2, IFN-gamma, IL-1-beta, and IL-4 cytokine expression. In the third article, Koziatek-Sadłowska and Sokół studied Dermanyssus gallinae, a parasitic poultry mite and aimed to determine its effect on the development of post-vaccination immune responses in layer hens. They found that D. gallinae infestation may inhibit humoral immune responses since the percentages of B cells and Th cells were negatively correlated with the number of mites. However, further research is needed to determine whether D. gallinae suppresses the production of vaccine-induced antibodies.

At least six original research articles in this collection study parasites, from intracellular protozoans, hookworms, roundworms, to tapeworms. Islam et al. (2021) conducts a systematic review on rodent helminths and potential threats of helminth parasites on public health in the Middle East using meta-analysis. These authors identified 44 rodent species as primary hosts for helminth infestation. The study detected 22 species of helminths that are of zoonotic importance, such as Capillaria hepatica, Cysticercus fasciolaris, Hymenolepis diminuta, and Hymenolepis nana. This study provides baseline data on rodent helminths at the human-animal interface in Middle East countries, which should facilitate the development of efficient rodent control programs in the region. The Metwally team surveyed two species of intracellular protozoan parasites, Sarcocystis cameli and S. camelicanis in Camelus dromedarius samples obtained in Saudi Arabia and for the first time identified the presence of S. camelicanis in this country. Slater and colleagues compared the intestinal microbiome and volatile organic compounds of colonic contents obtained from healthy horses and those infected with varying levels of the tapeworm, Anoplocephala perfoliate, an equine parasite causing abdominal pain. Their findings show that a general reduction in certain beneficial bacteria in horses with tapeworm infection, indicating a possible negative consequence of parasitic infection. The Hedberg-Alm team presents a case control study on parasite prevalence in Swedish horses and parasite management practice by horse owners. Their data demonstrate a high level of exposure of Swedish horses to Strongylus vulgaris, arguably the most pathogenic equine internal parasite. The findings also suggest an urgent need for education in the use of fecal parasite egg counts and pasture management, such as reducing stocking intensity and frequent removal of fecal matters. The teams led by Dr. Saruda Tiwananthagorn presents two original papers in this collection. They first examined geographical and spatial distribution of two nematode species in community dogs, Dirofilaria immitis and Brugia pahangi, which is a serious public health concern for humans, dogs, cats, and wildlife species, particularly in southeast Asian countries. They detected Brugia pahangi infection in dogs for the first time in Thailand. Spatial distribution patterns of these two species differ: at a higher altitude between 400 and 800m, B. pahangi infections significantly outnumber D. immitis infection while the overall prevalence rate of the former is lower than the latter. In the second paper, the prevalence of Ancylostoma ceylanicum, a zoonotic hookworm, was carefully examined in dogs and soil samples from Thailand and Asian countries. Genetics and diversity of this species was also evaluated; and nine haplotypes were identified from the Thai hookworm population. These authors conclude that increased public awareness and proper deworming programs are urgently needed to reduce the risk of the transmission of this important zoonotic disease. Dominika Mravčáková and colleagues tested anthelmintic activities of two medicinal plants, wormwood and mallow, against the barber's pole worm, Haemonchus contortus, in both in vitro and in vivo settings. The aqueous extracts of both plants exhibited strong parasiticidal effects in vitro. However, in sheep, the mean fecal egg counts did not differ significantly between the groups with and without plant supplements. While their study demonstrated the potential of medicinal plant-based alternatives to overcome the anthelmintic resistance, more research is warranted.

The Zaragoza-Bastida team evaluated the antibacterial and hemolytic activity of two rattlesnake venoms against Pseudomonas aeruginosa, an important bacterial pathogen that affects both animals and humans. These authors presented evidence for the first time that venoms from rattlesnakes contain bioactive compounds with bactericidal activity against Pseudomonas aeruginosa and can be developed as alternatives to antibiotics.

We strongly believe that this special collection has made a valuable contribution to the veterinary literature. The findings from these studies provide novel insights into host-pathogen interactions, which should facilitate the development of vaccines and alternative pathogen control strategies. We wish to express our gratitude to all contributing authors for their outstanding research efforts. Without their hard work, this collection will not come into existence. Finally, we want to thank the editorial board members and staff of Animals for their support to this special issue.

Robert W. Li
Editor

Review

Etiology, Epizootiology and Control of Maedi-Visna in Dairy Sheep: A Review

Aphrodite I. Kalogianni [1], Ioannis Bossis [1], Loukia V. Ekateriniadou [2] and Athanasios I. Gelasakis [1,*]

[1] Laboratory of Anatomy and Physiology of Farm Animals, Department of Animal Science, Agricultural University of Athens (AUA), Iera Odos 75 str., 11855 Athens, Greece; afrokalo@aua.gr (A.I.K.); bossisi@aua.gr (I.B.)

[2] Veterinary Research Institute, ELGO-Demeter, Thermi, 57001 Thessaloniki, Greece; ekateriniadou@vri.gr

* Correspondence: gelasakis@aua.gr

Received: 18 March 2020; Accepted: 31 March 2020; Published: 3 April 2020

Simple Summary: Maedi-visna is a chronic viral disease of sheep with worldwide prevalence. It is caused by a small ruminant lentivirus. Its clinical manifestation includes primarily pneumonia and mastitis, and secondarily, arthritis and encephalitis. It causes substantial economic losses associated with involuntary culling or death of infected animals and reduced productivity, challenging the sustainability of dairy sheep farms. The extensive spread of the disease and the lack of treatment or vaccines render surveillance and prevention strategies indispensable. Currently, the major obstacles in controlling the disease are (i) the absence of an effective protocol for the early and definitive diagnosis of infected animals, utilizing appropriate, universally accepted serological and molecular techniques, (ii) the long interval between infection and seroconversion, and (iii) lack of understanding whether horizontal or vertical transmission are the most important route of infection. Therefore, the most appropriate measures for the control of the disease should include more frequent serological testing with available diagnostics and isolation or culling of seropositive animals, incorporation of artificial suckling, and strengthening of hygiene and biosecurity protocols.

Abstract: Maedi-visna (MV) in sheep is caused by maedi-visna virus (MVV), a small ruminant lentivirus (SRLV) that causes chronic infection and inflammatory lesions in infected animals. Pneumonia and mastitis are its predominant clinical manifestations, and the tissues infected by MVV are mainly the lungs, the mammary gland, the nervous system and the joints. MV has a worldwide distribution with distinct MVV transmission patterns depending on circulating strains and regionally applied control/eradication schemes. Nevertheless, the prevalence rate of MV universally increases. Currently, gaps in understanding the epizootiology of MV, the continuous mutation of existing and the emergence of new small ruminant lentiviruses (SRLVs) strains, lack of an effective detection protocol and the inefficiency of currently applied preventive measures render elimination of MV an unrealistic target. Therefore, modifications on the existing MV surveillance and control schemes on an evidentiary basis are necessary. Updated control schemes require the development of diagnostic protocols for the early and definitive diagnosis of MVV infections. The objectives of this review are to summarize the current knowledge in the epizootiology and control of MV in dairy sheep, to describe the research framework and to cover existing gaps in understanding future challenges regarding MV.

Keywords: maedi-visna virus; ovine progressive pneumonia; small ruminant lentivirus; dairy sheep

1. Introduction

Maedi-visna (MV) (also known as ovine progressive pneumonia, OPP) is an incurable viral disease of sheep with very long incubation period that leads to life-long infection [1]. It is caused by non-oncogenic exogenous retroviruses [2], namely, maedi-visna virus (MVV) and caprine arthritis-encephalitis virus (CAEV), both belonging to a subgroup of viruses known as small ruminant lentiviruses (SRLVs). These viruses are structurally identical, display similar pathogenicity, and have genetic variants/strains that can infect both sheep and goats [3,4]. Species-specific categorization of SRLVs is not always valid since cross-species transmission of certain genotypes is evident with the direction of transmission not being always apparent [4]. Clinical symptoms of SRLVs infections are strain-dependent. Most prevalent are symptoms associated with the respiratory tract, such as dyspnea and abdominal breathing due to chronic interstitial pneumonia. Symptoms associated with encephalitis include instability, paresis, and paralysis. Other clinical manifestations such as mastitis (indurated udder), arthritis (lameness), progressive weakness and, in some cases, death of infected animals may also occur [5]. The primary route of infection for newborn lambs is the consumption of colostrum and milk from infected ewes [3]. Horizontal transmission via respiratory secretions is also significant [4], whereas, vertical transmission (transplacental) [6] and transmission via semen during mating or artificial insemination are also possible [6], but their significance and extent has not been thoroughly investigated.

Recognizing the socio-economic impact of MV, the World Organization for Animal Health (OIE) has included it in the list of notifiable terrestrial and aquatic animal diseases, with significant impact on international trade of animals and their products [7]. In dairy sheep, economic losses associated with the disease are rather extensive. During the last 5 years, the situation seems to have worsened as accumulating evidence suggests an increase in prevalence of the disease especially in the European continent. The scarcity however of epizootiological data regarding inter- and intra-farm transmission of the disease is a major obstacle for the development and implementation of evidence-based disease-control and eradication programs [8]. In particular, in the case of dairy sheep, there is a considerable shortage of certified MVV-free sheep breeding stocks in the international market. This is an obvious drawback in newly established intensive sheep farms, where the interest to buy MVV-free breeding stocks with high productive potential is large. Hence, the need to establish MVV-free dairy sheep farms is now more urgent than ever and forms a major issue for the conservation and evolution of the dairy sheep sector worldwide.

As there is no treatment against MV and all efforts for the development of vaccines did not produce any satisfactory results [6], the control of the disease and the reduction of its prevalence have been achieved to some degree through controlled eradication programs aiming at diagnosis of the infected animals at an early stage. However, these programs are heterogeneous in terms of their planning and effectiveness, which highly depends on the number of participating farms, the regional prevalence of the disease, and the diagnostic methods applied to detect the infected animals [9]. It has now been recognized that early diagnosis of viral infection in MV is more complicated than originally thought and constitutes a major barrier for the effective eradication of the disease [10,11]. Most of the programs used for the control of the disease are based on serological tests to detect antibodies against the virus [12]. This way, only the seropositive animals are diagnosed and removed from the flock. However, many animals infected with the virus remain undiagnosed due to the late (from a few months to three years) seroconversion [13,14]. In the past, several papers have reviewed the status of MV in sheep industry [5,6,15–22]; nevertheless, the objectives of this review paper are to summarize and provide an integrated overview of the current knowledge regarding the epizootiology and control strategies of MV in dairy sheep, to describe the research framework and to cover existing gaps in understanding future challenges regarding MV.

2. Current State of the Art in Epizootiology of the Disease

MV was initially described in Iceland in 1939, possibly originating from the importation of live animals from Germany. Although clinical signs were apparent, the situation was likely underestimated or misdiagnosed for many years [5,15]. Eventually, it was transmitted among several countries through the trading of breeding stocks (e.g., Denmark in 1968, Canada in 1970, Hungary in 1972, France in 1976, Norway in 1979, and Finland in 1994) [15].

Nowadays, MV has a worldwide spread, with the exception of Iceland, New Zealand, and Australia which are considered MVV-free regions, but not CAEV-free [23]. Despite the scarcity of MV epizootiological data in the Mediterranean countries with developed dairy sheep sector (Spain, Italy, and Greece), field observations suggest increased seroprevalence [16]. Based on these observations and having recognized the significance of the disease, several European countries (e.g., France, Germany, Netherlands, Italy, Switzerland, and Spain) have applied eradication programs. However, the effectiveness of these programs in controlling the disease has been questionable [6].

The majority of SRLVs epizootiological studies worldwide (Table 1) is limited to primarily small-scale estimation of seroprevalence; nationwide monitoring and active surveillance schemes are scarce as they require a substantial funding which is not available in most cases.

Table 1. Epizootiological studies regarding small ruminant lentiviruses (SRLVs).

Region	Species	Seropositivity at Animal Level (%)	Seroprevalence at Flock Level (%)	Number of Flocks	Number of Animals	Reference
Spain (Aragon)	s	52.8	100.0	554	274,048	[9]
Spain (Northwestern)	s	25.0	53.0	78	15,155	[10]
Italy (Southern)	g	18.64	51.69	1060	4800	[24]
Greece	s	41.96	N/A	6	143	[25]
Belgium	s	9.0	17.0	87	555	[11]
	g	6.0	13.0	76	401	
Switzerland	s	9.0	N/A	241	5084	[12]
Poland (Central Eastern)	s	10.2	N/A	98	2925	[26]
Turkey (Central Anatolia)	s	9.3	N/A	N/A	279	[14]
	g	7.5			146	
Turkey (Instanbul)	s	15.3	N/A	4	542	[14]
Germany (Mecklenburg Western-Pomerania)	s	28.8	51.2	41	2229	[8]
Canada (Manitoba)	s	2.5	25.1	77	2207	[27]
China (Yunnan, Guizhou, Gansu, Ningxia, Shandong, Sichuan, Hunan, Guangdong, Chongqing, Guangxi, Jilin, Anhui)	s	4.6–50.0	N/A	24	672	[28]

s: sheep, g: goats, N/A: relative data is not available.

3. Etiology

There is a continuous effort for the identification and classification of SRLVs strains due to large genetic diversity of the isolated viruses [17,29]. To date, five strain groups, namely, A, B, C, D, and E with their subtypes have been identified in sheep and goats [1,18]. Group A consists of MVV strains and has 15 subtypes, group B consists of CAEV strains with 3 subtypes, whereas, the other three groups refer to strains which have been isolated from Norway (group C), Spain and Switzerland (group D), and Italy (group E) [18]. Some of the classical strains with known sequences are the K1514 (Iceland) and its neurovirulent clones KV1772-kv72/67 and LV1-1KS1, SA-OMVV (South Africa), EV1 (UK), and P1OLV (Portugal). Many of the fore-mentioned strains have been found to infect both sheep and goats (e.g., all the strains of groups B, C, and D as well as group A subtypes A1–A6, A9, and A11–A13). The subtype A15 has been isolated only from sheep and subtypes A7, A8, A10, A14, E1, and E2 only from goats. The classification of SRLVs, the species infected and the countries first reported are summarized in Table 2.

Table 2. Classification of SRLVs strains.

Group	Subtype	Strain	Origin Country	Species	Reference
		P1OLV	Portugal	s/g	[30]
		KV1772	Iceland	s/g	[31]
	A1	LV1-1, KV1514	Iceland	s/g	[32,33]
		EV1	England	s/g	[34]
		SA-OMVV	South Africa	s/g	[35]
A	A2	85/34	USA	s	[36]
	A3	697	Spain	s/g	[37]
	A4	SRLV-A4	Switzerland	s/g	[29]
	A5, A6, A9, A11-A13			s/g	[17,18]
	A7, A8, A10, A14			g	[17,18]
	A15			S	[17,18]
	A16 and A17		Poland	g	[38]
		FESC-752	Mexico	s/g	[39]
	B1	CAEV-CO	USA	s/g	[40,41]
		GANSU	China	s/g	[42]
B		SHANXI	China	s/g	[43]
	B2	S-496	Spain	s/g	[44]
	B3	Volttera and Fonni	Italy	s/g	[45]
	B4		Canada	g	[46]
C		1GA	Norway	s/g	[47,48]
D			Spain, Switzerland	s/g	[17,18]
E	E1	Rocaverano	Italy	g	[49,50]
	E2	Seui	Italy	g	[49,50]

s: sheep, g: goats.

Recently, phylogenetic analysis revealed a new strain (CRF01_ABSRLV) which was isolated from Canadian goats and possibly originating from the combination of A2 and B1 subtypes [51]. The mechanism for generation of retroviral recombined strains is complicated and not very well understood. However, it could be considered among possible explanations for the emergence of new SRLV strains [51,52].

4. Clinical Signs and Gross Pathology

In the majority of cases, virus replication is slow and the number of infected blood cells in the circulation is very low [53]. Therefore, the clinical disease is latent or progressive and in many cases, the clinical signs are not evident or characteristic of the disease at its early stages [3]. Immunosuppression of animals due to aging, co-existing of diseases and environmental stressors accelerates virus replication and clinical evidence of the disease becomes apparent. At the flock-level, serum detection of antibodies, the severity of clinical disease and the number of deaths/culled animals, can be affected by management practices and the co-existence of other diseases [18].

Clinical manifestation of SRLVs infection depends on the virus strain, the host immune response and the host genetic profile regarding resistance or susceptibility to the virus [54,55]. Lesions of SRLVs infection in tissues and organs are caused both by the immune response to the viral antigens and the viral replication itself [3]. The tissues mainly infected by MVV are located at the lungs, the mammary gland, the nervous system, and the joints [3,19]. Pneumonia and mastitis are the predominant clinical manifestations of MVV in sheep [3]. Less frequently, lesions such as lymphoid tissue hyperplasia may be apparent in kidneys, liver, and heart, indicating them as possible target organs [56,57]. Multiple-organ infection may be observed in the progression of the disease, but the severity of lesions varies among the affected organs [19].

Respiratory clinical signs include dyspnea and increased respiratory rate, caused by the characteristic lymphocytic interstitial pneumonia; at necropsy, the lungs appear discolored, enlarged, and diffusely firm with grey spots on the pleural surface and the mediastinal lymph nodes are often enlarged [19,23].

Symptoms from the nervous system include ataxia, paresis, weakness in hind limbs, incoordination or, in heavier cases, total paralysis, due to meningoencephalitis, astrocytosis, microgliosis, and focal secondary demyelination in the brain and the spinal cord [3,19].

In mammary gland, MVV can cause an indurative non-suppurative interstitial mastitis [3,19]. The udder is hard but not painful, with decreased milk production mainly noticed the first days postpartum, a situation usually described as "hard udder syndrome". The lymphocytic inflammatory pattern, caused by the replication of virus in macrophages and mammary gland epithelial cells, provokes the destruction of the acinar structure and the reduction of milk production [19].

Arthritis can also be the outcome of MVV infection, although it is less common in sheep [3,16,18,19]. The affected joints are usually the carpal and tarsal, but metatarsal, metacarpal and vertebral joints can also be affected [16,18,19]. Infiltration of the synovial membrane by mononuclear cells is followed by villous hypertrophy, angiogenesis, and finally fibrosis, mineralization, and necrosis of synovium and joint capsule [3]. In advanced arthritis cases, the cartilage is destructed and the articular capsule is fibrotic [18,19]. In the majority of cases, arthritis is progressive causing lameness and involuntary culling of the animal, whereas less often it may regress [3].

5. Transmission

The mechanisms and the significance of horizontal and vertical transmission of the MVV have not been fully clarified yet [3]. The major routes of transmission have been described; however, their significance and extent remain unclear. This information is critical for the efficient designation of the eradication protocols, especially in intensively reared dairy sheep.

The vertical transmission of MVV refers to the transmission of the virus from the ewe to the lamb during pregnancy (transplacental), at lambing or during suckling [6]. Opinions regarding transplacental transmission are controversial [15,58]. Seropositivity of 4- to 9-months old lambs derived by caesarean section, isolated after birth and not allowed to suckle seropositive dams support the transplacental transmission [59]. However, the possibility of horizontal transmission post-lambing, complicates the assessment of the significance of either transplacental or horizontal transmission [6,15]. Vertical transmission at lambing refers to the transmission of virus while the lamb passes through the ewe's genital tract and it is exposed to maternal body fluids and blood. The exact significance of this route of transmission remains unknown [6,58].

The most significant route of vertical transmission is considered to be the lactogenic, through the ingestion of colostrum and milk from infected dams [6]. MVV shows tropism to the epithelial cells of the mammary gland and the resident macrophages, where it can replicate [60]. It has been found that isolated lambs fed colostrum or milk from infected ewes seroconverted a few months later and some of them were diagnosed with clinical disease later in their adult life [15]. There is evidence that the lactogenic transmission is more significant in small ruminants than in primates due to the higher permeability of the digestive tract of small ruminants in the first 24 h post-lambing [61,62], allowing virions and infected cells to be absorbed by the lamb's intestine [62]. However, not all the subgroup variants of MVV are efficiently transmitted via the lactogenic route as the envelope varies among the different subgroups of the MVV determining some of its physicochemical properties and facilitating or not the lactogenic transmission [62].

Horizontal transmission of MVV includes the environmental, mechanical and iatrogenic routes and mainly refers to the transmission through respiratory secretions [63]. Lungs are the main target organ of MVV in the respiratory tract. In lungs, the virus infects monocytes, macrophages, and dendritic cells, and it can be horizontally transmitted via respiratory secretions containing these cells [3,15]. In severe cases, MVV causes the characteristic lesions of interstitial pneumonia. In general, the lower respiratory tract, constitutes the main route of infection [15,64]. This route of transmission is of major importance in intensive and permanently housed sheep in sheds with inadequate ventilation and high stocking density [19]. In general, many researchers support that the airborne transmission can be an equally significant route of transmission as the vertical transmission. For this reason, the segregation of newborn lambs or non-infected animals from the infected ones is of major importance for the control of MVV transmission [3,15,65]. The significance of transmission through contaminated barns, sheds, feeding and water equipment and pastures or reusable veterinary equipment has not been fully clarified [6,15]. The presence of the virus in the water and air from pens with infected animals [63] indicates that waterborne and airborne transmission in the farms cannot be disregarded. Infection of dairy sheep via the teat canal during milking has also been reported [15].

Sexual transmission of the MVV is theoretically possible but not yet confirmed. However, there is evidence of virus proliferation in the genitals of infected rams and the virus has been found in the semen of rams with leucocytospermia and rams positive for *Brucella ovis* [3,66]. In another study using real-time PCR, proviral DNA of SRLV was found in semen (intermittent shedding) and the genital tract of rams suggesting possible sexual transmission [67].

6. Risk Factors

There are several risk factors that influence transmission of MVV between and within flocks. These factors determine the likelihood of infection, the prevalence, the incidence rate and other epizootiological characteristics of the disease. Identification and mitigation of risk factors at the farm level is therefore crucial when establishing a MVV control/eradication program. Flock size/stocking density, intensity of the farming system [9–12,26,27], and age distribution [9–11] affect the likelihood of seropositivity at flock level, indicating the significant role of horizontal transmission in the epizootiology of MV [9]. For example, lower prevalence in extensively reared sheep can be attributed to the reduced stocking rates and limited direct contact between animals [68,69], conditions that reduce the exposure to MVV and the possibility of airborne transmission through respiratory droplets during exhalation, sneezing, and coughing [3,6]. In flocks where MV co-exists with pulmonary adenomatosis, the transmission is favored by the increased quantities of respiratory secretions produced by infected sheep [9,15]. In these cases, late removal of clinical cases of MV and non-isolation of seropositive animals are significant risk factors for the transmission and increased seroprevalence of the disease.

Inappropriate cleaning and disinfection of milking equipment [15,19], reuse of infected needles and surgical equipment, inadequate hygiene conditions inside the barn and grazing at common pasturelands are also potential risk factors for the horizontal transmission of MVV.

Importation of breeding stocks from flocks of unknown MVV-status is associated with increased seroprevalence of MV [10,27]. The remarkable absence of certified MVV-free flocks to produce breeding stocks and the use of seropositive rams for mating or artificial insemination are the main causes. Surprisingly, despite the lactogenic transmission of the virus through colostrum/milk during suckling, a reduced seroprevalence in the replacement stocks has been observed in flocks with increased suckling period [9]. This is possibly the result of a confounding effect of farming system; increased weaning age is mainly observed in semi-extensive and extensive systems, where horizontal transmission is limited. On the other hand, early weaning is mainly practiced in intensive systems where virus transmission is facilitated mostly due to the permanent housing, the increased stocking density and the inappropriate ventilation [9,10,70]. Nevertheless, the use of colostrum/milk from seropositive dams and natural suckling of newborn animals constitute major risk factors. Also, in mixed-species flocks (sheep and goats) the seroprevalence has been found to be higher, possibly due to cross-species transmission of several SRLVs strains [10,70].

There is evidence of genetic resistance/susceptibility against SRLVs [71]. Different alleles of the cellular *TMEM154* (*Transmembrane protein 154*) gene have been found to be associated with the occurrence of MV. Haplotypes carrying nucleotide sequences that code for the amino acid glutamate at position 35 are associated with increased susceptibility to MV, whereas haplotypes carrying nucleotide sequences that code for lysine at the same position are associated with resistance to MV [72–74]. Also, the haplotype responsible for the susceptibility seems to be dominant against the "resistant" haplotype [74]. Although there is indication for association between *TMEM154* mutations and control of MVV infection, there is no proven association for all the haplotypes [73]. Other genes associated with virus susceptibility are the *DPPA2 (Developmental Pluripotency Associated 2)/DPPA4 (Developmental Pluripotency Associated 4), SYTL3 (Synaptotagmin-Like 3), CCR5 (Chemokine receptor 5),* MHC (Major Histocompatability Complex), *TLR7, TLR8, TLR9 (Toll-like receptors)* genes, and APOBEC3 (Apolipoprotein B mRNA-editing enzyme) proteins [75–77], whereas the zinc finger cluster, *C19orf42 (Chromosome 19 Open Reading Frame 19)/TMEM38A (Transmembrane Protein 38A)* and *DLGAP1 (Discs Large (Drosophila) Homolog-Associated Protein 1)* genes may be used in genetic selection programs to facilitate the control of the disease [78]. The tripartite motif-containing 5 (TRIM5) protein has been studied and has been proved to contribute to the restriction of MVV [79].

7. Diagnosis

Early and efficient diagnosis of MV is a critical parameter for the control and eradication of the disease. The diagnosis of SRLVs infections is based on the detection of antibodies against the virus proteins or the viral genome. Current control and eradication programs are based on serological tests (mainly enzyme-linked immunosorbent assays (ELISAs)) to detect antibodies against the virus [13]. Therefore, only the seropositive animals are considered infected and subsequently removed from the flock. This is a major drawback for the eradication of the disease as the immune response to the disease (seroconversion) requires a long period of time and thereby, many of the infected animals remain undiagnosed carriers of the virus [8,14]. The available assays for the detection of antibodies are the agar gel immunodiffusion (AGID) test, radioimmunoprecipitation assay (RIPA), Western blotting (WB) and ELISA, whereas, polymerase chain reaction (PCR) is used for the detection of proviral DNA as described below.

7.1. AGID Test

AGID test is commonly used as a diagnostic tool in MV control programs due to its simplicity. It is a highly specific diagnostic method but less sensitive than ELISA [6,19–21]. For this reason, it is supplementary used for the confirmation of the ELISA test [20].

7.2. RIPA

RIPA is as old as WB and both of them are considered as the reference standards. They have similar sensitivity and are mainly used as confirmatory assays [21,22]. RIPA is not frequently used due to its high cost and its difficult application.

7.3. WB

WB is a confirmatory laboratory test which has been used to detect antibodies in serum that recognize viral proteins. In general, WB is more sensitive than the ELISAs but more cumbersome and with lower throughput [6,21]. Cross-reactivity with non-specific cellular proteins is also a problem.

7.4. ELISA

ELISA is the most commonly used test in population screening and for the surveillance of SRLVs. It detects antiviral antibodies with satisfactory sensitivity and specificity, indicating the occurrence of infection and seroconversion at some point [6,20]. Seropositivity is not necessarily followed by clinical disease and a seronegative animal cannot be safely considered to be free of infection. In many cases, seroconversion requires several months even in early-infected animals (e.g., lambs infected via colostrum consumption) [17]. Except for this hurdle, the antigenic heterogeneity of SRLVs strains (especially among different subtypes, like the CAEV-like and the MVV-like) may limit the diagnostic performance of currently available ELISA [80]. Antibody titers present remarkable variations during an animal's life and in some cases, they are undetectable using an ELISA, which renders the test unreliable for a definitive diagnosis [17,21,22,81].

7.5. PCR

PCR can directly detect proviral DNA in fluids and tissues across the animal body (lungs, milk, peripheral blood, mammary gland, synovial membranes etc.). The most reliable cells for the detection of virus are the peripheral blood mononuclear cells [20,21]. The most significant advantage of PCR is its ability to detect infection before seroconversion. However, PCR is not a reference method and it is suggested to be combined with serological testing to overcome the problem of selective specificity associated with the lack of reliable universal primers [6,20,21]. PCR nested methods and Real-Time-PCR (RT-PCR) increase the sensitivity and specificity of the method, however, their use is less frequent [20,21,82].

8. Treatment–Vaccination

There is neither a treatment nor an effective vaccine against MV. In the past, there have been attempts for the development of attenuated and subunit vaccines but none of them proved to be effective in preventing viral infections [83–89]. The major obstacles for the development of an effective MV vaccine include the necessity for the induction of high antibody titers against the virus, the wide genetic variation of viral strains and its continuous mutations, the increased post-infection immunological reaction, the post-vaccination challenge on the immune system and the evidence that the production of SRLV-specific neutralizing antibodies, following vaccination with whole virus-, protein-, and live attenuated-vaccines, is not always protective and in some cases may favor persistent infection [3,54]. The last obstacle is correlated with the fact that the relationship of antibody production with the protection against the SRLVs infection is doubtful and the cellular immune response seems to be more critical [3,54]. Current research efforts for the development of an effective vaccine include pseudoviruses/viral particles, recombinant viruses carrying genes from MVV, and naked plasmids carrying MVV genes plus factors enhancing innate immune responses. However, the effectiveness of these alternative strategies has not been sufficiently validated and thus are considered inappropriate for commercial use [20,54].

9. Preventive and Eradication Measures

The preventive measures and management interventions that could aid in controlling or eradicating MV should be decided on case by case and include the following:

1. Annual, biannual, or more frequent blood sampling from the breeding stocks and serological and molecular testing for the diagnosis of the infected animals.

2. Post-lambing management primarily based on the application of artificial suckling and the use of colostrum and milk substitutes or pasteurized colostrum/milk. The management of colostrum involves the administration of bovine colostrum, commercial sheep colostrum or colostrum only from uninfected ewes or heat-treated colostrum (56 °C for 60 min) [6,20,90]. Artificial suckling should take place in an area isolated from adult animals, which has to be regularly cleaned and disinfected.

3. Immediate removal of animals with apparent clinical signs and positive laboratory diagnosis. The selective culling of these animals and their replacements with seronegative animals, or the grouping of animals according to their seroconversion status can be applied in areas with moderate seroprevalence [6,20,91]. The later strategy requires the spatial and managerial separation of seropositive and seronegative groups within the farm [20,91]. In flocks with high seroprevalence, the most efficient practice is the annual culling of the oldest and less productive seropositive animals and their exclusive replacement with seronegative breeding stocks. The selective culling of seropositive animals would not facilitate the rapid control of the disease, but can contribute to the reduction of seroprevalence and infection rate at flock-level, enhancing the potential of a more drastic elimination program in future time (i.e., culling of the remaining seropositive animals) [9,20].

4. Keeping the replacement animals, post-weaning, in separate housing facilities to avoid horizontal transmission of MVV through the contact with adult animals of the remaining flock.

5. Purchased animals should be from certified MVV-free farms. Imported animals need to remain on quarantine until the MV-status is determined using the most appropriate assays.

6. Regular cleaning and disinfection of facilities and equipment with appropriate disinfectants. The cleaning and disinfection schedule must include the barn (floor, walls, bedding), the milking machine, the feeders and the waterers.

7. Reduction of stocking density (sufficient area and volume) and adequate ventilation.

8. Implementation of general good hygiene practices. Use of disposable needles or sterilization of metal needles before their reuse is necessary. Similarly, the medical equipment should be sterilized after its use.

9. MV seronegative milking ewes should be grouped separately and machine-milked before the seropositive ones.

10. Grazing in communal pastures and sharing of infrastructures and equipment should be avoided when the MVV-status of the flocks is unknown.

11. Rams used either for mating or for semen collection need to be MVV-free [92]. Currently, attempts are being made to produce SRLVs-free breeding stocks via reproductive biotechnologies like artificial insemination and embryo transfer even from infected males and females, respectively. In the case of embryo transfer, this may be possible via the removal of cumulus oophorus cells [92].

12. Breeding for resistance could also be considered, but universally accepted resistant genotypes are yet to be developed [18].

The characterization of a flock as MVV-free demands two to five successive negative tests every 6 months, yearly or every 2 years (depending on the country). The trade of live animals is allowed when the animals have no clinical signs of MV, the adult animals are seronegative and MV has neither been clinically nor serologically diagnosed in the sheep flocks of origin during the last three years. Also, artificial insemination is allowed only using semen from seronegative rams. According to OIE, a country is considered MVV-free when <1.0% of herds are infected with 99.0% probability.

After the first implemented eradication program in Iceland, a lot of countries applied their own eradication programs (almost all the European countries and Canada) with variable results. The major obstacles for the successful implementation of eradication programs are (i) the possible refusal of farmers' participation [6,9], (ii) the breed variability, which, as previously mentioned associates with susceptibility and resistance against the disease, (iii) the genetic variability of the viral strains and the different epizootiological characteristics of the disease (virulence, transmission, seroconversion, seroprevalence at flock level, etc.) and (iv) the heterogeneous farming and herd health management systems. For this reason, the eradication program needs to be adjusted and optimized according to the fore-mentioned factors. An indicating classification of the flocks according to the observed seroprevalence could be: flocks with high (>70.0%), medium (40.0%–69.0%), low (10.0%–39.0%), very low (1.0%–9.0%) seroprevalence and the MVV-free flocks (<1.0%) [6,20].

The eradication measures when considered in a country-wide scale should include both species of small ruminants due to the fact that the cross transmission has been proven and is a significant risk factor in the spread of the virus [89]. Moreover, the existence of reference laboratories for the control of MV is of major importance [20]. These laboratories will be responsible for the surveillance of MV at national level and will coordinate all the efforts for the elimination of the disease [6].

Total replacement of seropositive flocks with breeding stocks from MVV-free flocks could be a sustainable option only in areas where the seroprevalence is very low and MVV-free flocks are available. Otherwise, there is a serious threat for significant monetary losses and restriction of genetic resources which may undermine the sustainability of the farms, particularly in areas with developed dairy sheep farming industry [20].

10. Conclusions and Future Challenges in Controlling MV

Epizootiological investigations regarding SRLVs infections are essential for the territorial mapping of virus distribution. The scarcity of detailed epizootiological data, in combination with the continuous mutations of SRLVs strains, render the creation of a universal, reliable and valid diagnostic protocol, and consequently the control of SRLVs, a rather complicated task.

For the early and definitive diagnosis of SRLVs infections, development of appropriate combinations of serological and molecular tests (ELISAs and PCR) on an evidential basis is suggested and forms a challenging research field in the future.

Determining the epizootiological traits of SRLVs infections and particularly the significance of the different routes of viral transmission will facilitate the decision-making process towards the determination of the most effective measures for the creation of MVV-free flocks. The global spread of SRLVs highlights the underestimation of the problem by the farmers and the insufficient implementation of the suggested preventive measures, which mainly focus on the lactogenic and airborne transmission of the disease, but may also be indicative of the failure in controlling the disease using these measures. Therefore, the significance of every other possible route of transmission (transplacental, sexual, and iatrogenic) needs to be carefully assessed and revised, and relevant preventive measures need to be integrated in the MV control protocols.

Exploitation of active surveillance programs and quantification of the consequences of SRLVs' infection on production [93], health and welfare traits are critical endeavors to determine the overall impact of the disease on the sustainability and resilience of the farms. For this reason, large-scale longitudinal and cohort studies are necessary to collect data for the determination of the most appropriate prediction models estimating the cost of the disease (production losses, involuntary culling/increased replacement rate, predisposition to other diseases and control measures), which remains unknown.

Breeding for resistance against SRLVs is an emerging research field. Specific genes have already been found to be associated with resistance or susceptibility to the disease and the discovery of additional genomic regions may increase the options for its control through marker-assisted selection (MAS) [76].

Despite the significant advances accomplished in SRLVs immunization studies, there are still important challenges to address, as vaccination against SRLVs may contribute to either controlling or enhancing MV [54]. Therefore, research efforts on the field of immunization have to be focused on the development of safe and effective vaccines, with the potential of universal application and mass production.

Funding: The research work was supported by the Hellenic Foundation for Research and Innovation (H.F.R.I.) under the "First Call for H.F.R.I. Research Projects to support Faculty members and Researchers and the procurement of high-cost research equipment grant" (Project Number: HFRI-FM17-1083).

Conflicts of Interest: None of the authors of this paper has a financial or personal relationship with other people or organizations that could inappropriately influence or bias the content of the paper.

References

1. De Boer, G.F. Zwoegerziekte virus, the causative agent for progressive interstitial pneumonia (maedi) and meningo-leucoencephalitis (visna) in sheep. *Res. Vet. Sci.* **1975**, *18*, 15–25. [CrossRef]
2. Pálsson, P.A. Maedi-Visna. History and Clinical Description. In *Maedi-Visna and Related Diseases. Developments in Veterinary Virology*; Pétursson, G., Hoff-Jørgensen, R., Eds.; Kluwer Academic Publishers: Norwell, MA, USA, 1990; Volume 10, pp. 3–74. ISBN 978-1-4612-8892-3.
3. Blacklaws, B.A. Small ruminant lentiviruses: Immunopathogenesis of Visna-Maedi and caprine arthritis and encephalitis virus. *Comp. Immunol. Microbiol. Infect. Dis.* **2012**, *35*, 259–269. [CrossRef] [PubMed]
4. Leroux, C.; Cruz, J.C.; Mornex, J.F. SRLVs: A genetic continuum of lentiviral species in sheep and goats with cumulative evidence of cross species transmission. *Curr. HIV Res.* **2010**, *8*, 94–100. [CrossRef] [PubMed]
5. Straub, O.C. Maedi–Visna virus infection in sheep. History and present knowledge. *Comp. Immunol. Microbiol. Infect. Dis.* **2004**, *27*, 1–5. [CrossRef]

6.	Peterhans, E.; Greenland, T.; Badiola, J.; Harkiss, G.; Bertoni, G.; Amorena, B.; Eliaszewicz, M.; Juste, R.A.; Kraβnig, R.; Lafont, J.P.; et al. Routes of transmission and consequences of small ruminant lentiviruses (SRLVs) infection and eradication schemes. *Vet. Res.* **2004**, *35*, 257–274. [CrossRef]

7.	*Terrestial Animal Health Code*, 28th ed.; World Organization for Animal Health: Paris, France, 2019; Volume 2, ISBN 978-92-95108-86-8.

8.	Hüttner, K.; Seelmann, M.; Feldhusen, F. Prevalence and risk factors for Maedi-Visna in sheep farms in Mecklenburg-Western-Pomerania. *Berl. Münch. Tierärztl.* **2010**, *123*, 463–467. [CrossRef]

9.	Pérez, M.; Biescas, E.; de Andrés, X.; Leginagoikoa, I.; Salazar, E.; Berriatua, E.; Reina, R.; Bolea, R.; de Andrés, D.; Juste, R.A.; et al. Visna/Maedi virus serology in sheep: Survey, risk factors and implementation of a successful control programme in Aragón (Spain). *Vet. J.* **2010**, *186*, 221–225. [CrossRef]

10.	Lago, N.; López, C.; Panadero, R.; Cienfuegos, J.; Pato, A.; Prieto, P.; Díaz, P.; Mourazos, N.; Fernández, G. Seroprevalence and risk factors associated with Visna/Maedi virus in semi-intensive lamb-producing flocks in northwestern Spain. *Prev. Vet. Med.* **2012**, *103*, 163–169. [CrossRef]

11.	Michiels, R.; van Mael, E.; Quinet, C.; Welby, S.; Cay, A.B.; de Regge, N. Seroprevalence and risk factors related to small ruminant lentivirus infections in Belgian sheep and goats. *Prev. Vet. Med.* **2018**, *151*, 13–20. [CrossRef]

12.	Schaller, P.; Vogt, H.R.; Strasser, M.; Nettleton, P.F.; Peterhans, E.; Zanoni, R. Seroprevalence of maedi-visna and border disease in Switzerland. *Schweiz. Arch. Tierh.* **2000**, *142*, 145–153.

13.	Kaba, J.; Czopowicz, M.; Ganter, M.; Nowicki, M.; Witkowski, L.; Nowicka, D.; Szaluś-Jordanow, O. Risk factors associated with seropositivity to small ruminant lentiviruses in goat herds. *Res. Vet. Sci.* **2013**, *94*, 225–227. [CrossRef]

14.	Azkur, A.K.; Aslan, M.E. Serological and Epidemiological Investigation of Bluetongue, Maedi-Visna and Caprine Arthritis-Encephalitis Viruses in Small Ruminant in Kirikkale District in Turkey. *Kafkas Univ. Vetr. Fak. Derg.* **2011**, *17*, 803–808. [CrossRef]

15.	Blacklaws, B.A.; Berriatua, E.; Torsteinsdottir, S.; Watt, N.J.; de Andres, D.; Klein, D.; Harkiss, G.D. Transmission of small ruminant lentiviruses. *Vet. Microbiol.* **2004**, *101*, 199–208. [CrossRef] [PubMed]

16.	Christodoulopoulos, G. Maedi–Visna: Clinical review and short reference on the disease status in Mediterranean countries. *Small Rumin. Res.* **2006**, *62*, 47–53. [CrossRef]

17.	Ramírez, H.; Reina, R.; Amorena, B.; de Andrés, D.; Martínez, H.A. Small Ruminant Lentiviruses: Genetic Variability, Tropism and Diagnosis. *Viruses* **2013**, *5*, 1175–1207. [CrossRef] [PubMed]

18.	Gomez-Lucia, E.; Barquero, N.; Domenech, A. Maedi-Visna virus: Current perspectives. *Vet. Med.* **2018**, *9*, 11–21. [CrossRef]

19.	Minguijón, E.; Reina, R.; Pérez, M.; Polledo, L.; Villoria, M.; Ramírez, H.; Leginagoikoa, I.; Badiola, J.J.; García-Marín, J.F.; de Andrés, D.; et al. Small ruminant lentivirus infections and diseases. *Vet. Microbiol.* **2015**, *181*, 75–89. [CrossRef]

20.	Reina, R.; Berriatua, E.; Lujan, L.; Juste, R.; Sanchez, A.; de Andres, D.; Amorena, B. Prevention strategies against small ruminant lentiviruses: An update. *Vet. J.* **2009**, *182*, 31–37. [CrossRef]

21.	Herrmann-Hoesing, L.M. Diagnostic assays used to control small ruminant lentiviruses. *J. Vet. Diagn. Investig.* **2010**, *22*, 843–855. [CrossRef]

22.	De Andres, D.; Klein, D.; Watt, N.J.; Berriatua, E.; Torsteinsdottir, S.; Blacklaws, B.A.; Harkiss, G.D. Diagnostic tests for small ruminant lentiviruses. *Vet. Microbiol.* **2005**, *107*, 49–62. [CrossRef]

23.	López, A.; Martinson, S.A. *Respiratory System, Mediastinum, and Pleurae, In Pathologic Basis of Veterinary Disease*, 6th ed.; Zachary, J.F., Ed.; Mosby: St. Louis, MO, USA, 2017; pp. 471–560. ISBN 9780323357753.

24.	Cirone, F.; Maggiolino, A.; Cirilli, M.; Sposato, A.; de Palo, P.; Ciappetta, G.; Pratelli, A. Small ruminant lentiviruses in goats in southern Italy: Serological evidence, risk factors and implementation of control programs. *Vet. Microbiol.* **2019**, *228*, 143–146. [CrossRef] [PubMed]

25.	Karanikolaou, K.; Angelopoulou, K.; Papanastasopoulou, M.; Koumpati-Artopiou, M.; Papadopoulos, O.; Koptopoulos, G. Detection of small ruminant lentiviruses by PCR and serology tests in field samples of animals from Greece. *Small Rumin. Res.* **2005**, *58*, 181–187. [CrossRef]

26.	Junkuszewa, A.; Dudko, P.; Bojar, W.; Olech, M.; Osinski, Z.; Gruszecki, T.M.; Kania, M.G.; Kuźmak, J.; Czerski, G. Risk factors associated with small ruminant lentivirus infection in eastern Poland sheep flocks. *Prev. Vet. Med.* **2016**, *127*, 44–49. [CrossRef] [PubMed]

27. Shuaib, M.; Green, C.; Rashid, M.; Duizer, G.; Whiting, T.L. Herd risk factors associated with sero-prevalence of Maedi-Visna in the Manitoba sheep population. *Can. Vet. J.* **2010**, *51*, 385 390.

28. Zhang, K.; He, J.; Liu, Y.; Shang, Y.; Liu, X. A Seroprevalence Survey of Maedi-Visna Among Twenty-Four Ovine Floks from Twelve Regions of China. *J. Integr. Agric.* **2013**, *12*, 2321–2323. [CrossRef]

29. Shah, C.; Boni, J.; Huder, J.B.; Vogt, H.R.; Muhlherr, J.; Zanoni, R.; Miserez, R.; Lutz, H.; Schüpbach, J. Phylogenetic analysis and reclassification of caprine and ovine lentiviruses based on 104 new isolates: Evidence for regular sheep-to-goat transmission and worldwide propagation through livestock trade. *Virology* **2004**, *319*, 12–26. [CrossRef]

30. Fevereiro, M.T.; Barros, S.S. Caracterização biológica e molecular de um lentivírus de ovino isolado em Portugal. *RPCV* **2004**, *99*, 27–39.

31. Andresson, O.S.; Elser, J.E.; Tobin, G.J.; Greenwood, J.D.; Gonda, M.A.; Georgsson, G.; Andrésdóttir, V.; Benediktsdóttir, E.; Carlsdóttir, H.M.; Mäntylä, E.O. Nucleotide sequence and biological properties of a pathogenic proviral molecular clone of neurovirulent visna virus. *Virology* **1993**, *193*, 89–105. [CrossRef]

32. Sonigo, P.; Alizon, M.; Staskus, K.; Klatzmann, D.; Cole, S.; Danos, O.; Retzel, E.; Tiollais, P.; Haase, A.; Wain-Hobson, S. Nucleotide sequence of the visna lentivirus: Relationship to the AIDS virus. *Cell* **1985**, *42*, 369–382. [CrossRef]

33. Staskus, K.A.; Retzel, E.F.; Lewis, E.D.; Silsby, J.L.; Sheila, C.S.T.; Rank, J.M.; Wietgrefe, S.W.; Haase, A.T.; Cook, R.; Fast, D.; et al. Isolation of replication-competent molecular clones of visna virus. *Virology* **1991**, *181*, 228–240. [CrossRef]

34. Sargan, D.R.; Bennet, I.D.; Cousens, C.; Roy, D.J.; Blacklaws, B.A.; Dalziel, R.G.; Watt, N.J.; Mc Connell, I. Nucleotide sequence of EV1, a British isolate of maedi-visna virus. *J. Gen. Virol.* **1991**, *72*, 1893–1903. [CrossRef] [PubMed]

35. Quérat, G.; Audoly, G.; Sonigo, P.; Vigne, R. Nucleotide sequence analysis of SA-OMVV, a visna-related ovine lentivirus: Phylogenetic history of lentiviruses. *Virology* **1990**, *175*, 434–447. [CrossRef]

36. Karr, B.M.; Chebloune, Y.; Leung, K.; Narayan, O. Genetic characterization of two phenotypically distinct North American ovine lentiviruses and their possible origin from caprine arthritis-encephalitis virus. *Virology* **1996**, *225*, 1–10. [CrossRef] [PubMed]

37. Glaria, I.; Reina, R.; Ramirez, H.; de Andres, X.; Crespo, H.; Jauregui, P.; Salazar, E.; Luján, L.; Pérez, M.M.; Benavides, J.; et al. Visna/Maedi virus genetic characterization and serological diagnosis of infection in sheep from a neurological outbreak. *Vet. Microbiol.* **2012**, *155*, 137–146. [CrossRef]

38. Olech, M.; Valas, S.; Kuźmak, J. Epidemiological survey in single-species flocks from Poland reveals expanded genetic and antigenic diversity of small ruminant lentiviruses. *PLoS ONE* **2018**, *13*, e0193892. [CrossRef]

39. Ramirez, H.; Glaria, I.; de Andres, X.; Martinez, H.A.; Hernandez, M.M.; Reina, R.; Iráizoz, E.; Crespo, H.; Berriatua, E.; Vázquez, J.; et al. Recombinant small ruminant lentivirus subtype B1 in goats and sheep of imported breeds in Mexico. *Vet. J.* **2011**, *190*, 169–172. [CrossRef]

40. Narayan, O.; Clements, J.E.; Strandberg, J.D.; Cork, L.C.; Griffin, D.E. Biological characterization of the virus causing leukoencephalitis and arthritis in goats. *J. Gen. Virol.* **1980**, *50*, 69–79. [CrossRef]

41. Saltarelli, M.; Querat, G.; Konings, D.A.; Vigne, R.; Clements, J.E. Nucleotide sequence and transcriptional analysis of molecular clones of CAEV which generate infectious virus. *Virology* **1990**, *179*, 347–364. [CrossRef]

42. Qu, J.; Guo, W.; Zhao, L.; Sheng, R.; Xiang, W. The gene cloning and sequence analysis of the whole genome of Caprine Arthritis Encephalitis virus (CAEV) GANSU strain. *Chin. J. Virol.* **2005**, *21*, 389–392.

43. Huang, J.; Sun, Y.; Liu, Y.; Xiao, H.; Zhuang, S. Development of a loop-mediated isothermal amplification method for rapid detection of caprine arthritis-encephalitis virus proviral DNA. *Arch. Virol.* **2012**, *157*, 1463–1469. [CrossRef]

44. Glaria, I.; Reina, R.; Crespo, H.; de Andres, X.; Ramirez, H.; Biescas, E.; Pérez, M.M.; Badiola, J.; Luján, L.; Amorena, B.; et al. Phylogenetic analysis of SRLV sequences from an arthritic sheep outbreak demonstrates the introduction of CAEV-like viruses among Spanish sheep. *Vet. Microbiol.* **2009**, *138*, 156–162. [CrossRef] [PubMed]

45. Bertolotti, L.; Mazzei, M.; Puggioni, G.; Carrozza, M.L.; Giudici, S.D.; Muz, D.; Juganaru, M.; Patta, C.; Tolari, F.; Rosati, S. Characterization of new small ruminant lentivirus subtype B3 suggests animal trade within the Mediterranean Basin. *J. Gen. Virol.* **2011**, *92*, 1923–1929. [CrossRef]

46. Santry, A.; de Jong, J.; Gold, A.C.; Walsh, S.R.; Menzies, P.I.; Wootton, S.K. Genetic characterization of small ruminant lentiviruses circulating in naturally infected sheep and goats in Ontario, Canada. *Virus Res.* **2013**, *175*, 30–44. [CrossRef] [PubMed]

47. Gjerset, B.; Jonassen, C.M.; Rimstad, E. Natural transmission and comparative analysis of small ruminant lentiviruses in the Norwegian sheep and goat populations. *Virus Res.* **2007**, *125*, 153–161. [CrossRef] [PubMed]

48. Gjerset, B.; Storset, A.K.; Rimstad, E. Genetic diversity of small-ruminant lentiviruses: Characterization of Norwegian isolates of Caprine arthritis encephalitis virus. *J. Gen. Virol.* **2006**, *87*, 573–580. [CrossRef]

49. Reina, R.; Grego, E.; Bertolotti, L.; de Meneghi, D.; Rosati, S. Genome analysis of small-ruminant lentivirus genotype E: A caprine lentivirus with natural deletions of the dUTPase subunit, vpr-like accessory gene, and 70-base-pair repeat of the U3 region. *J. Virol.* **2009**, *83*, 1152–1155. [CrossRef]

50. Reina, R.; Bertolotti, L.; Dei Giudici, S.; Puggioni, G.; Ponti, N.; Profiti, M.; Patta, C.; Rosati, S. Small Ruminant Lentivirus genotype E is widespread in Sarda Goat. *Vet. Microbiol.* **2010**, *144*, 24–31. [CrossRef]

51. L'Homme, Y.; Leboeuf, A.; Arsenault, J.; Fras, M. Identification and characterization of an emerging small ruminant lentivirus circulating recombinant form (CRF). *Virology* **2015**, *475*, 159–171. [CrossRef]

52. Pisoni, G.; Bertoni, G.; Puricelli, M.; Maccalli, M.; Moroni, P. Demonstration of coinfection with and recombination by caprine arthritis-encephalitis virus and maedi-visna virus in naturally infected goats. *J. Virol.* **2007**, *81*, 4948–4955. [CrossRef]

53. McNeilly, T.N.; Baker, A.; Brown, J.K.; Collie, D.; MacLachlan, G.; Rhind, S.M.; Harkiss, G.D. Role of Alveolar Macrophages in Respiratory Transmission of Visna/Maedi Virus. *J. Virol.* **2008**, *82*, 1526–1536. [CrossRef]

54. Reina, R.; de Andrés, D.; Amorena, B. Immunization against Small Ruminant Lentiviruses. *Viruses* **2013**, *5*, 1948–1963. [CrossRef] [PubMed]

55. Gayo, E.; Polledo, L.; Balseiro, A.; Martinez, C.P.; Garcia Iglesias, M.J.; Preziuso, S.; Rossi, G.; García Marín, J.F. Inflammatory Lesion Patterns in Target Organs of Visna/Maedi in Sheep and their Significance in the Pathogenesis and Diagnosis of the Infection. *J. Comp. Pathol.* **2018**, *159*, 49–56. [CrossRef] [PubMed]

56. Angelopoulou, K.; Brellou, G.D.; Vlemmas, I. Detection of Maedi-Visna Virus in the Kidneys of Naturally Infected Sheep. *J. Comp. Path.* **2006**, *134*, 329–335. [CrossRef] [PubMed]

57. Brellou, G.D.; Angelopoulou, K.; Poutahidis, T.; Vlemmas, I. Detection of Maedi-Visna Virus in the Liver and Heart of Naturally Infected Sheep. *J. Comp. Path.* **2007**, *136*, 27–35. [CrossRef]

58. Broughton-Neiswanger, L.E.; White, S.N.; Knowles, D.P.; Mousel, M.R.; Lewis, G.S.; Herndon, D.R.; Herrmann-Hoesing, L.M. Non-maternal transmission is the major mode of ovine lentivirus transmission in a ewe flock: A molecular epidemiology study. *Infect. Genet. Evol.* **2010**, *10*, 998–1007. [CrossRef]

59. Cutlip, R.C.; Lehmkuhl, H.D.; Jackson, T.A. Intrauterine transmission of ovine progressive pneumonia virus. *Am. J. Vet. Res.* **1981**, *42*, 1795–1797.

60. Bolea, R.; Monleón, E.; Carrasco, L.; Vargas, A.; de Andrés, D.; Amorena, B.; Badiola, J.J.; Luján, L. Maedi-visna virus infection of ovine mammary epithelial cells. *Vet. Res.* **2006**, *37*, 133–144. [CrossRef]

61. Preziuso, S.; Renzoni, G.; Allen, T.E.; Taccini, E.; Rossi, G.; DeMartini, J.C.; Braca, G. Colostral transmission of Maedi Visna virus: Sites of viral entry in lambs born from experimentally infected ewes. *Vet. Microbiol.* **2004**, *104*, 157–164. [CrossRef]

62. Pisoni, G.; Bertoni, G.; Manarolla, G.; Vogt, H.R.; Scaccabarozzi, L.; Locatelli, C.; Moroni, P. Genetic analysis of small ruminant lentiviruses following lactogenic transmission. *Virology* **2010**, *407*, 91–99. [CrossRef]

63. Radostits, O.M.; Gay, C.C.; Blood, D.C.; Hinchcliff, K.W. Ovine progressive pneumonia (Maedi, Maedi-Visna). In *Veterinary Medicine A Textbook of the Diseases of Cattle, Sheep, Pigs, Goats and Horses*, 9th ed.; W.B. Saunders Company Ltd.: London, UK, 2000; pp. 1186–1189.

64. McNeilly, T.N.; Tennant, P.; Lujan, L.; Perez, M.; Harkiss, G.D. Differential infection efficiencies of peripheral lung and tracheal tissues in sheep infected with Visna/maedi virus via the respiratory tract. *J. Gen. Virol.* **2007**, *88*, 670–679. [CrossRef] [PubMed]

65. Villoria, M.; Leginagoikoa, I.; Luján, L.; Pérez, M.; Salazar, E.; Berriatua, E.; Juste, R.A.; Minguijón, E. Detection of small ruminant lentivirus in environmental samples of air and water. *Small Rumin. Res.* **2013**, *110*, 155–160. [CrossRef]

66. Preziuso, S.; Sanna, E.; Sanna, M.P.; Loddo, C.; Cerri, D.; Taccini, E.; Mariotti, F.; Braca, G.; Rossi, G.; Renzoni, G. Association of maedi visna virus with Brucella ovis infectionin rams. *Eur. J. Histochem.* **2003**, *47*, 151–1588. [CrossRef] [PubMed]

67. Peterson, K.; Brinkhof, J.; Houwers, D.J.; Colenbrander, B.; Gadella, B.M. Presence of pro-lentiviral DNA in male sexual organs and ejaculates of small ruminants. *Theriogenology* **2008**, *69*, 433–442. [CrossRef] [PubMed]

68. Leginagoiko, I.; Daltabuit-Test, M.; Álvarez, V.; Arranz, J.; Juste, R.A.; Amorena, B.; de Andrés, D.; Luján, L.L.; Badiola, J.J.; Berriatua, E. Horizontal Maedi-Visna virus (MVV) infection in adult dairy-sheep raised under varying MVV-infection pressures investigated by ELISA and PCR. *Res. Vet. Sci.* **2006**, *80*, 235–241. [CrossRef]

69. Leginagoikoa, I.; Juste, R.A.; Barandika, J.; Amorena, B.; de Andrés, D.; Lujánc, L.; Badiola, J.; Beriattua, E. Extensive rearing hinders Maedi-Visna Virus (MVV) infection in sheep. *Vet. Res.* **2006**, *37*, 767–778. [CrossRef] [PubMed]

70. Alba, A.; Allepuz, A.; Serrano, E.; Casal, J. Seroprevalence and spatial distribution of Maedi–Visna virus and pestiviruses in Catalonia (Spain). *Small Rumin. Res.* **2008**, *78*, 80–86. [CrossRef]

71. Leginagoikoa, I.; Minguijon, E.; Juste, R.A.; Barandika, J.; Amorena, B.; de Andres, D.; Badiola, J.J.; Luján, L.; Berriatua, E. Effects of housing on the incidence of visna/maedi virus infection in sheep flocks. *Res. Vet. Sci.* **2010**, *88*, 415–421. [CrossRef]

72. Heaton, M.; Clawson, M.; Chitko-Mckown, C.; Leymaster, K.; Smith, T.; Harhay, G.P.; White, S.N.; Herrmann-Hoesing, L.M.; Mousel, M.R.; Lewis, G.S.; et al. Reduced Lentivirus Susceptibility in Sheep with TMEM154 Mutations. *PLoS Genet.* **2012**, *8*, e1002467. [CrossRef]

73. Alshanbari, F.A.; Mousel, M.R.; Reynolds, J.O.; Herrmann-Hoesing, L.M.; Highland, M.A.; Lewis, G.S.; White, N. Mutations in Ovis aries TMEM154 are associated with lower small ruminant lentivirus proviral concentration in one sheep flock. *Anim. Genet.* **2014**, *45*, 565–571. [CrossRef]

74. Leymaster, K.A.; Chitko-McKown, C.G.; Clawson, M.L.; Harhay, G.P.; Heaton, M.P. Effects of TMEM154 haplotypes 1 and 3 on susceptibility to ovine progressive pneumonia virus following natural exposure in sheep. *J. Anim. Sci.* **2013**, *91*, 5114–5121. [CrossRef]

75. Larruskain, A.; Jugo, B.M. Retroviral infections in sheep and goats: Small ruminant lentiviruses and host interaction. *Viruses* **2013**, *5*, 2043–2061. [CrossRef]

76. White, S.N.; Knowles, D.P. Expanding possibilities for intervention against small ruminant lentiviruses through genetic marker-assisted selective breeding. *Viruses* **2013**, *5*, 1466–1499. [CrossRef] [PubMed]

77. Stonos, N.; Wootton, S.K.; Karrow, N. Immunogenetics of small ruminant lentiviral infections. *Viruses* **2014**, *6*, 3311–3333. [CrossRef] [PubMed]

78. White, S.N.; Mousel, M.R.; Herrmann-Hoesing, L.M.; Reynolds, J.O.; Leymaster, K.A.; Neibergs, H.L.; Lewis, G.S.; Knowles, D.P. Genome-Wide Association Identifies Multiple Genomic Regions Associated with Susceptibility to and Control of Ovine Lentivirus. *PLoS ONE* **2012**, *7*, e47829. [CrossRef] [PubMed]

79. Jáuregui, P.; Crespo, H.; Glaria, I.; Luján, L.; Contreras, A.; Rosati, S.; de Andrés, D.; Amorena, B.; Towers, G.J.; Reina, R. Ovine TRIM5a Can Restrict Visna/Maedi Virus. *J. Virol.* **2012**, *86*, 9504–9509. [CrossRef]

80. Lacerenza, D.; Giammarioli, M.; Grego, E.; Marini, C.; Profiti, M.; Rutili, D.; Rosati, S. Antibody response in sheep experimentally infected with different small ruminant lentivirus genotypes. *Vet. Immunol. Immunopathol.* **2006**, *112*, 264–271. [CrossRef]

81. De Andrés, D.; Ramírez, H.; Bertolotti, L.; San Román, B.; Glaria, I.; Crespo, H.; Jáuregui, P.; Minguijón, E.; Juste, R.; Leginagoikoa, I.; et al. An insight into a combination of ELISA strategies to diagnose small ruminant lentivirus infections. *Vet. Immunol. Immunopathol.* **2013**, *152*, 277–288. [CrossRef]

82. Carrozza, M.L.; Mazzei, M.; Bandecchi, P.; Fraisier, C.; Pérez, M.; Suzan-Monti, M.; de Andrés, D.; Amorena, B.; Rosati, S.; Andrésdottir, V.; et al. Development and comparison of strain specific gag and pol real-time PCR assays for the detection of Visna/maedi virus. *J. Virol. Methods.* **2010**, *165*, 161–167. [CrossRef]

83. Cheevers, W.P.; Knowles, D.P.; McGuire, T.C.; Baszler, T.V.; Hullinger, G.A. Caprine arthritis–encephalitis lentivirus (CAEV) challenge of goats immunized with recombinant vaccinia virus expressing CAEV surface and transmembrane envelope glycoproteins. *Vet. Immunol. Immunopathol.* **1994**, *42*, 237–251. [CrossRef]

84. Petursson, G.; Matthiasdottir, S.; Svansson, V.; Andresdottir, V.; Georgsson, G.; Martin, A.H.; Agnarsdóttir, G.; Gísladóttir, E.; Arnadóttir, S.; Högnadóttir, S.; et al. Mucosal vaccination with an attenuated maedi-visna virus clone. *Vaccine* **2005**, *23*, 3223–3228. [CrossRef]

85. Gonzalez, B.; Reina, R.; Garcia, I.; Andres, S.; Glaria, I.; Alzueta, M.; Mora, M.I.; Jugo, B.M.; Arrieta-Aguirre, I.; de la Lastra, J.M.; et al. Mucosal immunization of sheep with a Maedi-Visna virus (MVV) env DNA vaccine protects against early MVV productive infection. *Vaccine* **2005**, *23*, 4342–4352. [CrossRef] [PubMed]

86. Torsteinsdottir, S.; Carlsdottir, H.M.; Svansson, V.; Matthiasdottir, S.; Martin, A.H.; Petursson, G. Vaccination of sheep with Maedi-visna virus gag gene and protein, beneficial or harmful? *Vaccine* **2007**, *25*, 6713–6720. [CrossRef] [PubMed]

87. de Andres, X.; Reina, R.; Ciriza, J.; Crespo, H.; Glaria, I.; Ramirez, H.; Grilló, M.J.; Pérez, M.M.; Andrésdóttir, V.; Rosati, S.; et al. Use of B7 costimulatory molecules as adjuvants in a prime-boost vaccination against Visna/Maedi ovine lentivirus. *Vaccine* **2009**, *27*, 591–600. [CrossRef]

88. Lin, Y.Z.; Shen, R.X.; Zhu, Z.Y.; Deng, X.L.; Cao, X.Z.; Wang, X.F.; Ma, J.; Jianga, C.G.; Zhao, L.P.; Lv, X.L.; et al. An attenuated EIAV vaccine strain induces significantly different immune responses from its pathogenic parental strain although with similar in vivo replication pattern. *Antiviral Res.* **2011**, *92*, 292–304. [CrossRef] [PubMed]

89. Brülisauer, F.; Vogt, H.; Perler, L.; Rüfenacht, J. Risk factors for the infection of Swiss goat herds with small ruminant lentivirus: A case-control study. *Vet. Rec.* **2005**, *157*, 229–233. [CrossRef] [PubMed]

90. Polledo, L.; González, J.; Fernández, C.; Miguélez, J.; Martínez-Fernández, B.; Morales, S.; Ferreras, M.C.; García Marín, J.F. Simple control strategy to reduce the level of Maedi-Visna infection in sheep flocks with high prevalence values (>90%). *Small Rumin. Res.* **2013**, *112*, 224–229. [CrossRef]

91. Pérez, M.; Munoz, J.A.; Biescas, E.; Salazar, E.; Bolea, R.; de Andrés, D.; Amorena, B.; Badiola, J.J.; Reina, R.; Luján, L. Successful Visna/maedi control in a highly infected ovine dairy flock using serologic segregation and management strategies. *Prev. Vet. Med.* **2013**, *112*, 423–427. [CrossRef]

92. Cortez-Romero, C.; Pellerin, J.L.; Ali-Al-Ahmad, M.Z.; Chebloune, Y.; Gallegos-Sánchez, J.; Lamara, A.; Pépin, M.; Fieni, F. The risk of small ruminant lentivirus (SRLV) transmission with reproductive biotechnologies: State-of-the-art review. *Theriogenology* **2013**, *79*, 1–9. [CrossRef]

93. Martínez-Navalón, B.; Peris, C.; Gómez, E.A.; Peris, B.; Roche, M.L.; Caballero, C.; Goyena, E.; Berriatua, E. Quantitative estimation of the impact of caprine arthritis encephalitis virus infection on milk production by dairy goats. *Vet. J.* **2013**, *197*, 311–317. [CrossRef]

Article

Pathological, Bacteriological and Virological Findings in Sudden and Unexpected Deaths in Young Dogs

Giuseppe Piegari [1,*], Lorena Cardillo [2], Flora Alfano [2], Lucia Vangone [2], Valentina Iovane [3] and Giovanna Fusco [2]

1   Department of Veterinary Medicine and Animal Production, Unit of Pathology, University of Naples Federico II, 80137 Naples, Italy
2   Istituto Zooprofilattico Sperimentale del Mezzogiorno, Portici, 81100 Naples, Italy; lorena.cardillo85@gmail.com (L.C.); flora.alfano@cert.izsmportici.it (F.A.); lucia.vangone@izsmportici.it (L.V.); giovanna.fusco@izsmportici.it (G.F.)
3   Department of Pharmacy, University of Salerno, 84084 Fisciano, Italy; viovane@unisa.it
*   Correspondence: giuseppe.piegari@unina.it

Received: 6 May 2020; Accepted: 2 July 2020; Published: 3 July 2020

Simple Summary: "Sudden death" has been defined by the World Health Organization as a non-violent, unexpected death occurring less than 24 h from the onset of symptoms. The causes of sudden death have been widely investigated in human forensic medicine. In contrast, few studies have been reported in the veterinary literature. This study aimed to investigate the frequency of sudden deaths in young dogs in different age ranges. A secondary aim was to collect information regarding clinical symptoms, and pathological and microbiological findings related to sudden death in young dogs. The results of the present study demonstrate that the highest frequency of sudden death occurs in animals in an age range from 10 days to 1 month and from 6 to 12 months. The most frequently observed clinical symptoms in cases of sudden death were acute respiratory symptoms. Furthermore, *Canine parvovirus type 2*, *E. coli*, *Canine Distemper Virus*, *Clostridium perfringens type A, and Pasteurella* spp. were the main causes of death observed in the present study. The results reported in the present study could provide a reference basis to better investigate sudden death in veterinary clinical practice.

Abstract: In human medicine, "sudden death" has been defined by the World Health Organization (WHO) as a non-violent, unexpected death occurring less than 24 h from the onset of symptoms. The aims of this study were: (1) to estimate the proportional mortality ratio for "sudden and unexpected death" (SUD) in young dogs; (2) to investigate the pathological and microbiological findings in SUD cases in young dogs. For these purposes, a retrospective study of a total of 145 cases of young dead dogs was performed. For each case, we collected information about the age, medical history and the gross and microbiological findings of the animals. The results of this study found 21 cases of SUD. The most frequently observed clinical symptoms in the cases of sudden death were acute respiratory symptoms, followed by acute gastroenteric symptoms, non-specific symptoms and neurological symptoms. The evaluation of necropsy reports allowed us to observe enteritis in 18 out of 21 cases and pneumonia in seven out of 21 cases. Viral infection with *Canine parvovirus type 2* was the most common cause of SUD observed. These results could provide a valuable tool for the investigation of sudden death in young dogs.

Keywords: post-mortem microbiology; veterinary forensic pathology; sudden death; young dogs

1. Introduction

In human medicine, "sudden death" has been defined by the World Health Organization (WHO) as a non-violent, unexpected death occurring less than 24 h from the onset of symptoms [1]; in particular, the term "sudden and unexpected infant death" (SUID) is used to describe deaths that occur relatively suddenly and unexpectedly in children less than 1 years old [2,3]. Infections are reported in the literature as an important cause of SUID, followed by metabolic or molecular disorders [2–5]. The main pathogens reported in SUID cases are as follows: *Staphylococcus aureus*, *Escherichia coli*, *Streptococcus pyogenes*, *Streptococcus agalactiae*, *Streptococcus pneumoniae*, *Group B Streptococci (GBS)*, *Respiratory Syncytial Virus (RSV)*, *Cytomegalovirus (CMV) and Adenovirus* [2,6,7]. However, a broad range of pathogens has been reported in the literature as being causes or a co-factors in SUID, such as *Parvovirus B19, Epstein-Barr virus, Influenza A virus* and *Mycobacterium tuberculosis* [2,6,8,9]. Furthermore, recent studies reported the relatively benign Coxsackie virus A16 as a possible contributing factor in SUID in humans [10]. For these reasons, the current SUID autopsy protocol in the UK and the international guidelines advocate for a multidisciplinary approach to the investigation of all cases of SUID, which should be based not only on the findings of the macroscopic examination, but also on a broad range of ancillary investigations, such as bacteriological and virological analyses [11,12]. Sudden infant death syndrome (SIDS) is considered to be a sub-class of SUID, in which the cause of death remains unexplained even after the forensic necropsy, ancillary tests and evaluation of the anamnestic data and crime scene analysis [1,2]. Indeed, among the cases of SUID, only 20% have a clear cause, while most cases remain unexplained and are categorized as SIDS [1,5,13]. Although the cause is unknown, specific genetic mutations or mild infections could be involved in the genesis of the syndrome [5]. Mild infections have been suggested to play a key role, as demonstrated by altered levels of immunoglobulin or cytokine and the high frequency of mild tracheal infections commonly observed during post-mortem examinations of the subjects with a final diagnosis of SIDS [5]. Although, in human medicine, the concept of sudden death, SUID and SIDS has been well defined by the WHO, in veterinary medicine, a universal definition is lacking. Some authors have defined sudden death in animals as death that occurs in a few minutes or several hours, due to pre-existing disease or a functional disorder [14]. However, in the opinion of the authors, this definition should be avoided, because it lacks a well-defined temporal reference range. In contrast, even if not yet validated in veterinary medicine, the WHO definition provides an important temporal reference range useful for the identification of cases of sudden death in veterinary clinical practice.

Over the last few years, many studies have investigated the cause of death in animals. In particular, infectious diseases that affect the gastrointestinal system are reported to be the main cause of death in puppies and young dogs [15,16]. In contrast, neoplastic diseases appear to be the prevalent cause of death in adult dogs [15]. Among the infections, *canine parvovirus type 2* (CPV-2) is reported to be one of the most common and important causes of morbidity and mortality in young dogs [17,18]. Moreover, this virus is considered to be an important pathogen responsible for acute gastroenteritis and myocarditis in dogs [17,18]. However, with regards to sudden and unexpected deaths, despite the underlying causes having been sporadically investigated in dogs [19], to the best of our knowledge, no study has evaluated the microbiological findings in cases of sudden death in young dogs. In light of these observations, the aims of this study were as follows: (1) to estimate the proportional mortality ratio (PMR) for "sudden and unexpected death" in puppies and young dogs; (2) to investigate the pathological, bacteriological, and virological findings in sudden and unexpected death in young dogs; and (3) to introduce a standardized microbiological protocol for the diagnostic investigation of cases of sudden death in veterinary medicine

2. Materials and Methods

2.1. Study Design

An observational retrospective study of a total of 145 cases of young dead dogs, consecutively presented by veterinary practitioners, owners, or law enforcement to the "Istituto Zooprofilattico del Mezzogiorno" (IZSM) of Portici city, Southern Italy, was performed over a 3-year period (2015–2017). The submission forms were collected to obtain information about the medical history and age of the animals. On the basis of the medical history, the animals were divided into groups as follows:

-   (Sudden and unexpected death group—SUD group): dogs with a clinical diagnosis of sudden and unexpected death. According to the WHO, sudden and unexpected death (SUD) cases were considered to be a non-violent and unexpected death that occurs less than 24 h after the onset of symptoms;
-   (Expected death group—ED group): dogs without a clinical history of sudden and unexpected death

On the basis of age, the available data were categorized as follows: (Group 1) 10 days–4 weeks; (Group 2) 4 weeks–6 weeks; (Group 3) 6 weeks–2 months; (Group 4) 2–3 months; (Group 5) 3–6 months; (Group 6) 6–12 months. Each examined case was subjected to a complete necropsy and bacteriological and virological investigations; however, for the purposes of this study, only the necropsy and microbiological reports from animals in the SUD group were included. Furthermore, the microbiological investigations were restricted to molecular tests for the virological analysis, and microbiological cultures for the bacteriological examinations. In all assessed cases, molecular tests were performed using a real-time polymerase chain reaction assay (RT-PCR) for *canine parvovirus (CPV), canine coronavirus, canine adenovirus, herpesvirus, and canine distemper virus.* Furthermore, in all cases positive for *canine parvovirus type 2,* multiplexed PCR panels were used to distinguish between wild-type and vaccine CPV-2 and to identify the pathogen subtype (CPV-2a; CPV2b; CPV 2c) [20]. The analyzed samples included the liver, lung, kidney, spleen, heart, brain and intestine. Microbiological results and necropsy reports were both extracted from the IZSM information system (SIGLA). Animals that tested positive in the toxicological investigations, or that had died from trauma, were excluded from the study.

2.2. Necropsy Protocol

All necropsies were performed in the necropsy room of the "Istituto Zooprofilattico Sperimentale del Mezzogiorno" (IZSM), Portici, Italy, with a standard necropsy protocol [21]. All SUD cases were stored at 4 °C before necropsy. The period between death and necropsy was between 12 and 36 h. During the necropsy, according to internal institute protocol, all samples were taken in rigorous asepsis conditions using sterile instruments and transported to the laboratory of microbiology. Furthermore, to obtain uncontaminated specimens, a sterilization of the body and organs surfaces was performed before sampling. Finally, the mean time between sample collection and transport to the reference laboratory was under 3 h.

2.3. Analytical Validation of the Results

For each case of sudden and unexpected death, the clinical history, necropsy report and microbiological findings were reviewed, and the final cause of death was categorized as "explained" or "unexplained". However, since determining the pathological significance of the microorganisms isolated during necropsy is often difficult, as has been frequently reported in the literature [22,23]. For the purposes of this study, viruses detected by PCR were considered to be the cause of death, only when associated with the typical macroscopic changes observed during the anatomopathological examination. In addition, the bacteriological and virological findings were interpreted considering a broad range of variables, such as the location of pathogen detection, the capacity of pathogens for virulence, the correlation with injuries observed during the necropsy, the multisite location of the pathogens, the age of the dog and the composition of the normal flora.

2.4. Statistical Analysis

The frequencies of sudden and unexpected death (SUD), expected death (ED), and total deaths (SUD + ED) were evaluated and stratified by age classes. Furthermore, we estimated the proportional mortality ratio (PMR) for "SUD" in each assessed age group. The Chi-square test was used to assess differences in the distributions of ED and SUD among age groups.

3. Results

Out of the 145 examined reports, we found 21 cases of SUD and 124 cases of ED during the 3-year study period. The PMR of SUD was therefore 14.48%, while the ED was 85.52%. Furthermore, the Chi-square test showed a significant difference in the frequencies of ED and SUD among the assessed age groups ($p < 0.05$). All SUD cases were submitted by Italian veterinary practitioners. The highest frequencies of expected death were observed in animals in Group 2 (100% of the cases), Group 3 (87.7% vs. 12.5%), Group 4 (97.2% vs. 12.8%) and Group 5 (93.3% vs. 6.7%). In contrast, the highest frequencies of SUD were found in animals in Group 1 (58.8% vs. 41.2%) and Group 6 (37.5% vs. 62.5%). Table 1 summarizes the frequencies and percentages of SUD and ED and the frequency of total deaths (SUD + ED) stratified by age classes. Overall, of the 21 SUD cases, 10 out of 21 (47.61%) dogs were less than 4 weeks old (Group 1), 0 out of 21 were between 4 weeks and 6 weeks old (Group 2), five out of 21 (23.8%) were between 6 weeks and 2 months old (Group 3), one out of 21 (4.76%) was between 2 and 3 months old (Group 4), two out of 21 (9.51%) were between 3 and 6 months old (Group 5), and three out of 21 (14.28%) were between 6 months and 1 year old (Group 6).

Table 1. Frequency and percentage of sudden and unexpected deaths, expected deaths and the frequency of total deaths stratified by age groups.

Age Group	Sudden Death	Expected Death	Total Deaths
Group 1	10 (58.8%)	7 (41.2%)	17
Group 2	0	14 (100%)	14
Group 3	5 (12.5%)	35 (87.5%)	40
Group 4	1 (2.8%)	35 (97.2%)	36
Group 5	2 (6.7%)	28 (93.3%)	30
Group 6	3 (37.5%)	5 (62.5%)	8
Total	21	124	145

3.1. Clinical Background and Gross Findings

The most frequently observed clinical symptoms in cases of sudden death were as follows: acute respiratory symptoms in 12 out of 21 cases, followed by acute gastroenteric symptoms (a single or few episodes of vomiting or diarrhea) in six out of 21 cases, neurological symptoms in one case, and finally, non-specific symptoms in two out of 21 cases. The evaluation of necropsy reports allowed us to observe haemorrhagic gastroenteritis in 12 out of 21 cases, pneumonia in seven out of 21 cases, and catarrhal enteritis in five out of 21 cases (Figure 1). Pulmonary oedema or multiorgan congestion were also observed in 19 out of 21 cases.

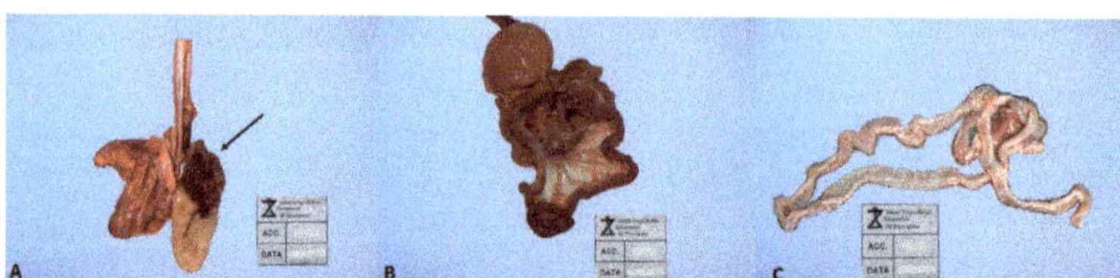

Figure 1. Pathological lesions in cases of sudden death. (**A**): pneumonia (arrows) (**B**): haemorrhagic gastroenteritis (**C**): catarrhal enteritis.

3.2. Microbiological Examination

In all animals dead for sudden and unexpected death, virological investigations were performed with a panel of viruses tested by PCR (*Canine parvovirus, Canine coronavirus, Canine adenovirus, Canine herpesvirus* and *Canine distemper virus*), and a bacteriological examination was performed with microbiological cultures. The retrospective analysis showed positive microbiological results in 18 out of 21 cases (Table 2).

Table 2. Viruses or bacteria detected in cases of sudden and unexpected death.

Pathogen	NO. of Cases
Canine Parvovirus	10
E. Coli	7
Clostridium perfringens type A	6
Adenovirus	3
Canine Distemper Virus	2
Streptococcus sanguinis	2
Pasteurella spp.	2
Streptococcus dysgalactiae	1
Rotavirus	1

However, for the purposes of this study, the microbiological findings were interpreted considering a broad range of variables. In particular, the location of pathogen detection, the capacity of pathogens for virulence and the correlation of that microorganisms with the observed macroscopic injuries were the most important parameters assessed in this study. Therefore, after the review of the necropsies and microbiological reports, the detected pathogens were considered the main cause of death in only 14 out of 21 cases. In particular, among the evaluated cases, the main cause of death was viral infection with *Canine parvovirus type 2* (8/21), followed by viral infection with *Canine parvovirus type 2*, and co-infection with *E. coli* (2/21), bacterial co-infection with *Clostridium perfringens* type A and *E. coli* (2/21) and viral and bacterial co-infection with the *Canine distemper virus* and *Pasteurella* spp. (2/21). Finally, in seven out of 21 cases, the microbiological results did not explain the injuries observed during the necropsy. Therefore, the causative agent of infection was considered undetermined after the microbiological examination. Table 3 summarizes the clinical backgrounds, pathological findings, microbiological results and causes of death of the cases of sudden death.

Table 3. Microbiological and anatomopathological findings of the studied dogs.

Group	Sex	Clinical Background	Pathological Findings	Virological Examination	Bacteriological Examination	Cause of Death
1	M	Acute respiratory insufficiency	Visceral congestion, pulmonary edema, hemorrhagic gastroenteritis	*Canine parvovirus type 2a* (wild type)—detected in the lung, liver, heart, brain, and intestine	*Streptococcus dysgalactiae*—isolated in the lung	Viral infection
	M	Acute respiratory insufficiency	Visceral congestion, pulmonary edema, severe broncho-pneumonia, catarrhal enteritis	No viruses detected	*Streptococcus sanguinis*—isolated in the intestine	Undetermined: severe pneumonia due to unexplained causes
	M	Acute respiratory insufficiency	Visceral congestion, pulmonary edema, severe broncho-pneumonia, catarrhal enteritis	No viruses detected	*Streptococcus sanguinis*—isolated in the intestine	Undetermined: severe pneumonia due to unexplained causes
	F	Acute respiratory insufficiency	Visceral congestion, multifocal pulmonary hemorrhages, hemorrhagic enteritis	*Canine parvovirus type 2b and 2c* (wild type)—detected in the lung, liver, spleen, heart, and intestine	*Clostridium perfringens*, Detection of *Clostridium perfringens* alpha toxin—isolated in the intestine	Viral infection
	M	Acute respiratory insufficiency	Lobar pneumonia, pulmonary edema, catarrhal enteritis	*Canine distemper virus*—detected in the lung, liver, and brain	Pasteurella spp.—detected in the lung	Viral and bacterial infection
	F	Acute respiratory insufficiency	Visceral congestion, multifocal pulmonary hemorrhages, hemorrhagic enteritis	*Canine parvovirus type 2b and 2c* (wild type)—detected in the lung, liver, brain, heart, and intestine	*E. coli, Clostridium perfringens*, Detection of *Clostridium perfringens* alpha toxin—isolated in the intestine	Viral infection
	M	Acute respiratory insufficiency	Visceral congestion, pulmonary edema, focal broncho-pneumonia, hemorrhagic gastroenteritis	*Canine distemper virus*—detected in the lung, liver, and brain	*Pasteurella* spp.—detected in the lung	Viral and bacterial co-infection
	M	Acute respiratory insufficiency	Visceral congestion, pulmonary edema, hemorrhagic gastroenteritis	No viruses detected	No bacteria detected	Undetermined
	M	Acute respiratory insufficiency	Visceral congestion, pulmonary edema, severe broncho-pneumonia	No viruses detected	No bacteria detected	Undetermined
	M	Acute respiratory insufficiency	Visceral congestion, hemorrhagic enteritis	*Canine parvovirus type 2a* (wild type)—detected in the heart, spleen, and intestine	No bacteria detected	Viral infection

Table 3. *Cont.*

Group	Sex	Clinical Background	Pathological Findings	Virological Examination	Bacteriological Examination	Cause of Death
	M	Sialorrhea, unilateral eye swelling, muscle stiffness, a single episode of vomiting	Visceral congestion, pulmonary edema, hemorrhagic enteritis	No virus detected	*E. coli; Clostridium perfringens* Detection of *Clostridium perfringens* alpha toxin—detected in the intestine and lung	Bacterial infection
	M	Neurological symptoms	Visceral congestion, bi-lateral pneumonia, pulmonary edema segmental catarrhal enteritis	Rotavirus (detected in the intestine)	*E. coli*—isolated in the intestine	Undetermined: severe pneumonia due to unexplained causes
3	F	Acute gastrointestinal symptoms	Visceral congestion, hemorrhagic enteritis, focal pneumonia	*Canine parvovirus type 2a*, (wild type)—detected in the lung, liver, and intestine; *Adenovirus*—detected in intestine	*Clostridium perfringens* Detection of *Clostridium perfringens* alpha toxin—isolated in the intestine	Viral infection
	F	Vomiting	Visceral congestion, hemorrhagic enteritis, focal pneumonia	*Canine parvovirus type 2a* (wild type)—detected in the lung, liver, heart, and intestine	*E. coli*—isolated in intestine and lung	Viral and bacterial co-infection
	M	Vomiting	Visceral congestion, pulmonary congestion, enteritis, abdominal, thoracic and pericardial effusion, multifocal pulmonary hemorrhage	No virus detected	*E. coli*—isolated in intestine	Undetermined: insufficient findings to explain death
4	F	Acute respiratory insufficiency	Multifocal hemorrhage, abdominal, thoracic and pericardial effusion, hemorrhagic enteritis	No virus detected	*E. coli*—detected in the liver, lung, and intestine *Clostridium perfringens* Detection of *Clostridium perfringens* alpha toxin—detected in the intestine and lung	Bacterial infection
	M	Acute respiratory insufficiency	Pulmonary congestion, segmental catarrhal enteritis	*Canine parvovirus type 2a* (wild type)—detected in lung, liver, and intestine	*E. coli*—isolated in the intestine	Viral infection
5	F	A single episode of diarrhea	Pulmonary congestion, pulmonary edema, segmental hemorrhagic enteritis	*Canine parvovirus type 2a* (wild type)—detected in lung, liver, intestine, and spleen *Adenovirus*—detected in the lung	No bacteria detected	Viral infection

Table 3. *Cont.*

Group	Sex	Clinical Background	Pathological Findings	Virological Examination	Bacteriological Examination	Cause of Death
6	F	Lack of appetite	Thoracic effusion, visceral congestion, multifocal hemorrhage, severe hemorrhagic enteritis	*Canine parvovirus type 2a* (vaccinal) and *2c* (wild type)—detected in lung, liver, intestine, and spleen *Canine distemper virus-detected in lung Adenovirus—detected in lung*	No bacteria detected	Viral infection
	F	Lack of appetite and fever for 12 h	Multifocal hepatic necrosis, hemorrhagic enteritis	*Canine parvovirus type 2b* (wild type)—detected in the lung, liver, brain, and intestine	*E. coli. Clostridium perfringens,* Detection of *Clostridium perfringens* alpha toxin (isolated in lung, liver, and intestine)	Viral and bacterial infection
	F	Single episode of diarrhea	Congestion of the spleen, abdominal effusion	No viruses detected	No bacteria detected	Undetermined: insufficient findings to explain death

4. Discussion

In human forensic pathology, the autopsy for cases of SUID are primarily performed according to the "Kennedy Report" [11]. This protocol and the published international guidelines advocate a multidisciplinary approach to investigations of all cases of SUID, which should be based not only on the findings of the post-mortem macroscopic examination, but also on a broad range of ancillary investigations, such as bacteriological and virological analyses. Although a broad range of tests have been proposed in cases of SUID in human forensic medicine, in the present study, we focused on the pathological and post-mortem microbiology findings in cases of sudden death in young dogs. The results of this study show a low frequency of sudden deaths in young dogs, accounting for 14.48% of the total observed deaths. Furthermore, the Chi-squared test showed a significant difference in the frequencies of ED and SUD among the assessed age groups ($p < 0.05$). In particular, the highest frequency of sudden death was observed in dogs younger than 4 weeks old. In contrast, the highest frequency of ED was observed in animals in Groups 2–5. This difference could be due to the immaturity of the immune system of puppies younger than 6–12 weeks of age [24]. Indeed, the endotheliochorial placentation of this species is relatively impenetrable to the transfer of maternal immunoglobulin [24]. Thus, the immune protection of the puppies during the first weeks of life depends on the ingestion of maternal colostrum antibodies (MCA) [24]. In the absence of the passive transfer of MCA, newborn puppies are only able to develop an immune response to antigens at 2–3 weeks of age. Therefore, any delay in colostrum intake or reduction of colostrum ingestion leads to a reduction in the immune protection of the animals [24–26]. Under these conditions, viruses or bacteria can replicate and spread quickly, leading to the death of the puppies, without the development of characteristic symptoms. Furthermore, congenital malformation or maternal malnutrition could be considered additional causes of sudden death in this age range. In contrast, after 2–3 weeks of age, the immune system of puppies, although immature, is able to develop a mild immune response against pathogens, avoiding the rapid spread of the pathogens and allowing the development of characteristic symptoms of the pathology. In addition, in our study, the most frequently observed gross injuries in cases of sudden death were haemorrhagic gastroenteritis in 12 out of 21 cases, pneumonia in seven out of 21 cases, and catarrhal enteritis in five out of 21 cases. Pulmonary oedema or multiorgan congestion were also observed in 19 out of 21 cases. Intra-abdominal and respiratory lesions have been previously reported in the literature as two important causes of sudden and unexpected infant death in human forensic pathology [27]. Indeed, respiratory tract lesions, although mild, can easily lead to serious complications and sudden death of the subjects [27]. Similarly, acute gastroenteric lesions can cause severe dehydration and serum electrolyte disturbance, which have the potential to cause sudden and unexpected death in children [27]. Overall, in our study, positive microbiological results were observed in 18 out of 21 cases. However, as frequently highlighted in the human literature, the isolation of pathogens in cases of sudden death does not necessarily imply a correlation between those pathogens and the death. In particular, this correlation must be confirmed by the observation of severe and specific injuries during the anatomopathological examination [6]. Therefore, after the review of the necropsy findings, detected pathogens were considered the main cause of death in only 14 out of 21 cases, while in the remaining seven cases, the microbiological results did not explain the injuries observed during the necropsy. The negative findings observed in our study could suggest: (1) a non-infectious cause of death of the assessed animals; or (2) a death due to viruses or other pathogens not detected by the virological panel in use in this study. Indeed, there are a wide range of viruses that are potentially pathogenic in young dogs, including both DNA and RNA viruses. However, our virological panel was limited to the detection of the following five specific viruses: *canine parvovirus, canine coronavirus, canine adenovirus, canine herpesvirus* and *canine distemper virus*. With regard to the positive results, viral infection due to *canine parvovirus type 2* (wild type) was the most common cause of death observed in our study. Overall, *CPV-2* is a causative agent of acute gastroenteritis and myocarditis [17,18]. Furthermore, it is reported in the literature to be one of the most common and important causes of morbidity and mortality in young dogs [17,18]. Usually, the clinical symptoms of the infection are as follows: anorexia, depression, lethargy and fever, followed by vomiting and diarrhea [17,18,28]. However, it is also reported to be a cause of sudden cardiac death in puppies between 4 and 8 weeks of age [17,29,30]. Interestingly, we also observed two cases of sudden death due to *Canine distemper virus* and *Pasteurella* spp. co-infection. *Canine distemper virus* is a member of the genus *Morbillivirus*, which can cause a large variety of disorders in dogs including rhinitis, pneumonia, demyelinating leukoencephalitis, necrotizing bronchiolitis and enteritis [31]. *Canine distemper virus* is not reported in the veterinary literature as a cause of sudden death in animals. However, *Canine distemper virus* causes immunosuppression by targeting cells that express the CD150 protein (signaling lymphocyte activation molecule (SLAM)) [31]. Previous studies showed that this immunosuppression favors secondary infections caused by pathogens, such as *Bordetella bronchiseptica* or *C. piliforme* [32]. Therefore, it is possible to suppose that opportunistic pathogens could complicate a sub-clinical *Canine distemper virus* infection, replicating and spreading quickly, and leading to the death of the puppies without the development of characteristic symptoms. With regards to the bacteriological examination, the most common isolated bacteria were *C. perfringens* type A (6/19) and *E. coli* (6/19). However, they were considered the cause of death in only two cases. Indeed, *C. perfrigens* and *E. coli* are considered normal components of canine intestinal flora [33,34]. Similarly, the alpha toxin gene of *C. perfringens* may be found in asymptomatic dogs as part of the normal intestinal microflora [33]. However, in some cases, *E. coli* can cause pleuro-pneumonitis [35], gastroenteritis [36], urogenital infections, cholangitis, cholangiohepatitis and septicaemia [37] in both humans and other animals. Similarly, *C. perfringens* type A has been associated with gastro-enteric disease, such as haemorrhagic enteritis in dogs and abomasitis in ruminants [38–40]. Furthermore, this bacterium has been reported in the literature as a cause of sudden and unexpected death in dogs [39]. Unfortunately, no specific test for the diagnosis of enteritis due to *C. perfringens* is described in the literature [40]. Thus, generally, the clinical signs, the pathological findings, the microbiological analysis, and the absence of other pathogens must be examined before confirming the diagnosis [40]. In our case, the multisite locations of the pathogens, the absence of other viruses or bacteria and the specific anatomopathological findings of haemorrhagic enteritis supported the diagnosis of enteritis due to *C. perfringens* and *E. coli* as the final cause of death.

Finally, this study allowed the detection of a wide range of pathogens that, after the review of the necropsy and microbiological reports, were not considered the main cause of death of the animals, such as *Canine Adenovirus*, *Rotavirus*, *Streptococcus sanguinis*, *Streptococcus dysgalactiae* and, in some cases, *E. coli* and *C. perfringens*. Therefore, further studies will be needed to evaluate the possible contributions of these pathogens to cases of sudden and unexpected death in young dogs.

5. Conclusions

Sudden death is an uncommon cause of death in young dogs. However, the high frequencies of viruses and bacteria detected in our study highlights the importance of performing complete bacteriological and virological analyses in all cases of sudden death in young dogs. The results of this study suggest that our PCR panel combined with a bacteriological analysis could facilitate the rapid detection and type-specific identification of the pathogenic causes or co-factors of death in most cases of sudden death in young dogs. Finally, these results could provide a valuable epidemiological tool for the investigation of sudden death in young dogs.

Author Contributions: The individual contributions in the present study were as follows: conceptualization and methodology G.P., G.F., L.C.; validation F.A., L.V., V.I.; investigation G.P., L.C.; data curation G.P.; writing—original draft preparation G.P.; project administration G.F.; funding acquisition G.F. All authors have read and agreed to the final version of the manuscript.

Funding: This research did not receive any specific grant from funding agencies in the public, commercial, or not-for-profit sectors.

Conflicts of Interest: The authors declare no conflict of interest.

References

1. Byard, R.W. Sudden death—An overview. In *Sudden Death in Infancy, Childhood, and Adolescence*, 1st ed.; Byard, R.W., Cohle, S.D., Eds.; Cambridge University Press: Cambridge, UK, 1994; pp. 1–4.
2. Kruger, M.M.; Martin, L.J.; Maistry, S.; Heathfield, L.J. A systematic review exploring the relationship between infection and sudden unexpected death between 2000 and 2016: A forensic perspective. *Forensic Sci. Int.* **2018**, *289*, 108–119. [CrossRef]
3. Prtak, L.; Al-Adnani, M.; Fenton, P.; Kudesia, G.; Cohen, M.C. Contribution of bacteriology and virology in sudden unexpected death in infancy. *Arch. Dis. Child.* **2010**, *95*, 371–376.
4. Tajiri, T.; Tate, G.; Miura, K.; Masuda, S.; Ohike, N.; Kunimura, T.; Mitsuya, T.; Morohoshi, T. Sudden death caused by fulminant bacterial infection: Background and pathogenesis of Japanese adult cases. *Intern. Med.* **2008**, *47*, 1499–1504. [CrossRef] [PubMed]
5. Kinney, H.C.; Thach, B.T. The Sudden Infant Death Syndrome. *N. Engl. J. Med.* **2012**, *8*, 795–805.
6. Fernández-Rodríguez, A.; Ballesteros, S.; De Ory, F.; Echevarría, J.E.; Alvarez-Lafuente, R.; Vallejo, G.; Gómez, J. Virological analysis in the diagnosis of sudden children death: A medico-legal approach. *Forensic Sci. Int.* **2006**, *161*, 8–14. [CrossRef]
7. Yagmur, G.; Ziyade, N.; Elgormus, N.; Das, T.; Sahin, M.F.; Yildirim, M.; Ozgun, A.; Akcay, A.; Karayel, F.; Koc, S. Postmortem diagnosis of cytomegalovirus and accompanying other infection agents by real-time PCR in cases of sudden unexpected death in infancy (SUDI). *J. Forensic Leg. Med.* **2016**, *38*, 18–23. [CrossRef]
8. Dettmeyer, R.; Baasner, A.; Schlamann, M.; Padosch, S.A.; Haag, C.; Kandolf, R.; Madea, B. Role of virus-induced myocardial affections in sudden infant death syndrome: A prospective postmortem study. *Pediatr. Res.* **2004**, *55*, 947–952. [CrossRef]
9. Dempers, J.; Ann, S.M.; Wadee, S.A.; Kinney, H.C.; Odendaal, H.J.; Wright, C.A. Progressive Primary Pulmonary Tuberculosis Presenting as the Sudden Unexpected Death in Infancy: A Case Report. *Forensic Sci. Int.* **2011**, *206*, 27–30. [CrossRef]
10. Astrup, B.S.; Johnsen, I.B.G.; Engsbro, A.L. The role of Coxsackievirus A16 in a case of sudden unexplained death in an infant—A SUDI case. *Forensic Sci. Int.* **2016**, *259*, 9–13. [CrossRef]
11. Kennedy, H. *Sudden Unexpected Death in Infancy. A Multi-Agency Protocol for Care and Investigation*, 1st ed.; The Report of a Working Group Convened by the Royal College of Pathologists and the Royal College of Paediatrics and Child Health; The Royal College of Pathologists/The Royal College of Paediatrics and Child Health: London, UK, 2014.
12. Bajanowski, T.; Vege, A.; Byard, R.W.; Krous, H.F.; Arnestad, M.; Bachs, L.; Banner, J.; Blair, P.S.; Borthne, A.; Dettmeyer, R.; et al. Sudden infant death syndrome (SIDS)—Standardized investigations and classification: Recommendations. *Forensic Sci. Int.* **2007**, *165*, 129–143. [CrossRef]
13. Vennemann, M.; Bajanowski, T.; Butterfass-Bahloul, T.; Jorch, G.; Brinkmann, B.; Mitchell, E.A. Do risk factors differ between explained sudden unexpected death in infancy and sudden infant death syndrome? *Arch. Dis. Child.* **2007**, *92*, 133–136. [CrossRef] [PubMed]
14. Parry, N.M. Medicine: Sudden and unexpected death in small animal patients: Part 1. *Companion Anim.* **2008**, *13*, 23–30. [CrossRef]
15. Fleming, J.M.; Creevy, K.E.; Promislow, D.E.L. Mortality in North American Dogs from 1984 to 2004: An Investigation into Age-, Size-, and Breed-Related Causes of Death. *J. Vet. Intern. Med.* **2011**, *25*, 187–198. [CrossRef]
16. Eleni, C.; Panetta, V.; Scholl, F.; Scaramozzino, P. Causes of death in dogs in the province of Rome (Italy). *Vet. Ital.* **2014**, *50*, 137–143. [PubMed]
17. Nandi, S.; Kumar, M. *Canine Parvovirus*: Current Perspective. *Indian J. Virol.* **2010**, *21*, 31–44. [CrossRef]
18. Mylonakis, M.E.; Kalli, I.; Rallis, T.S. Canine parvoviral enteritis: An update on the clinical diagnosis, treatment, and prevention. *Vet. Med.* **2016**, *11*, 91–100. [CrossRef]
19. Olsen, T.F.; Allen, A.L. Causes of sudden and unexpected death in dogs: A 10-year retrospective study. *Can. Vet. J.* **2000**, *41*, 873–875.
20. Kaur, G.; Chandra, M.; Dwivedi, P.N.; Narang, D. Multiplex real-time PCR for identification of canine parvovirus antigenic types. *J. Virol. Methods* **2016**, *233*, 1–5. [CrossRef]
21. Brownlie, H.W.; Munro, R. The Veterinary Forensic Necropsy: A Review of Procedures and Protocols. *Vet. Pathol.* **2016**, *53*, 919–928. [CrossRef]

22. Wilson, S.J.; Wilson, M.L.; Reller, L.B. Diagnostic utility of postmortem blood cultures. *Arch. Pathol. Lab.* **1993**, *117*, 986–988.

23. Morris, J.A.; Harrison, L.M.; Partridge, S.M. Postmortem bacteriology: A re-evaluation. *J. Clin. Pathol.* **2006**, *59*, 1–9. [CrossRef] [PubMed]

24. Day, M.J. Immune System Development in the Dog and Cat. *J. Comp. Path.* **2007**, *137*, 10–15. [CrossRef] [PubMed]

25. Decaro, N.; Buonavoglia, C. *Canine parvovirus*—A review of epidemiological and diagnostic aspects, with emphasis on type 2c. *Vet. Microbiol.* **2012**, *155*, 1–12. [CrossRef] [PubMed]

26. Mila, H.; Grellet, A.; Desario, C.; Feugier, A.; Decaro, N.; Buonavoglia, C.; Chastant-Maillard, S. Protection against *canine parvovirus type 2* infection in puppies by colostrum-derived antibodies. *J. Nutr. Sci.* **2014**, *3*, 54.

27. Byarnt, V.A. Natural Diseases Causing Sudden Death in Infancy and Early Childhood. In *SIDS Sudden Infant and Early Childhood Death: The Past, the Present and the Future*, 1st ed.; Duncan, J.R., Byard, W.R., Eds.; The University of Adelaide: Adelaide, Australia, 2018; pp. 300–312.

28. Stann, S.E.; DiGiacomo, R.F.; Giddens, W.E.; Evermann, J.F. Clinical and pathological features of parvoviral diarrhoea in dogs. *J. Am. Vet. Med. Assoc.* **1984**, *185*, 651–654. [PubMed]

29. Hayes, M.A.; Russell, R.G.; Babiuk, L.A. Sudden death in young dogs with myocarditis caused by *parvovirus*. *J. Am. Vet. Med. Assoc.* **1979**, *174*, 1197–1203. [PubMed]

30. Ridgway, E.J.; Harvey, D. The hospital autopsy: A manual of fundamental autopsy practice. In *The Hospital Autopsy*, 3rd ed.; Burton, J., Rutty, G., Eds.; Hodder Education: London, UK, 2009.

31. Beineke, A.; Puff, C.; Seehusen, F.; Baumgärtner, W. Pathogenesis and immunopathology of systemic and nervous *canine distemper*. *Vet. Immunol. Immunopathol.* **2009**, *127*, 1–18. [CrossRef]

32. Headley, S.A.; Oliveira, T.E.S.; Pereira, A.H.T.; Moreira, J.R.; Michelazzo, M.M.Z.; Pires, B.G.; Marutani, V.H.B.; Xavier, A.A.C.; Di Santis, W.G.; Garcia, J.L.; et al. *Canine morbillivirus (canine distemper virus)* with concomitant *canine adenovirus, canine parvovirus-2*, and *Neospora caninum* in puppies: A retrospective immunohistochemical study. *Sci. Rep.* **2018**, *8*, 13477. [CrossRef]

33. Goldstein, M.R.; Kruth, S.A.; Bersenas, A.M.E.; Holowaychuk, M.K.; Weese, J.S. Detection and characterization of *Clostridium perfringens* in the feces of healthy and diarrheic dogs. *Can. J. Vet. Res.* **2012**, *76*, 161–165.

34. Bettelheim, K.A.; Goldwater, P.N. *Escherichia coli* and Sudden Infant Death Syndrome. *Front. Immunol.* **2015**, *6*, 343. [CrossRef]

35. Handt, L.K.; Stoffregen, D.A.; Prescott, J.S.; Pouch, W.J.; Ngai, D.T.; Anderson, C.A.; Gatto, N.T.; DebRoy, C.; Fairbrother, J.M.; Motzel, S.L.; et al. Clinical and microbiologic characterization of hemorrhagic pneumonia due to extraintestinal pathogenic *Escherichia coli* in four young dogs. *Comp. Med.* **2003**, *53*, 663–670. [PubMed]

36. Astrid, B.K.; Anthony, P.C.; Casey, M. Gaunt Enteropathogenic *Escherichia coli* (EPEC) infection in association with acute gastroenteritis in 7 dogs from Saskatchewan. *Can. Vet. J.* **2016**, *57*, 964–968.

37. Beutin, L. *Escherichia coli* as a pathogen in dogs and cats. *Vet. Res.* **1999**, *30*, 285–298. [PubMed]

38. Sasaki, J.; Goryo, M.; Asahina, M.; Makara, M.; Shishido, S.; Okada, K. Hemorrhagic enteritis associated with *Clostridium perfringens type A* in a dog. *J. Vet. Med. Sci.* **1999**, *61*, 175–177. [CrossRef]

39. Schlegel, J.B.; Dreumel, T.D.; Slavić, D.; Prescott, J.F. *Clostridium perfringens type A* fatal acute hemorrhagic gastroenteritis in a dog. *Can. Vet. J.* **2012**, *53*, 555–557.

40. Marks, S.L.; Kather, E.J. *Clostridium perfringens*- and *Clostridium difficile*-associated diarrhea. In *Infectious Diseases of the Dog and Cat*, 3rd ed.; Greene, C.E., Ed.; Saunders Elsevier: St Louis, MO, USA, 2016; pp. 363–366.

 *animals*

Brief Report

Microbiome Shift, Diversity, and Overabundance of Opportunistic Pathogens in Bovine Digital Dermatitis Revealed by 16S rRNA Amplicon Sequencing

Hector M. Espiritu [1], Lovelia L. Mamuad [1], Seon-ho Kim [1], Su-jeong Jin [1], Sang-suk Lee [1], Seok-won Kwon [2] and Yong-il Cho [1,*]

[1]   Department of Animal Science and Technology, Sunchon National University, Suncheon, Jeonnam 57922, Korea; 1193001@s.scnu.ac.kr (H.M.E.); loveliamamuad2306@gmail.com (L.L.M.); mhs0425@daum.net (S.-h.K.); herodias@scnu.ac.kr (S.-j.J.); rumen@scnu.ac.kr (S.-s.L.)

[2]   Woosarang Animal Hospital, Yongin, Gyeonggi 17026, Korea; gastra00@naver.com

*   Correspondence: ycho@scnu.ac.kr; Tel.: +82-61-750-3234

Received: 27 August 2020; Accepted: 30 September 2020; Published: 3 October 2020

Simple Summary: Bovine digital dermatitis (BDD) is a foot infection known as the primary cause of lameness in cattle due to painful lesions, posing serious impacts on the productivity and welfare of affected animals. Members of the bacterial group *Treponema* have long been considered as the main causative agents because previous investigations by bacterial isolation, tissue analyses, and high molecular sequencing have persistently identified this group in BDD. However, other studies indicated that the presence of several bacteria on the lesion due to the slurry environment the cattle foot are exposed to, suggests an interdependent polybacterial nature which could also play a role in disease development and progression. Therefore, we analyzed the diversity and relationship of the diverse microbiome in BDD lesions compared to normal skin from cattle foot by using next-generation high throughput sequencing. Based on the results obtained, we concluded that the shift in microbial composition which leads to richer diversity in BDD, and the overabundance of opportunistic bacterial pathogens could be associated with BDD pathogenesis.

Abstract: This study analyzed the diversity and phylogenetic relationship of the microbiome of bovine digital dermatitis (BDD) lesions and normal skin from cattle foot by using 16S rRNA amplicon sequencing. Three BDD samples and a normal skin sample were pre-assessed for analysis. The Illumina Miseq platform was used for sequencing and sequences were assembled and were categorized to operational taxonomic units (OTUs) based on similarity, then the core microbiome was visualized. The phylogeny was inferred using MEGA7 (Molecular evolutionary genetics analysis version 7.0). A total of 129 and 185 OTUs were uniquely observed in normal and in BDD samples, respectively. Of the 47 shared OTUs, 15 species presented increased abundance in BDD. In BDD and normal samples, Spirochetes and Proteobacteria showed the most abundant phyla, respectively, suggesting the close association of observed species in each sample group. The phylogeny revealed the evolutionary relationship of OTUs and the Euclidean distance suggested a high sequence divergence between OTUs. We concluded that a shift in the microbiome leads to richer diversity in BDD lesions, and the overabundance of opportunistic pathogens and its synergistic relationship with commensal bacteria could serve as factors in disease development. The influence of these factors should be thoroughly investigated in future studies to provide deeper insights on the pathogenesis of BDD.

Keywords: bovine digital dermatitis; cattle lameness; microbiome; *Treponema* spp

1. Introduction

Bovine digital dermatitis (BDD) is known as the most important foot infection causing lameness in cattle [1]. This severe lameness, which is the primary clinical manifestation of BDD caused by painful hyperkeratotic lesions [2], poses serious concerns on the welfare of the affected animals [3]. Serious economic impacts have also been implicated with BDD due to significant milk production losses, poor reproductive performance [4,5], and most extremely, premature culling of the affected animals [6].

The first reported case of BDD was in Italy in 1974 by Chelli and Mortellaro [7], and since then, it has been globally reported reaching an endemic state in many countries [8]. Etiological investigations have identified a variety of bacteria in BDD lesions [9], but advancements in sequencing technology have provided essential information on the identity of associated causal agents [8]. In previous investigations, Spirochetes are considered to be the major pathogen in BDD [1,10]. However, studies indicated that the presence of several bacteria on the lesion due to the slurry environment the cattle foot are exposed to [2], suggests a synergistic polybacterial nature [11] which plays a role in disease development and progression [12].

Many reports have attempted to elucidate the association of the microbiome of BDD towards disease development [11,13–15]. Surprisingly, for more than 20 years since it emerged as a major problem in the cattle industry in certain countries, studies on BDD have just recently started in Korea. In a recent report, *Treponema* spp. was established as the dominant pathogen in BDD lesions in Korea [16], but an emphasis on the plethora of bacteria that was observed is still lacking. Although the majority of studies considered that the main pathogen of BDD is *Treponema* spp., its polybacterial nature must also be considered in order to understand its pathogenesis [15]. Therefore, this study analyzed the diversity and phylogenetic relationship of the microbiome of BDD lesions and normal skin from the interdigital space of the cattle using 16S rRNA amplicon sequencing.

2. Materials and Methods

2.1. Sample Collection

This study was performed following general ethical principles and with the consent of farm owners. The collection of samples was done in Holstein–Friesian cattle in housed dairy farm, and was conducted during hoof cleaning, trimming and treatment performed by a professional veterinarian with years of expertise in the field. Inspection was done by anatomic pathological observation of grossly visible active BDD lesions on the proximal border of interdigital space characterized by the presence of ulceration, with hyperkeratosis and proliferative growth with hair-like projections as described by Zinicola et al. (2015) [17]. For normal skin sample, healthy animals with no sign of lameness and no history of BDD were inspected. After thorough cleaning of the foot surface, lidocaine (2%) was subcutaneously injected around the lesion and a 5-mm punch biopsy was taken from the center of the lesion for BDD, while for normal skin, biopsy sample was taken where BDD most often take place. Samples were washed thoroughly with buffered phosphate saline (pH 7.4) and delivered to the laboratory with ice. A total of 66 pre-assessed active BDD samples were subjected to detection of *Treponema* spp. by genus-specific PCR as described in the results of our previous study [16]. Since 100% of the active BDD lesions are positive in PCR, three samples were randomly selected for metagenomics sequencing. A normal sample was provided for BDD negative control. Samples were categorized as Group A for normal skin sample, and Group B for BDD-infected samples.

2.2. Metagenomics Sequencing and Diversity Analysis

The four randomly pre-assessed samples were submitted to Macrogen (Korea) for high-throughput sequencing. DNA samples extracted from lesion and normal sample biopsies were subjected to quality control by Picogreen method before library construction. By targeting the V3-V4 region of the 16S rRNA gene, libraries were constructed and were purified. Sequencing was carried out using

Illumina Miseq platform (Illumina, San Diego, CA, USA). The base call binary data produced by Real-Time Analysis (RTA) were converted to FASTQ files by bcl2fastq package (Illumina, San Diego, CA, USA) and were filtered using Scythe (v0.994) (https://github.com/vsbuffalo/scythe) and Sickle (https://github.com/najoshi/sickle) programs to remove adapter sequences. The obtained 16S rRNA sequences were binned into Operational Taxonomic Units (OTUs) based on 97% identity using Quantitative Insights Into Microbial Ecology (QIIME) [18]. The microbiome was visualized using Metagenomics Core Microbiome Exploration Tool (MetaCoMET) [19] using a Biological Observation Matrix format (BIOM) [20] generated using Mothur [21].

2.3. Divergence and Phylogenetic Analysis

The phylogeny of the microbiome was inferred by aligning the obtained 16S rRNA sequences of the operational taxonomic units identified using ClustalW [22], and the Newick tree data were obtained from MEGA7 [23] by maximum likelihood method following the general time-reversible model as the fit model for calculating the rate of nucleotide base substitution. The final dendrogram was constructed using Iroki [24]. The divergence distance between OTUs was calculated by Euclidian distance method based on the rate of nucleotide base substitution between OTUs, and presented as a heatmap matrix generated by Heatmapper [25] for the aligned sequences of OTUs using the Euclidean distance method.

3. Results

In total, 66 samples pre-assessed from our previous research [16] were used in this study. From these, one normal skin sample (Group A) and three BDD lesions (Group B) were subjected for analysis.

The Chao1 diversity index shows that Group B has richer species diversity compared to Group A (Figure 1A). The size of the microbiome presented in Figure 1B shows a higher observed OTUs in BDD lesions compared to the normal skin sample. Out of 267 OTUs observed, 138 were uniquely observed in Group B, 82 in Group A, while 47 were overlapping in both groups. Fifteen of these shared OTUs increased their abundance from normal to BDD and includes T. pedis (20.93%), a group of unclassified species (12.4%), Treponema denticola (9.8%), T. medium (6.48%), Porphyromonas levii (1.56%), P. somerae (1.22%), and Acholeplasma vituli (0.89%). OTUs absent in Group A with increased abundance ratio in Group B were as follows in decreasing order: Carboxylicivirga mesophila (5.89%), T. lecithinolyticum (5.35%), A. morum (5.03%), Spirochaeta africana (3.74%), Mycoplasma feliminutum (3.65%), A. modicum (2.77%), Pelobacter propionicus (2.04%), H. sueciensis (1.64%), P. uenonis (1.30%), Falcatimonas natans (1.25%), Devosia confluentis (1.17), Christensenella minuta (1.08%), and M. fermentans (1.30%). Additional information can be accessed in Supplementary Table S1. The shift in abundance ratio of each species from normal to BDD (Figure 1C) shows the drop in abundance of dominant bacteria, Psychrobacter fulvigenes and Pseudomonas caeni from the normal sample, and the increased abundance of Treponema spp., and other bacteria in BDD. In addition, the Euclidian distance heat-map based on OTU abundance between samples and representative phyla (Figure 1D) illustrated the high association of Spirochetes with BDD. Moraxellaceae and Pseudomonadaceae, both under phylum Proteobacteria presented a close relationship with normal tissue. Moreover, Bacteroidetes and Firmicutes are considerably associated in both groups, while Tenericutes is associated with BDD only.

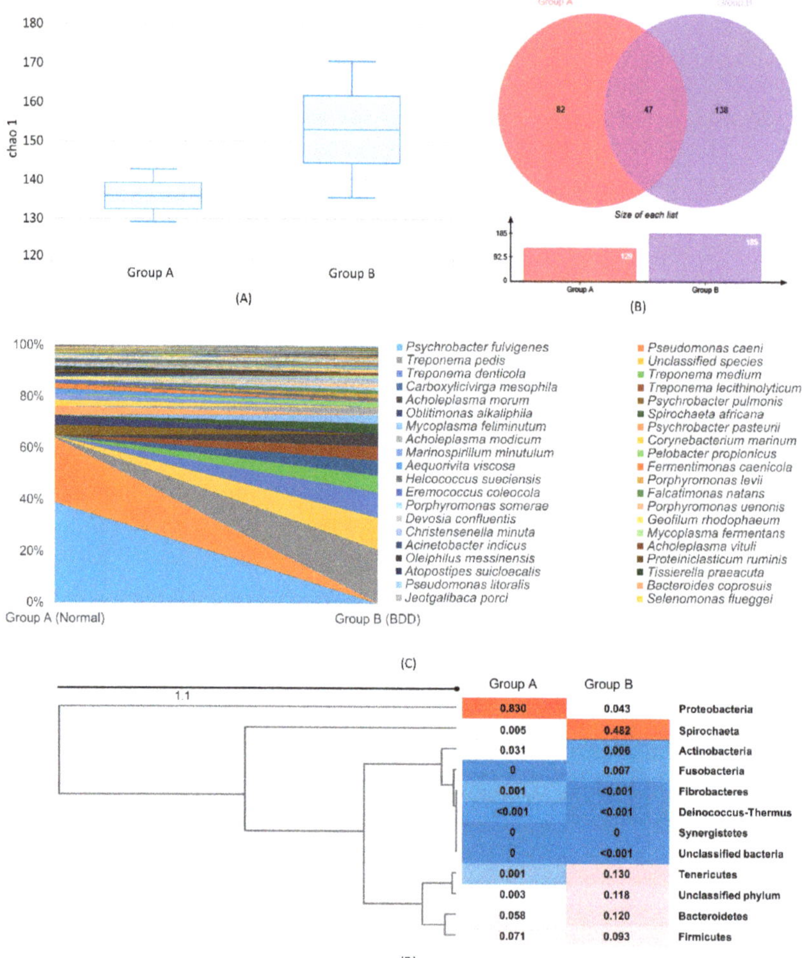

Figure 1. (**A**) Box plot of Chao1 diversity index for bovine digital dermatitis (BDD) (Group A) and normal (Group B), (**B**) the microbiome presenting the number of operational taxonomic units (OTUs) in Group A and Group B showing the unique and shared OTUs between groups, and (**C**) relative abundance of operational taxonomic units from Group A and Group B (legend: 22 most abundant). (**D**) Euclidian distance heatmap presenting the association between phylum and sample groups. Red denotes closer association while blue denotes a lesser association with normal sample and BDD.

The maximum likelihood tree constructed using the general time-reversible model illustrated the phylogenetic relationship of all the OTUs (Figure 2A). The tree showed the phylum classification of the OTUs, and the abundance ratio and OTU count (Figure 2B). Firmicutes was the most diverse phylum representing 41.85% and 37.21% of the observed OTUs in BDD and normal skin, respectively. Spirochetes (which is the most abundant) only represented 3.75% of the over-all observed OTUs. In addition, the Euclidean distance heat map matrix showed the estimates of the evolutionary divergence between OTUs in lesions and normal tissue (Figure 2C), and the frequency distribution of the computed distance of the pairwise comparisons was graphed in Figure 2D. Overall, 63.89% of the pairwise comparisons were above the median distance.

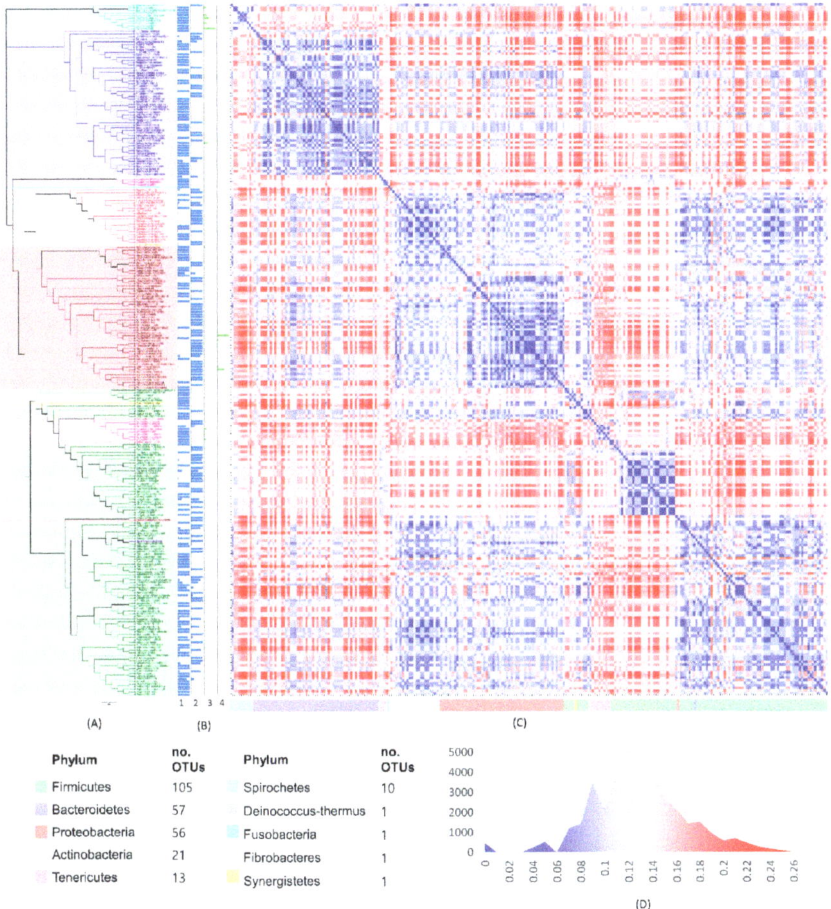

Figure 2. General time-reversible tree representing all observed OTUs in BDD-infected and normal skin samples color-coded based on phylum classification (**A**). The scale value is 2.57. The blue bar graph denotes the abundance ratio of each species on BDD (**B1**) and normal sample (**B2**), while the green bar graph shows the actual count per species (OTU) for BDD (**B3**) and normal (**B4**) sample. Euclidean distance heat map matrix is based on the nucleotide base substitution rate presenting the divergence between all OTUs in BDD and normal sample (**C**). The graph (**D**) shows the frequency distribution of pairwise comparison between OTUs in each calculated distance from low (blue) to high (red).

4. Discussion

This study analyzed the diversity and phylogenetic relationship of the microbiome of BDD lesions compared to normal skin from the interdigital space of the cattle. From our previous study [16], we elucidated the relative abundance of Spirochetes, specifically *Treponema* spp. in BDD lesions and concluded that *Treponema* spp., was the dominant pathogen involved in BDD in Korea. Although the major abundance of *Treponema* spp. in lesions based on several investigations suggest that BDD is polytreponemal, the presence of other bacteria implies that they may also play certain roles in its pathogenesis.

The data analyzed exhibited a change in the microbiome in BDD lesions from the normal skin sample. This altered microbiome has been previously observed by Krull et al., (2014) [13] and Zinicola et al., (2015) [11] by investigating the microbiome of each stage and layer of BDD lesions, respectively. They observed the replacement of species from the normal sample by other species along with the disease progression and the alteration of the microbiome in each layer of lesions. In the current data, the 82 OTUs from the normal sample were replaced in BDD-infected sample by 138 OTUs. This alteration also led to a richer diversity in BDD as opposed to the findings of Krull et al., (2014) where disease progression leads to a drop in diversity [13]. Although samples subjected in our study is categorized only as active BDD, it is recommended to use several normal samples and BDD samples categorized in varying lesion stages to gain more conclusive data. Another interesting finding of this study is the shared OTUs between normal and BDD found in the overlap comprised of three *Treponema* spp. (*T. pedis*, *T. medium*, *T. denticola*), along with two *Phorphyromonas* spp. as the five most abundant bacteria in BDD. The increased abundance of these species from normal to BDD, along with others could imply that these bacteria are commensals which later progresses as opportunistic pathological agents which initiates disease development when triggered under favorable conditions.

The abundance data previously presented by Mamuad et al., (2020) [16], to some extent, agree with the reports of Moreira et al., (2018) [14], Zinicola et al., (2015) [11], Beninger et al., (2016) [26], Krull et al., (2014) [13], and Nielsen et al., (2016) [12], that BDD is polybacterial, with *Treponema* spp., as the most abundant genus. However, the dominance and diversity of species under this genus still varies between reports from different geographical locations. Moreira et al., (2018) reported that *Treponema pedis* is the most abundant species in BDD in Brazil, which is in accordance with Mamuad et al., (2020). However, in one study in the USA, *T. denticola* and *T. phagedenis* were the most abundant species in active and inactive lesions, respectively [11]. Additionally, the results of investigation of Beninger et al., (2015) [26] and Krull et al., (2014) [13] were in agreement with each other that the most abundant species in BDD is *T. phagedenis*, which was absent in all samples in our data. This suggests that BDD microbiome has geographical and/or sample-to-sample variation.

Given that certain members of genus *Treponema* are recognized as the abundant bacteria, and considered as major contributors in the development and progression of the disease, other bacterial genera have also been reported in BDD from other countries as also in this study such as *Porphyromonas* [14], *Campylobacter*, *Acholeplasma* [11], *Peptoniphilus* and *Romboutsia* [27]. In this study, a major abundance of *Carboxylicivirga mesophila* and *Treponema lecithinolyticum* was observed as the fourth and fifth most abundant identified bacteria in BDD, respectively, which were not observed from previous studies. Further investigation is required to verify if these bacteria have actual involvement with the disease, since *C. mesophila* was first isolated from tidal flat sediment in Korea [28], while *T. lecithinolyticum* was commonly found in human oral microbiome [29].

The distance heat map between phyla and samples based on OTU count and taxonomy showed the close relationship of the observed species under phylum Spirochetes and Proteobacteria in BDD and normal skin, respectively. Phylum Bacteroidetes and Firmicutes can be associated in both groups, while Tenericutes is associated only with BDD which agrees with Nielsen et al., (2016) [12]. In a study by Bay et al., (2018) Firmicutes was the most abundant phylum in other polybacterial foot infections in bovine models such as interdigital hyperplasia, interdigital phlegmon, sole ulcer, toe necrosis, and white line disease, with Spirochetes being the fifth most abundant phylum overall, after Bacteroidetes, Actinobacteria, and Proteobacteria [27]. This suggests that there are variations on the major pathogens between these diseases.

The GTR tree shows the phylogenetic relationship of all OTUs present in both BDD-infected and normal skin samples, classified under 10 phyla, with Firmicutes representing the highest diversity. This large number of Firmicutes was supported by the findings of Yano et al., (2010) [30] and Santos et al., (2012) [31] in both normal and BDD-infected samples, regardless of its relative abundance. In this study, Spirochetes with the highest abundance in BDD have lower diversity both before and during infection. Compared with Firmicutes with highest diversity in normal and BDD, there was no increase in its abundance before and during infection. This suggests that diversity richness of a certain phylum may be irrelative with the abundance or its pathogenic involvement in BDD. The microbiome of BDD in this study was verified to be highly diverse, thus, we hypothesize that synergism between overabundant opportunistic pathogens and the diversity of commensals makes the disease more complicated. The pairwise distances of OTUs based on the computed rate of nucleotide substitution show higher frequencies for above-median distances between species found in all samples. This supports the idea that bacterial diversity is evolutionarily diverged.

5. Conclusions

We concluded that a shift in the microbiome leads to richer diversity in BDD lesions, and the overabundance of opportunistic pathogens and its possible synergistic relationship between less abundant commensal bacteria could serve as factors in disease development and progression. Spirochetes is the most abundant phylum associated with BDD in other previous studies, and in this study, we deliberated that the abundance of species on each of the observed phylum varied between reports, suggesting either geographical or sample-to-sample variation. The influence of the overabundance of opportunistic species and the synergistic interaction of the plethora of commensal bacteria should be thoroughly investigated in future studies by including additional samples categorized to varying degree of the severity of infection, not just in BDD, but also in other lameness-related foot diseases, to provide deeper insights on the pathogenesis and microbiome relationship of these debilitating diseases.

Supplementary Materials: The following are available online at http://www.mdpi.com/2076-2615/10/10/1798/s1, Table S1: Abundance ratio of operational taxonomic units in normal and BDD samples.

Author Contributions: Conceptualization, Y.-i.C.; methodology, H.M.E., L.L.M., S.-j.J., S.-h.K., and S.-w.K. software, H.M.E. and L.L.M.; validation, Y.-i.C.; formal analysis, H.M.E. and L.L.M.; investigation, H.M.E. and L.L.M.; resources, Y.-i.C. and S.-w.K.; data curation, H.M.E. and L.L.M.; writing—original draft preparation, H.M.E. and L.L.M.; writing—review and editing, Y.-i.C., and S.-s.L.; visualization, H.M.E. and L.L.M.; supervision, Y.-i.C., and S.-s.L.; project administration, S.-j.J.; funding acquisition, Y.-i.C. All authors have read and agreed to the published version of the manuscript.

Funding: This study was supported by the National Research Foundation of Korea (NRF-2018R1C1B6004589).

Conflicts of Interest: The authors declare no conflict of interest.

References

1. Evans, N.J.; Murray, R.D.; Carter, S.D. Bovine digital dermatitis: Current concepts from laboratory to farm. *Vet. J.* **2016**, *211*, 3–13. [CrossRef]
2. Demirkan, I.; Evans, N.J.; Singh, P.; Brown, J.M.; Getty, B.; Carter, S.D.; Timofte, D.; Hart, C.A.; Vink, W.D.; Birtles, R.J.; et al. Association of unique, isolated treponemes with bovine digital dermatitis lesions. *J. Clin. Microbiol.* **2009**, *47*, 689–696. [CrossRef]
3. Klitgaard, K.; Nielsen, M.W.; Ingerslev, H.C.; Boye, M.; Jensen, T.K. Discovery of bovine digital dermatitis-associated *Treponema* spp. in the dairy herd environment by a targeted deep-sequencing approach. *Appl. Environ. Microbiol.* **2014**, *80*, 4427–4432. [CrossRef] [PubMed]
4. De Jesús Argáez-Rodríguez, F.; Hird, D.W.; Hernández De Anda, J.; Read, D.H.; Rodríguez-Lainz, A. Papillomatous digital dermatitis on a commercial dairy farm in Mexicali, Mexico: Incidence and effect on reproduction and milk production. *Prev. Vet. Med.* **1997**, *32*, 275–286. [CrossRef]
5. Read, D.H.; Walker, R.L. Papillomatous digital dermatitis (footwarts) in California dairy cattle: Clinical and gross pathologic findings. *J. Vet. Diagn. Investig.* **1998**, *10*, 67–76. [CrossRef]
6. Elliott, M.K.; Alt, D.P.; Zuerner, R.L. Lesion formation and antibody response induced by papillomatous digital dermatitis-associated spirochetes in a murine abscess model. *Infect. Immun.* **2007**, *75*, 4400–4408. [CrossRef] [PubMed]
7. Cheli, R.; Mortellaro, C. Digital dermatitis in cattle. In Proceedings of the 8th International Conference Diseases of Cattle, Milan, Italy, 9–13 September 1974; pp. 208–213.

8. Palmer, M.A.; O'Connell, N.E. Digital dermatitis in dairy cows: A review of risk factors and potential sources of between-animal variation in susceptibility. *Animals* **2015**, *5*, 512–535. [CrossRef]

9. Nally, J.E.; Hornsby, R.L.; Alt, D.P.; Whitelegge, J.P. Phenotypic and proteomic characterization of treponemes associated with bovine digital dermatitis. *Vet. Microbiol.* **2019**, *235*, 35–42. [CrossRef]

10. Wilson-Welder, J.H.; Alt, D.P.; Nally, J.E. Digital dermatitis in cattle: Current bacterial and immunological findings. *Animals* **2015**, *5*, 1114–1135. [CrossRef]

11. Zinicola, M.; Lima, F.; Lima, S.; Machado, V.; Gomez, M. Altered microbiomes in bovine digital dermatitis lesions, and the gut as a pathogen reservoir. *PLoS ONE* **2015**, 1–23. [CrossRef]

12. Nielsen, M.W.; Strube, M.L.; Isbrand, A.; Al-Medrasi, W.D.H.M.; Boye, M.; Jensen, T.K.; Klitgaard, K. Potential bacterial core species associated with digital dermatitis in cattle herds identified by molecular profiling of interdigital skin samples. *Vet. Microbiol.* **2016**, *186*, 139–149. [CrossRef] [PubMed]

13. Krull, A.C.; Shearer, J.K.; Gorden, P.J.; Cooper, V.L.; Phillips, G.J.; Plummera, P.J. Deep sequencing analysis reveals temporal microbiota changes associated with development of bovine digital dermatitis. *Infect. Immun.* **2014**, *82*, 3359–3373. [CrossRef] [PubMed]

14. Moreira, T.F.; Facury Filho, E.J.; Carvalho, A.U.; Strube, M.L.; Nielsen, M.W.; Klitgaard, K.; Jensen, T.K. Pathology and bacteria related to digital dermatitis in dairy cattle in all year round grazing system in Brazil. *PLoS ONE* **2018**, *13*, 1–15. [CrossRef] [PubMed]

15. Hesseling, J.; Legione, A.R.; Stevenson, M.A.; McCowan, C.I.; Pyman, M.F.; Finochio, C.; Nguyen, D.; Roic, C.L.; Thiris, O.L.; Zhang, A.J.; et al. Bovine digital dermatitis in Victoria, Australia. *Aust. Vet. J.* **2019**, *97*, 404–413. [CrossRef]

16. Mamuad, L.L.; Joo, B.; Al, S.; Espiritu, H.M.; Jeong, S.; Kim, W.; Lee, S.; Cho, Y. *Treponema* spp., the dominant pathogen in the lesion of bovine digital dermatitis and its characterization in dairy cattle. *Vet. Microbiol.* **2020**, *245*, 108696. [CrossRef]

17. Zinicola, M.; Higgins, H.; Lima, S.; Machado, V.; Guard, C.; Bicalho, R. Shotgun metagenomic sequencing reveals functional genes and microbiome associated with bovine digital dermatitis. *PLoS ONE* **2015**, *10*, 1–17. [CrossRef]

18. Caporaso, J.G.; Kuczynski, J.; Stombaugh, J.; Bittinger, K.; Bushman, F.D.; Costello, E.K.; Fierer, N.; Peña, A.G.; Goodrich, J.K.; Gordon, J.I.; et al. QIIME allows analysis of high- throughput community sequencing data intensity normalization improves color calling in SOLiD sequencing. *Nat. Methods* **2010**, *7*, 335–336. [CrossRef]

19. Wang, Y.; Xu, L.; Gu, Y.Q.; Coleman-Derr, D. MetaCoMET: A web platform for discovery and visualization of the core microbiome. *Bioinformatics* **2016**, *32*, 3469–3470. [CrossRef]

20. McDonald, D.; Clemente, J.C.; Kuczynski, J.; Rideout, J.R.; Stombaugh, J.; Wendel, D.; Wilke, A.; Huse, S.; Hufnagle, J.; Meyer, F.; et al. The Biological Observation Matrix (BIOM) format or: How I learned to stop worrying and love the ome-ome. *Gigascience* **2012**, *464*, 1–6. [CrossRef]

21. Schloss, P.D.; Westcott, S.L.; Ryabin, T.; Hall, J.R.; Hartmann, M.; Hollister, E.B.; Lesniewski, R.A.; Oakley, B.B.; Parks, D.H.; Robinson, C.J.; et al. Introducing mothur: Open-source, platform-independent, community-supported software for describing and comparing microbial communities. *Appl. Environ. Microbiol.* **2009**, *75*, 7537–7541. [CrossRef]

22. Thompson, J.D.; Higgins, D.G.; Gibson, T.J. CLUSTAL W: Improving the sensitivity of progressive multiple sequence alignment through sequence weighting, position-specific gap penalties and weight matrix choice. *Nucl. Acids Res.* **1994**, *22*, 4673–4680. [CrossRef] [PubMed]

23. Kumar, S.; Stecher, G.; Tamura, K. MEGA7: Molecular evolutionary genetics analysis version 7.0 for bigger datasets. *Mol. Biol. Evol.* **2016**, *33*, 1870–1874. [CrossRef] [PubMed]

24. Moore, R.M.; Harrison, A.O.; McAllister, S.M.; Polson, S.W.; Wommack, K.E. Iroki: Automatic customization and visualization of phylogenetic trees. *PeerJ* **2020**, *8*, e8584. [CrossRef]

25. Babicki, S.; Arndt, D.; Marcu, A.; Liang, Y.; Grant, J.R.; Maciejewski, A.; Wishart, D.S. Heatmapper: Web-enabled heat mapping for all. *Nucl. Acids Res.* **2016**, *44*, W147–W153. [CrossRef] [PubMed]

26. Beninger, C.; Naqvi, S.A.; Naushad, S.; Orsel, K.; Luby, C.; Derakhshani, H.; Khafipour, E.; De Buck, J. Associations between digital dermatitis lesion grades in dairy cattle and the quantities of four *Treponema* species. *Vet. Res.* **2018**, *49*, 10–12. [CrossRef] [PubMed]

27. Bay, V.; Griffiths, B.; Carter, S.; Evans, N.J.; Lenzi, L.; Bicalho, R.C.; Oikonomou, G. 16S rRNA amplicon sequencing reveals a polymicrobial nature of complicated claw horn disruption lesions and interdigital phlegmon in dairy cattle. *Sci. Rep.* **2018**, *8*, 1–12. [CrossRef]

28. Yang, S.H.; Seo, H.S.; Woo, J.H.; Oh, H.M.; Jang, H.; Lee, J.H.; Kim, S.J.; Kwon, K.K. *Carboxylicivirga* gen. nov. in the family Marinilabiliaceae with two novel species, *Carboxylicivirga mesophila* sp. nov. and *Carboxylicivirga taeanensis* sp. nov., and reclassification of *Cytophaga fermentans* as *Saccharicrinis fermentans* gen. nov., comb. nov. *Int. J. Syst. Evol. Microbiol.* **2014**, *64*, 1351–1358. [CrossRef]

29. Belkacemi, S.; Khalil, J.B.; Ominami, Y.; Hisada, A.; Fontanini, A.; Caputo, A.; Levasseur, A.; Lagier, J.C.; Khelaifia, S.; Raoult, D. Passive filtration, rapid scanning electron microscopy, and matrix-assisted laser desorption ionization–time of flight mass spectrometry for *Treponema* culture and identification from the oral cavity. *J. Clin. Microbiol.* **2019**, *57*. [CrossRef]

30. Yano, T.; Moe, K.K.; Yamazaki, K.; Ooka, T.; Hayashi, T.; Misawa, N. Identification of candidate pathogens of papillomatous digital dermatitis in dairy cattle from quantitative 16S rRNA clonal analysis. *Vet. Microbiol.* **2010**, *143*, 352–362. [CrossRef]

31. Santos, T.M.A.; Pereira, R.V.; Caixeta, L.S.; Guard, C.L.; Bicalho, R.C. Microbial diversity in bovine Papillomatous digital dermatitis in Holstein dairy cows from upstate New York. *FEMS Microbiol. Ecol.* **2012**, *79*, 518–529. [CrossRef]

Article

Molecular Detection and Phylogeny of *Anaplasma phagocytophilum* and Related Variants in Small Ruminants from Turkey

Münir Aktaş *, Sezayi Özübek and Mehmet Can Uluçeşme

Department of Parasitology, Faculty of Veterinary Medicine, University of Firat, 23119 Elazig, Turkey;
sozubek@firat.edu.tr (S.Ö.); mculucesme@firat.edu.tr (M.C.U.)
* Correspondence: maktas@firat.edu.tr; Tel.: +90-(424)-237-0000

Simple Summary: We explored the existence of *Anaplasma phagocytophilum* and related variant in samples of goats and sheep obtained from Antalya and Mersin provinces, representative of Mediterranean region of Turkey. Based on *16S rRNA* and *groEL* genes of *A. phagocytophilum* and related variants, we examined blood samples by polymerase chain reaction (PCR) followed by sequencing. The results showed that the prevalence of *A. phagocytophilum* and *A. phagocytophilum*-like 1 infection was 1.4% and 26.5%, respectively. Sequencing confirmed molecular data and showed the presence of *A. phagocytophilum* and *A. phagocytophilum*-like-1 variant in the sampled animals.

Abstract: *Anaplasma phagocytophilum* causes tick-borne fever in small ruminants. Recently, novel *Anaplasma* variants related to *A. phagocytophilum* have been reported in ruminants from Tunisia, Italy, South Korea, Japan, and China. Based on *16S rRNA* and *groEL* genes and sequencing, we screened the frequency of *A. phagocytophilum* and related variants in 433 apparently healthy small ruminants in Turkey. *Anaplasma* spp. overall infection rates were 27.9% (121/433 analyzed samples). The frequency of *A. phagocytophilum* and *A. phagocytophilum*-like 1 infections was 1.4% and 26.5%, respectively. No *A. phagocytophilum*-like 2 was detected in the tested animals. The prevalence of *Anaplasma* spp. was comparable in species, and no significant difference was detected between sheep and goats, whereas the prevalence significantly increased with tick infestation. Sequencing confirmed PCR-RFLP data and showed the presence of *A. phagocytophilum* and *A. phagocytophilum*-like-1 variant in the sampled animals. Phylogeny-based on *16S rRNA* gene revealed the *A. phagocytophilum*-like 1 in a separate clade together with the previous isolates detected in small ruminants and ticks. In this work, *A. phagocytophilum*-like 1 has been detected for the first time in sheep and goats from Turkey. This finding revealed that the variant should be considered in the diagnosis of caprine and ovine anaplasmosis.

Keywords: tick-borne fever; *Anaplasma phagocytophilum*-like 1; PCR-RFLP; small ruminant

Citation: Aktaş, M.; Özübek, S.; Uluçeşme, M.C. Molecular Detection and Phylogeny of *Anaplasma phagocytophilum* and Related Variants in Small Ruminants from Turkey. *Animals* **2021**, *11*, 814. https://doi.org/10.3390/ani11030814

Academic Editor: Robert W. Li

Received: 25 February 2021
Accepted: 11 March 2021
Published: 14 March 2021

1. Introduction

Anaplasma phagocytophilum is the agent of tick-borne fever (TBF) or pasture fever, a disease affecting some species of domestic ruminants (cattle, sheep, goats). The bacterium is a pathogenic species for livestock such as ruminants as well as humans in temperate and tropical countries [1–4]. *Anaplasma phagocytophilum* is transmitted by *Ixodes* spp. and infects host neutrophils and monocytes, where reproduction occurs [1,5]. *Anaplasma phagocytophilum* infection is known as pasture fever and characterized by fever, anorexia, lateral recumbency, dullness, and loss of milk yield in affected hosts [2,4,6].

Increased attention to *A. phagocytophilum* reveals new information about the genetic diversity of the pathogen. Recently two *Anaplasma* variants related to *A. phagocytophilum* have been documented in cattle, sheep, goats, and ticks [7–9]. In Japan, *A. phagocytophilum*-like 1 has been detected in deer and *Hemaphysalis longicornis* [10], cattle [11], *Ixodes* spp. [12], and *Haemaphysalis megaspinosa* [13]. *A. phagocytophilum*-like 2 has been identified in *Hyalomma asiaticum* [14], sheep and goats from China [15]. Recently those *Anaplasma* variants have been documented in ruminants from Tunisia [7,8], South Korea [16], and Italy [17].

Various *Anaplasma* species including *A. phagocytophilum* have been documented in ruminants and ticks in Turkey [5,6,18–21]. However, until now no data on *A. phagocytophilum* variants is available in Turkey. In the current study, 16S rRNA, groEL (heat shock protein) PCR and sequencing were performed to identity *A. phagocytophilum* and *A. phagocytophilum*-like variants in small ruminants from sampling sites in Antalya and Mersin provinces, where the representative Mediterranean area of Turkey.

2. Materials and Methods

2.1. Study Region and Sample Collection

This survey was conducted in small ruminants farmed in three districts (Alanya, Akseki, Manavgat) from Antalya (latitude 36° 53′ N, longitude 30° 42′ E) and two districts (Anamur, Bozyazı) from Mersin (latitude 36° 47′ N, longitude 34° 37′ E) provinces of Turkey (Figure 1). This area has a Mediterranean climate, with hot humid summers and warm rainy winters. The goats and sheep are kept in closed areas in villages near to the coast during the winter months, and they are taken to the plateaus in the Taurus Mountains in the early spring and grazed in the pastures here until autumn.

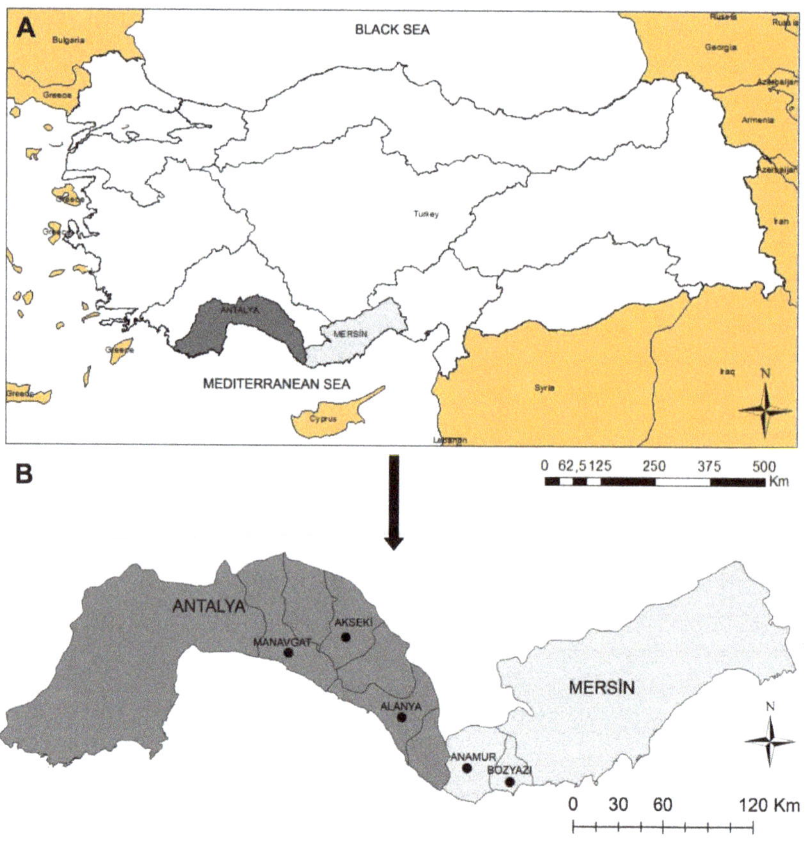

Figure 1. Map of Turkish provinces, indicating the localities studied in the study. (**A**) Geographical position of the provinces of Antalya and Mersin in Turkey. (**B**) Position of localities sampled in the provinces of Antalya and Mersin.

The sample size was calculated using the online tool Sample Size Calculator (www. calculator.net/sample-sizecalculator.html, accessed on 1 February 2019), for a confidence level (CL) of 95%, an error margin of 5%. According to this, during April–July 2019, a total of 433 apparently healthy small ruminants (296 goats, 137 sheep) were included in the survey. Blood samples were drawn from the punctured jugular vein into anticoagulated (K3-EDTA) vacutainer tubes and stored at −20 °C freezer until DNA extraction. The goats and sheep were also checked for tick infestations, and a total of 1475 ticks were removed. The collected ticks were preserved in 70% ethanol in Eppendorf tubes. They were identified using taxonomic keys [22]. The animals were grouped into categories according to species (goat and sheep) and the presence of ticks (yes/no). This study secured the approval of the Elazig Veterinary Control Institute (number: 2018/02).

2.2. DNA Extraction and Amplification of 16S rRNA Gene

DNA was isolated from 200 µL volumes of whole blood using a DNAeasy Blood Minikit according to the vendor's recommendations. Genomic DNA from blood of clinically infected cattle with *A. phagocytophilum* [6] was used for positive control. *Anaplasma phagocytophilum*-like variants DNAs, received from Alberto Alberti (University of Sassari, Sassari, Italy) were used as positive controls.

To detect *A. phagocytophilum* and *A. phagocytophilum*-like variants, a nested 16S rRNA PCR was carried out described by Kawahara et al. [10]. The PCR reaction conditions were made according to the previously described studies [10,21]. The nested amplicons were examined by 1.5% agarose gel electrophoresis and visualized using the gel Documentation System (Vilber Lourmat, Marne La Vallee Ceedex, France).

2.3. Restriction Fragment Length Polymorphism (RFLP)

XcmI and *BsaI* restriction enzymes allow the specific discrimination amongst *A. phagocytophilum* and related variants [8,17]. For differentiation of *A. phagocytophilum* and related variants, the nested amplicons obtained in this study were digested with the *XcmI* and *BsaI* restriction enzymes as previously described [8,17].

2.4. GroEL PCR

To confirm the results of the RFLP assay, the positive samples were screened by a groEL nested PCR for the amplification of *A. phagocytophilum* [23]. The semi-nested PCR reported by Ybañez et al. [24] with the primers EEGro1F/AnaGroe712R and AnaGroe240F was utilized for amplifying of *A. phagocytophilum*-like 1 *groEL* gene. Oligonucleotide primers used in this study were presented in Table 1.

Table 1. Oligonucleotide primers used in this study (* Degenerate primer: Y = C or T).

Target Gene	Specificity	Primer Name	Oligonucleotide Dequence (5′-3′)	Annealing	Amplicon Size (bp)	Reference
-	All *Anaplasma/Ehrlichia*	EC9	TACCTTGTTACGACTT	54	1462	
-	-	EC12a	TGATCCTGGCTCAGAACGAACG			
-	-	-	-	-	-	[10]
16S rRNA	*A. phagocytophilum* and related variants	SSAP2f	GCTGAATGTGGGGATAATTTAT	53	641–642	
		SSAP2r	ATGGCTGCTTCCTTTCGGTTA			
		EphplgroEL(569)F	ATGGTATGCAGTTTGATCGC	54	624	
		EphgroEL(1142)R	TTGAGTACAGCAACACCACCGGAA			
-	-	-	-	-	-	[23]
groEL	*A. phagocytophilum*	EphplgroEL(569)F	ATGGTATGCAGTTTGATCGC	54	573	
		EphgroEL(1142)R	TTGAGTACAGCAACACCACCGGAA			
	A. phagocytophilum-like	EEGro1F	GAGTTCGACGGTAAGAAGTTCA	52	670	
groEL	1	AnaGroe712R	ATTAGY *AAGCCTTATGGGTC	-	-	[25]
		AnaGroe240F	CCGCGATCAAACTGCATACC	57	432	

2.5. Sequencing and Phylogenetic Analyses

Anaplasma phagocytophilum (*n* = 6) and *A. phagocytophilum*-like 1 (*n* = 10) positive PCR amplicons were purified using the QIAquick PCR Purification Kit (Qiagen, Hilden, Germany). The purified amplicons were sent to BM Labosis (Ankara, Turkey) for Sanger sequencing to determine DNA sequences of the *16S rRNA* gene. Multiple alignments were performed with the CLUSTAL Omega ver. 1.2.1 (https://www.ebi.ac.uk, accessed on 1 February 2019). The representative sequences have been submitted to the GenBank (MT881655 and MT881656 for *16S rRNA* gene of *A. phagocytophilum*-like 1 and *A. phagocytophilum*, respectively). The sequence alignment was performed using MUSCLE in Geneious prime [25].

Phylogenetic analyses of the 16S rRNA sequences obtained in this work and the other sequences submitted to GenBank were carried out. The maximum likelihood analysis (ML) carried out in Mega X [26] was utilized to determine the phylogenetic relationship of the *Anaplasma* spp. To sequence evolution, best-fit model was assessed as TN93+G+I by using the jModel test v.0.1.1 [27]. Reliability of internal branches of the tree was evaluated by the bootstrapping method with 1000 iterations [28].

2.6. Statistical Analysis

Association of the presence of *Anaplasma* spp. with host species and presence of tick was performed with Epi Info 6.01 (CDC, Atlanta), using the $\chi 2$ test and Fisher's exact test.

3. Results

3.1. Tick Infestation

Of the 433 small ruminants examined, 190 (43.9%) were infested with at least one tick species. A total of 1475 adult ticks (449 females, 1026 males) were collected from goats (1409/1475, 95.5%) and sheep (66/1475, 4.5%). Six tick species were identified among all collected ticks. *Rhipicephalus bursa* (1269/1475, 86%) was the dominant tick species, followed by *R. turanicus* (98/1475, 6.6%), *Dermacentor marginatus* (94/1475, 6.4%), *Hyalomma marginatum* (8/1475, 0.5%), *R. sanguineus* s.l. (5/1475, 0.3%), and *Ixodes ricinus* (0.06%, only one specimen). The goats were infested with all the identified tick species, whereas sheep were infested with *R. bursa* and *R. turanicus*.

3.2. Prevalence and Distribution of Anaplasma spp.

The prevalence of *A. phagocytophilum* and related variants in sampled goats and sheep is presented in Table 2. Overall, 121/433 (27.9%) samples collected in studied regions tested positive for *Anaplasma* spp. by 16S rRNA PCR. The infection rate in goats and sheep was determined as 28% and 27.7%, respectively. RFLP revealed the prevalence of *A. phagocytophilum* and *A. phagocytophilum*-like 1 as 1.4% and 26.5%, respectively. No PCR amplicons derived from goats and sheep were digested by the *Bsa*I enzyme, confirming the absence of Chinese variant (*A. phagocytophilum*-like 2). Of the 121 positive samples with 16S rRNA PCR, 110 (95.6%) were positive with groEL nested PCR. Six of them (6/110, 5.4%) were positive for *A. phagocytophilum* and 104 (94.5%) were positive for *A. phagocytophilum*-like 1 (Table 2).

Association of the frequency of *A. phagocytophilum* and *A. phagocytophilum*-like 1 variant in small ruminants with species and tick infestation is documented in Table 3. The prevalence of *Anaplasma* spp. was comparable in species, and no difference was detected between infection rates in sheep and goats (*p* = 0.9603). However, the prevalence significantly increased with tick infestation in small ruminants (*p* = 0.0003) (Table 3).

Table 2. Samples origin, 16S rRNA PCR, RFLP and groEL PCR.

Host	District/Province	16S rRNA PCR+/No. of Samples	16S rRNA PCR + RFLP			groEL+/16S+	groEL PCR	
			AP	AP-like 1	AP-like 2		AP	AP-like 1
Goat	Akseki/Antalya	13/56 (23.2%)	2	11	0	12/13	2	10
-	Manavgat/Antalya	26/111 (23.4%)	3	23	0	23/26	3	20
-	Alanya/Antalya	24/55 (43.6%)	0	24	0	24/24		24
-	Anamur/Mersin	15/44 (34.1%)	0	15	0	12/15		12
-	Bozyazı/Mersin	5/30 (16.7%)	0	5	0	5/5		5
Goat Total	-	83/296 (28%)	5	78	0	76/83 (91.5%)	5	71
Sheep	Akseki/Antalya	5/9 (55.6%)	1	4	0	3/5	1	2
-	Manavgat/Antalya	18/103 (17.5%)	0	18	0	17/18	-	17
-	Alanya/Antalya	9/9 (100%)	0	9	0	9/9	-	9
-	Anamur/Mersin	6/16 (37.5%)	0	6	0	5/6	-	5
Sheep Total	-	38/137 (27.7%)	1	37	0	34/38 (89.4%)	1	33
Grand Total	-	121/433 (27.9%)	6 (1.4%)	115 (26.5%)	0	110/115 (95.6%)	6	104

Table 3. Association of the frequency (16S rRNA PCR) of *Anaplasma phagocytophilum* and related variants in small ruminants with species and tick infestation.

	Species		Presence of Ticks on the Animals	
	Goats *n* (%)	**Sheep** *n* (%)	**No** *n* (%)	**Yes** *n* (%)
Number	296	137	243	190
Positive	83 (28)	38 (27.7)	51 (20.9)	70 (36.8)
Negative	213 (72)	99 (72.3)	192 (79.1)	120 (63.2)
p-Value	0.9603		0.0003	

3.3. Molecular and Phylogenetic Analyses

To validate the RFLP results and identify genetic variants of *A. phagocytophilum*-like 1, randomly selected 10 representative samples were sequenced. The sequences shared 100% identity to each other. Therefore, one representative sequence for *A. phagocytophilum*-like 1 was submitted to the NCBI GenBank database, and deposited with accession number MT881655. This finding indicated that one variant was identified, and named as Aplike1OvineCaprine in this work. BlastN analysis demonstrated that the Aplike1OvineCaprine variant indicated high similarity (99–100%) to those *Anaplasma* isolates deposited in the GenBank as uncultured *Anaplasma* sp. and *A. phagocytophilum*. Moreover, the Aplike1OvineCaprine variant was 100% identical to those of *A. phagocytophilum*-like 1 detected in sheep (Aplike1Ov1, KX702978) and goat (Aplike1GGo2, KM285227) from Tunisia, and cattle from Turkey (Aplike1Bv, MT338494) (Table 4). The *A. phagocytophilum* Akseki11 Sheep Turkey isolate obtained in this study shared 99.3–99.6% identity isolated from *Niviventer confucianus* (*A. phagocytophilum* ZJ-HGA strain, DQ458805) and human (*A. phagocytophilum* HZ strain, NR_074113), respectively.

Phylogenetic analysis using the *16S rRNA* gene showed that our variant (Aplike1OvineCaprine) clustered a distinct group with those of *A. phagocytophilum*-like 1 previously published sequences reported in sheep, goats, cattle, deer, and *Haemaphysalis ginghaiensis* (Figure 2).

Table 4. Nucleotide differences among 16S rRNA sequences from *Anaplasma* variants related to *A. phagocytophilum* (598–599 bp).

Host	Genetic Variant [a]	Country	GenBank	Nucleotide Positions [b]														Reference
-	-	-	-	823	830	1011	1109	1111	1113	1120	1137	1148	1237	1239	1240	1260	1291	-
Human	Webster	USA	NR_044762	T	T	A	G	T	A	C	A	T	T	T	C	G	C	Unpublished
Horse	Camawi	USA	AF172167	*	*	*	*	*	*	*	*	*	*	*	*	*	*	Unpublished
Dog	Dog2	USA	CP006618	*	*	*	*	*	*	*	*	*	*	*	*	*	*	Unpublished
Deer	Clone 1	Japan	JN055357	C	*	*	*	A	-	*	G	C	*	*	*	*	*	[24]
Goat	Aplike1GGo1	Tunisia	KM285226	C	*	*	*	A	-	*	G	C	*	*	*	*	*	[7]
Goat	Aplike1GGo2	Tunisia	KM285227	C	*	*	*	A	T	*	G	*	*	*	*	*	*	[7]
Sheep	Aplike1GOv1	Tunisia	KM285230	C	*	*	*	A	-	*	G	C	*	*	*	*	*	[7]
Cattle and Goat	Aplike1BvCp1	Tunisia	KX702974	C	*	*	*	A	-	*	G	C	*	*	*	*	*	[8]
Sheep	Aplike1Ov1	Tunisia	KX702978	C	*	*	*	A	T	*	G	*	*	*	*	*	*	[8]
Sheep and Goat	Aplike2OvCp1	Tunisia	KX702980	C	A	G	*	A	T	T	G	C	C	C	T	A	T	[8]
Cattle	Aplike1Bv	Turkey	MT338494	C	*	*	*	A	T	*	G	*	*	*	*	*	*	Unpublished
Sheep and Goat	Aplike1OvineCaprine	Turkey	MT881655	C	*	*	*	A	T	*	G	*	*	*	*	*	*	Present study
Sheep	Aphaakseki11	Turkey	MT881656	*	*	*	*	*	*	*	*	*	*	*	T	*	*	Present study

* Asterisks show the conserved nucleotide positions. [a] Aplike1OvineCaprine variant has been registered with GenBank under accession number MT881655. [b] Numbers indicate the nucleotide position (*A. phagocytophilum*, NC 007797). The position of nucleotide 1011 indicates the substitution of A by G allowing differentiation between variants of *A. phagocytophilum* (like 1 and 2) by *Bsa*I enzyme, while the position of 1137 nucleotide indicates the substitution of A by G allowing for the distinction between *A. phagocytophilum* and related variants (like 1 and 2) by *Xcm*I enzyme [8].

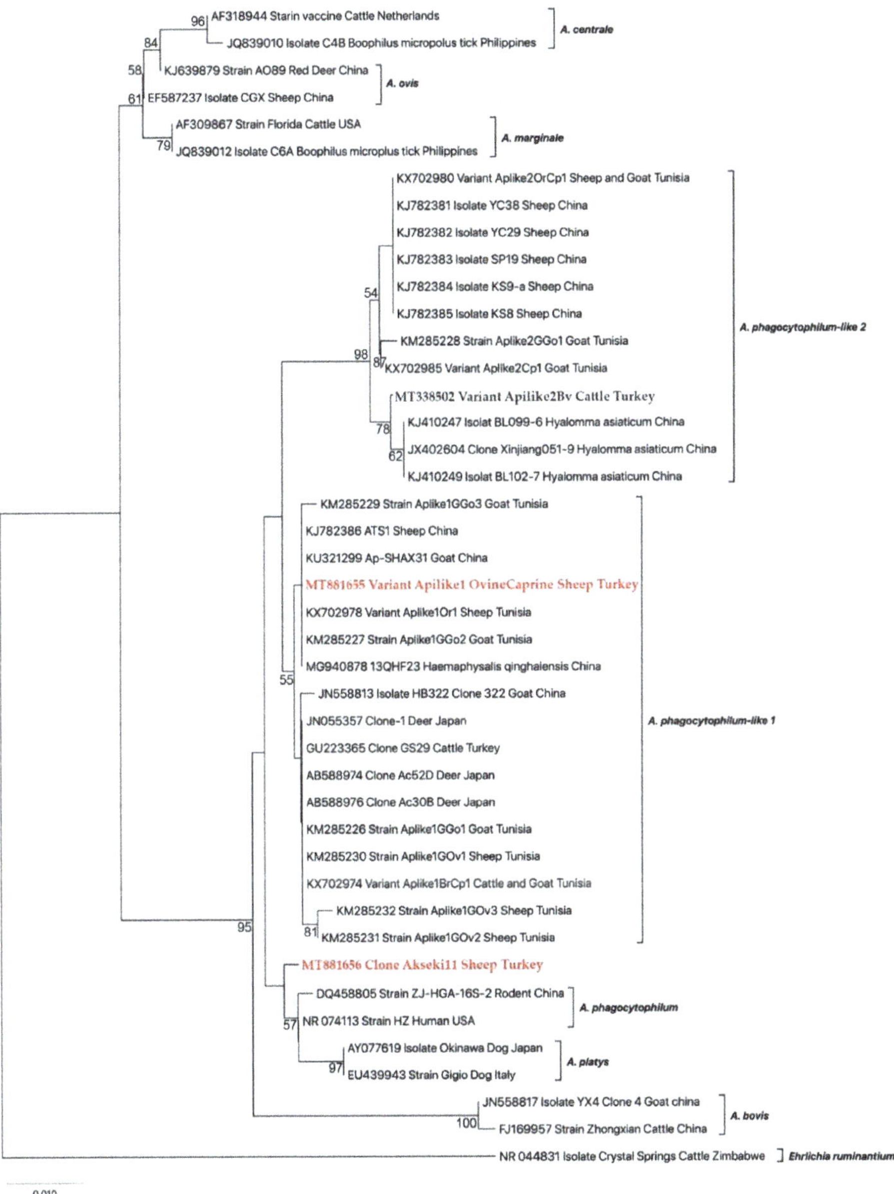

Figure 2. Maximum likelihood phylogenetic tree was inferred with partial sequences (598–599 bp) of the 16S rRNA gene of *Anaplasma* sp. related to *A. phagocytophilum* isolated from sheep in Turkey (highlighted in red) with other *Anaplasma* spp. retrieved from GenBank. Numbers at the nodes refer percentage occurrence in 1000 the bootstrap replication. The new sequences of *A. phagocytophilum*-like 1 and *A. phagocytophilum* from this study were highlighted in red. *Ehrlichia ruminantium* was used as an outgroup.

4. Discussion

Anaplasma phagocytophilum causes tick-borne fever in small ruminants and granulocytic anaplasmosis in horses and dogs [1,2]. It is an emerging tick-borne pathogen for humans as well [3]. The genetic diversity of *A. phagocytophilum* is much greater than expected. Indeed, recent studies have revealed the existence of two distinct *Anaplasma* species or variants related to *A. phagocytophilum*, one in Japan and the other in China [11–14,24]. Then, these pathogens were designated as *A. phagocytophilum*-like 1 and *A. phagocytophilum*-like 2 variants [7,8]. More recently, both genotypes have been documented in ruminants and *R. turanicus* in Tunisia [7,8,17,29], cattle in South Korea [16], and small ruminants in Italy [17]. In the present study, a survey was carried out to detect and identify *A. phagocytophilum* and *A. phagocytophilum*-like variants in small ruminants from the Mediterranean region of Turkey. Our findings provide molecular evidence for the presence of *A. phagocytophilum* and *A. phagocytophilum*-like 1 in sampled sheep and goats. In the previous studies carried out in Turkey, *A. phagocytophilum* has been reported in small ruminants [18,21,30]. However, this is the first time that *A. phagocytophilum*-like 1 variant in sheep and goats have been reported in the country.

Contrary to *A. phagocytophilum*, it has been suggested that both Japanese and Chinese variants do not cause clinical infection in ruminants [8,17]. In this study, a high prevalence for *A. phagocytophilum*-like 1 variant was determined (26.5%), but no clinical infection for tick-borne fever was observed in sheep and goats during sample collection. This result is consistent with the previous suggestions that *A. phagocytophilum*-like variants are considered non-pathogenic for ruminants [8,16,17]. The prevalence of *A. phagocytophilum*-like 1 (26.5%) in small ruminants obtained in this study was higher than that observed in Tunisian sheep (7%) and goats (13.1%) [8], however, it was lower than that observed in other studies conducted in Mediterranean small ruminants (122/203, 60%) from Tunisia and Italy [17].

It has been previously suggested that serological cross-reactions occur between *A. phagocytophilum* and other *Anaplasma* species [31,32]. The same situation may be true in some circumstances for molecular markers, for example a pair of primers (SSAP2f/SSAP2r) based on the *16S rRNA* gene of *A. phagocytophilum* were designed for the specific amplification [10]. However, it has been shown that these primers also detect distinct *Anaplasma* variants related to *A. phagocytophilum* [7,8,24]. In this work, the frequency of pathogenic *A. phagocytophilum* was 1.4%, which is not consistent with the previous studies in Turkey that reported values of 66.7% in Central Anatolia [30] and 19.7% in Eastern Anatolia [21]. The high infection rates obtained in the previous studies may be due to the selected primers for the amplification of *A. phagocytophilum*. EE1/EE2 and SSAP2f/SSAP2r primers have been selected to detect *A. phagocytophilum* in the studies conducted in Central Anatolia [30] and Eastern Anatolia [21], respectively. However, the EE1/EE2 primers are universal for the detection of all *Anaplasma* spp. including *A. phagocytophilum*-like variants [33]. It has been also reported that the SSAP2f/SSAP2r can amplify not only *A. phagocytophilum*, but also *A. phagocytophilum*-like variants [7,8,24]. This study provides molecular data for the circulation of *A. phagocytophilum* and *A. phagocytophilum*-like 1 Turkish small ruminants. Therefore, cross-reactivity between *A. phagocytophilum* and related variant should be considered in interpreting the findings of surveys to be carried out in the area, where *A. phagocytophilum* and *A. phagocytophilum*-like variant co-exist.

As several domestic and wild mammals are hosts or reservoirs for *A. phagocytophilum* [1,2], abundance and intensity of the tick vector, *I. ricinus* in Europe including Turkey are considered a major factor affecting the distribution of the pathogen in a specific area. It is well known that there is no *I. ricinus* in the Eastern and Central Anatolian regions of Turkey [34]. It has been reported that *A. phagocytophilum* is transmitted by *Ixodes* spp. (*I. persulcatus, I. scapularis* and *I. ricinus*) in some parts of the world including in Europe [1,35]. In Turkey, *I. ricinus* collected from humans were positive for *A. phagocytophilum* [5]. So far, data on the transmission of *A. phagocytophilum*-like variants by ticks are lacking. A recent study reported that *R. turanicus* was common in sampled sheep and goats in Tunisia, and one *R. turanicus* tick feeding on the goat was found to be infected with *A. phagocytophilum*-like 2 [28]. In the present study, potential vectors of *A. phagocytophilum*-like 1 was not studied, but we found that the sampled sheep and goats were commonly infested with *R. bursa* (86%), *R. turanicus* (6.6%), *D. marginatus* (6.4%), and very rarely *I. ricinus* (0.06%, only one specimen). Our finding also showed a correlation between *Anaplasma* positivity and the presence of ticks ($p = 0.0003$), compatible with the finding that the prevalence of *A. phagocytophilum*-like 1 was higher in goats infested by ticks than in not infested [7]. Based on the abundance of *Rhipicepahlus* and *Dermacentor* ticks and the very rarity of *I. ricinus*, we can assume that *Rhipicephalus* and *Dermacentor* may play an important role in the transmission of *A. phagocytophilum*-like 1 rather than *I. ricinus*. This assumption is supported by the previous findings that a high prevalence of *A. phagocytophilum*-like variants have been reported in ruminants in the higher semi-arid area of Tunisia, where *I. ricinus* is not present [8]. However, more detailed studies are needed to validate this assumption and to establish what tick species may play a role in the transmission of *A. phagocytophilum*-like 1 in Turkey.

Our sequencing validated RFLP findings, and showed that the sampled small ruminants were found to be infected with *A. phagocytophilum*-like 1. Phylogenetic analysis indicated two main separate branches. The Aplike1OvineCaprine (MT881655) variant obtained in this study, as well as those previously reported from sheep, goats, cattle, and ticks, formed a monophylogenic clade distinct from *A. phagocytophilum* and *A. phagocytophilum*-like 2, and other members of *Anaplasma* spp. [7,8,24].

5. Conclusions

This work provides molecular data for the circulation of *A. phagocytophilum*-like 1 for the first time in Turkey. The novel strain is widespread in small ruminants in the Mediterranean area of Turkey with an overall prevalence of 26.5%. This finding revealed that the variant should be considered in the diagnosis of caprine and ovine anaplasmosis.

Author Contributions: Conceptualization, project administration, methodology, validation, formal analysis, investigation, writing—review and editing, M.A.; methodology, validation, formal analysis, S.Ö.; validation, formal analysis, M.C.U. All authors have read and agreed to the published version of the manuscript.

Funding: This work was supported by founding from the Scientific and Technological Research Council of Turkey (TUBITAK) (project no. 118O871).

Institutional Review Board Statement: The study was conducted according to the guidelines of the Declaration of Helsinki and approved by Firat University Animal Experiment Local Ethics Committee (2018/100, 109. 11. 2018).

Data Availability Statement: Data available in a publicly accessible repository.

Acknowledgments: The authors are grateful to Alberto Alberti (University of Sassari, Sassari, Italy) for positive control DNAs from *A. phagocytophilum*-like variants. We are also thankful to the farmers for their cooperation and the veterinary clinician's Ali Asar for helping with sample collection.

Conflicts of Interest: The authors declare no conflict of interest.

References

1. Woldehiwet, Z. The natural history of *Anaplasma phagocytophilum*. *Vet. Parasitol.* **2010**, *167*, 108–112. [CrossRef]
2. Stuen, S.; Granquist, E.G.; Silaghi, C. *Anaplasma phagocytophilum*-a widespread multi-host pathogen with highly adaptive strategies. *Front. Cell. Infect. Microbiol.* **2013**, *3*, 31. [CrossRef]
3. Kocan, K.M.; de la Fuente, J.; Cabezas-Cruz, A. The genus *Anaplasma*: New challenges after reclassification. *Rev. Sci. Technol.* **2015**, *34*, 577–586. [CrossRef]
4. Langenwalder, D.B.; Schmidt, S.; Gilli, U.; Pantchev, N.; Ganter, M.; Silaghi, C.; Aardema, M.L.; von Loewenich, F.D. Genetic characterization of *Anaplasma phagocytophilum* strains from goats (*Capra aegagrus hircus*) and water buffalo (*Bubalus bubalis*) by 16S rRNA gene, ankA gene and multilocus sequence typing. *Ticks Tick Borne Dis.* **2019**, *10*, 101267. [CrossRef] [PubMed]
5. Aktas, M.; Vatansever, Z.; Altay, K.; Aydin, M.F.; Dumanli, N. Molecular evidence for *Anaplasma phagocytophilum* in *Ixodes ricinus* from Turkey. *Trans. R. Soc. Trop. Med. Hyg.* **2010**, *104*, 10–15. [CrossRef] [PubMed]

6. Aktas, M.; Ozubek, S. Bovine anaplasmosis in Turkey: First laboratory confirmed clinical cases caused by *Anaplasma phagocytophilum*. *Vet. Microbiol.* **2015**, *178*, 246–251. [CrossRef]

7. Ben Said, M.; Belkahia, H.; Alberti, A.; Zobba, R.; Bousrih, M.; Yahiaoui, M.; Daaloul-Jedidi, M.; Mamlouk, A.; Gharbi, M.; Messadi, L. Molecular survey of *Anaplasma* species in small ruminants reveals the presence of novel strains closely related to *A. phagocytophilum* in Tunisia. *Vector Borne Zoonotic Dis.* **2015**, *15*, 580–590. [CrossRef]

8. Ben Said, M.; Belkahia, H.; El Mabrouk, N.; Saidani, M.; Ben Hassen, M.; Alberti, A.; Zobba, R.; Bouattour, S.; Bouattour, A.; Messadi, L. Molecular typing and diagnosis of *Anaplasma* spp. closely related to *Anaplasma phagocytophilum* in ruminants from Tunisia. *Ticks Tick Borne Dis.* **2017**, *8*, 412–422. [CrossRef] [PubMed]

9. Ben Said, M.; Belkahia, H.; Messadi, L. *Anaplasma* spp. in North Africa: A review on molecular epidemiology, associated risk factors and genetic characteristics. *Ticks Tick Borne Dis.* **2018**, *9*, 543–555. [CrossRef] [PubMed]

10. Kawahara, M.; Rikihisa, Y.; Lin, Q.; Isogai, E.; Tahara, K.; Itagaki, A.; Hiramitsu, Y.; Tajima, T. Novel genetic variants of *Anaplasma phagocytophilum*, *Anaplasma bovis*, *Anaplasma centrale*, and a novel *Ehrlichia* sp. in wild deer and ticks on two major islands in Japan. *Appl. Environ. Microbiol.* **2006**, *72*, 1102–1109. [CrossRef]

11. Jilintai, S.N.; Hayakawa, D.; Suzuki, M.; Hata, H.; Kondo, S.; Matsumoto, K.; Yokoyam, N.; Inokuma, H. Molecular survey for *Anaplasma bovis* and *Anaplasma phagocytophilum* infection in cattle in pastureland where sika deer appear in Hokkaido, Japan. *Jpn. J. Infect. Dis.* **2009**, *62*, 73–75.

12. Ohashi, N.; Inayoshi, M.; Kitamura, K.; Kawamori, F.; Kawaguchi, D.; Nishimura, Y.; Naitou, H.; Hiroi, M.; Masuzawa, T. *Anaplasma phagocytophilum*-infected ticks, Japan. *Emerg. Infect. Dis.* **2005**, *11*, 1780–1783. [CrossRef] [PubMed]

13. Yoshimoto, K.; Matsuyama, Y.; Matsuda, H.; Sakamoto, L.; Matsumoto, K.; Yokoyama, N.; Inokuma, H. Detection of *Anaplasma bovis* and *Anaplasma phagocytophilum* DNA from *Haemaphysalis megaspinosa* in Hokkaido, Japan. *Vet. Parasitol.* **2010**, *168*, 170–172. [CrossRef] [PubMed]

14. Kang, Y.J.; Diao, X.N.; Zhao, G.Y.; Chen, M.H.; Xiong, Y.; Shi, M.; Fu, W.M.; Guo, Y.J.; Pan, B.; Chen, X.P.; et al. Extensive diversity of Rickettsiales bacteria in two species of ticks from China and the evolution of the Rickettsiales. *BMC Evol. Biol.* **2014**, *14*, 167. [CrossRef]

15. Yang, J.; Li, Y.; Liu, Z.; Liu, J.; Niu, Q.; Ren, Q.; Chen, Z.; Guan, G.; Luo, J.; Yin, H. Molecular detection and characterization of *Anaplasma* spp. in sheep and cattle from Xinjiang, northwest China. *Parasit. Vectors* **2015**, *19*, 108. [CrossRef] [PubMed]

16. Seo, M.G.; Ouh, I.-O.; Kwon, O.-D.; Kwak, D. Molecular detection of *Anaplasma phagocytophilum*-like *Anaplasma* spp. and pathogenic *A. Phagocytophilum* in cattle from South Korea. *Mol. Phyl. Evol.* **2018**, *126*, 23–30. [CrossRef]

17. Zobba, R.; Ben Said, M.; Belkahia, H.; Pittaua, M.; Cacciotto, C.; Parpagliaa, M.L.P.; Messadic, L.; Alberto, A. Molecular epidemiology of *Anaplasma* spp. related to *A. phagocytophilum* in Mediterranean small ruminants. *Acta Tropica*. **2020**, *202*, 105286. [CrossRef]

18. Gokce, H.I.; Genc, O.; Akca, A.; Vatansever, Z.; Unver, A.; Erdogan, H.M. Molecular and serological evidence of *Anaplasma phagocytophilum* infection of farm animals in the Black Sea Region of Turkey. *Acta Vet. Hung.* **2008**, *56*, 281–292. [CrossRef] [PubMed]

19. Aktas, M.; Altay, K.; Ozubek, S.; Dumanli, N. A survey of ixodid ticks feeding on cattle and prevalence of tick-borne pathogens in the BlackSea Region of Turkey. *Vet. Parasitol.* **2012**, *187*, 567–571. [CrossRef]

20. Aktas, M. A survey of ixodid tick species and molecular identification of tick-borne pathogens. *Vet. Parasitol.* **2014**, *200*, 276–283. [CrossRef]

21. Altay, K.; Dumanli, N.; Aktas, M.; Ozubek, S. Survey of *Anaplasma* infections in small ruminants from East part of Turkey. *Kafkas Univ. Vet. Fak. Derg.* **2014**, *20*, 1–4. [CrossRef]

22. Estrada-Peña, A.; Bouattour, A.; Camicas, J.L.; Walker, A.R. *Ticks of Domestic Animals in the Mediterranean Region. A Guide of Identification of Species*; University of Zaragoza Press: Zaragoza, Spain, 2004.

23. Alberti, A.; Zobba, R.; Chessa, B.; Addis, M.F.; Sparagano, O.; Pinna Parpaglia, M.L.; Cubeddu, T.; Pintori, G.; Pittau, M. Equine and canine *Anaplasma phagocytophilum* strains isolated on the island of Sardinia (Italy) are phylogenetically related to pathogenic strains from the United States. *Appl. Environ. Microbiol.* **2005**, *71*, 6418–6422. [CrossRef]

24. Ybañez, A.P.; Matsumoto, K.; Kishimoto, T.; Inokuma, H. Molecular analyses of a potentially novel *Anaplasma* species closely related to *Anaplasma phagocytophilum* detected in sika deer (Cervus nippon yesoensis) in Japan. *Vet. Microbiol.* **2012**, *157*, 232–236. [CrossRef] [PubMed]

25. Edgar, R.C. MUSCLE: A multiple sequence alignment method with reduced time and space complexity. *BMC Bioinform.* **2004**, *5*, 113. [CrossRef] [PubMed]

26. Kumar, S.; Stecher, G.; Li, M.; Knyaz, C.; Tamura, K. MEGA X: Molecular Evolutionary Genetics Analysis across computing platforms. *Mol. Biol. Evol.* **2018**, *35*, 1547–1549. [CrossRef]

27. Posada, D. jModelTest: Phylogenetic model averaging. *Mol. Biol. Evol.* **2008**, *25*, 1253–1256. [CrossRef] [PubMed]

28. Felsenstein, J. Confidence limits on phylogenies: An approach using the bootstrap. *Evolution* **1985**, *39*, 783–791. [CrossRef]

29. Belkahia, H.; Ben Said, M.; Ghribi, R.; Selmi, R.; Ben Asker, A.; Yahiaoui, M.; Bousrih, M.; Daaloul-Jedidi, M.; Messadi, L. Molecular detection, genotyping and phylogeny of *Anaplasma* spp. in *Rhipicephalus* ticks from Tunisia. *Acta Trop.* **2019**, *191*, 38–49. [CrossRef]

30. Benedicto, B.; Ceylan, O.; Moumouni, P.F.A.; Lee, S.H.; Tumwebaze, M.A.; Li, X.; Galoni, E.M.; Li, Y.; Ji, S.; Ringo, A.; et al. Molecular detection and assessment of risk factors for tick-borne diseases in Sheep and Goats from Turkey. *Acta Parasit.* **2020**, *65*, 723–732. [CrossRef] [PubMed]
31. Movilla, R.; García, C.; Siebert, S.; Roura, X. Countrywide serological evaluation of canine prevalence for *Anaplasma* spp., *Borrelia burgdorferi* (sensu lato), *Dirofilaria immitis* and *Ehrlichia canis* in Mexico. *Parasit. Vectors* **2016**, *9*, 421. [CrossRef]
32. Dreher, U.M.; de la Fuente, J.; Hofmann-Lehmann, R.; Meli, M.L.; Pusterla, N.; Kocan, K.M.; Woldehiwet, Z.; Braun, U.; Regula, G.; Staerk, K.D.; et al. Serologic cross-reactivity between *Anaplasma marginale* and *Anaplasma phagocytophilum*. *Clin. Diagn. Lab. Immunol.* **2005**, *12*, 1177–1183. [CrossRef] [PubMed]
33. Yang, J.; Liu, Z.; Niu, Q.; Liu, J.; Xie, J.; Chen, Q.; Chen, Z.; Guan, G.; Liu, G.; Luo, J.; et al. Evaluation of different nested PCRs for detection of *Anaplasma phagocytophilum* in ruminants and ticks. *BMC Vet. Res.* **2016**, *12*, 35. [CrossRef] [PubMed]
34. Aydin, L.; Bakırcı, S. Geographical distribution of ticks in Turkey. *Parasitol. Res.* **2007**, *101*, 163–166. [CrossRef] [PubMed]
35. Sarih, M.; M'ghirbi, Y.; Bouattour, A.; Gern, L.; Baranton, G.; Postic, D. Detection and identification of *Ehrlichia* spp. in ticks collected in Tunisia andMorocco. *J. Clin. Microbiol.* **2005**, *43*, 1127–1132. [CrossRef] [PubMed]

Article

Host-Parasite Interaction in *Sarcoptes scabiei* Infestation in Porcine Model with a Preliminary Note on Its Genetic Lineage from India

Arun Kumar De [1,*,†], Sneha Sawhney [1,†], Samiran Mondal [2], Perumal Ponraj [1], Sanjay Kumar Ravi [1], Gopal Sarkar [2], Santanu Banik [3], Dhruba Malakar [4], Kangayan Muniswamy [1], Ashish Kumar [5], Arvind Kumar Tripathi [1], Asit Kumar Bera [6] and Debasis Bhattacharya [1]

[1] Animal Science Division, ICAR-Central Island Agricultural Research Institute, Port Blair 744101, India; snehasawhney88@gmail.com (S.S.); perumalponraj@gmail.com (P.P.); skravivet@gmail.com (S.K.R.); swamy02_vet@yahoo.co.in (K.M.); akt9@rediffmail.com (A.K.T.); debasis63@rediffmail.com (D.B.)

[2] Department of Veterinary Pathology, West Bengal University of Animal and Fishery Sciences, Kolkata 700037, India; vetsamiran@gmail.com (S.M.); gsarkar999@gmail.com (G.S.)

[3] Department of Animal Genetics and Breeding, ICAR-National Research Centre on Pig, Guwahati 781131, India; sbanik2000@gmail.com

[4] Animal Biotechnology Centre, National Dairy Research Institute, Karnal 132001, India; dhrubamalakar@gmail.com

[5] CTARA, IIT Bombay, Mumbai 400076, India; ashish7@iitb.ac.in

[6] Reservoir and Wetland Fisheries Division, ICAR-Central Inland Fishery Research Institute, Barrackpore, Kolkata 700120, India; asitmed2000@yahoo.com

[*] Correspondence: biotech.cari@gmail.com; Tel.: +91-967-951-5260

[†] These authors contributed equally to this paper.

Received: 4 November 2020; Accepted: 1 December 2020; Published: 7 December 2020

Simple Summary: Scabies or mange caused by *Sarcoptess cabiei* is the latest addition of WHO's list oftropical neglected diseases. It causes severe itching to the host. It has a wide host range including humans, farm animals, companion animals, and wild animals. It is anemerging/re-emerging disease with high prevalence in underdeveloped and developing countries. The disease has zoonotic importance and is of significant public health concern as cross-transmission or species jumping is very common. To date, fifteen *Sarcoptes* varieties have been reported as per host origin. Differential diagnosis at variety level is very crucial for epidemiological study and scratching future eradication program of the disease. As morphotaxonomy fails to differentiate varieties, use of molecular markers is crucial. Moreover, it is very important to understand the host-parasite interaction at the systemic level for a better understanding on the pathogenicity of the disease. Here, we report the genetic characterization of *S. scabiei* from India and host-parasite interaction in a porcine model.

Abstract: The burrowing mite *Sarcoptes scabiei* causes scabies in humans or mange in animals. It infests a wide range of mammalian species including livestock, companion animals, wild animals, and humans. Differential diagnosis of *Sarcoptes* varieties is key for epidemiological studies and for formulation of an eradication program. Host-parasite interaction at the systemic level is very important to understand the pathogenicity of the mite. This communication deals with the preliminary report on the genetic characterization of *S. scabiei* from India. Moreover, the effect of *S. scabiei* infestation on host physiology with special emphasis on serum biochemical parameters, lipid profile, oxidant/antioxidant balance, stress parameters, and immune responses were evaluated in a porcine model. Cytochrome C oxidase 1 and voltage-sensitive sodium channel based phylogenetic study could distinguish human and animals isolates but could not distinguish host or geographical specific isolates belonging to animal origin. An absence of host-specific cluster among animal isolates argues against the hypothesis of delineating *S. scabiei* as per host origin. Elevated levels of markers of liver function such as albumin, AST, ALT, ALP, and LDH in infested animals indicated impaired liver function in infested animals. *S. scabiei* infestation induced atherogenic dyslipidemia indicated by elevated levels of total cholesterol, low-density lipoprotein cholesterol and triglycerides, and a decreased level of high-density lipoprotein cholesterol. Oxidative stress in infested animals was indicated by a high level of nitric oxide and serum MDA as oxidative stress markers and low antioxidant capacity. *S. scabiei* triggered stress response and elevated levels of serum cortisol and heat shock proteins were recorded in infested animals. *S. scabiei* infestation increased the serum concentration of immunoglobulins and was associated with up-regulation of IL-2, IFN-γ, IL-1β, and IL-4 indicating both Th1 and Th2 response. The results of the study will be helpful for a better understanding of host-parasite interaction at the systemic level in crusted scabies in pigs.

Keywords: *Sarcoptes scabiei*; host-parasite interaction; molecular characterization; lipid profile; antioxidant

---

1. Introduction

Scabies in human, or mange in animals, caused by the burrowing mite *Sarcoptes scabiei*, is a neglected tropical infectious disease and is prevalent worldwide especially in developing and under-developed countries [1,2]. It infests more than 100 mammalian species including livestock, companion animals, wild animals, and humans [3–5]. Mange is considered as an emerging/re-emerging infectious disease and imparts huge economic loss to the livestock industry due to devastating morbidity and reduced production [3,5]. It is regarded as a significant public health concern in underdeveloped and developing countries as it is frequently associated with bacterial super-infection, which may sometimes be fatal especially in children if not treated in time [6]. Worldwide, around 300 million people are estimated to be infested with scabies every year [7] and its prevalence among children in Africa and indigenous communities of Northern Australia is astoundingly high ranging from 25 to 80% [8–10]. Scabies is broadly categorized into two forms, ordinary scabies with low mite burden (<20) or crusted scabies with high mite load and hyper-keratosis of the skin [11]. The mite *S. scabiei* causes inflammation of the skin which is associated with an exudate which forms crusts on the surface [12].

Controversy and conflicting opinion exist regarding speciation of *S. scabiei*; some believe that *S. scabiei* infesting different hosts are monospecific whereas others claim that they belong to different species or subspecies [13,14]. Currently, there is a broad consensus that the species *Sarcoptes* is divided into different varieties as per its host origin [13,15] and to date, more than 15 diverse varieties have been reported [2,16]. Of recent, this hypothesis has been challenged as cross-transmission or species jumping is very common in some varieties [17]. Therefore, differential diagnosis of *Sarcoptes* varieties is crucial for epidemiological studies [18] and formulating eradication strategies [15]. Morphotaxonomy has failed to differentiate between varieties as they share similar morphology [19]. Molecular diagnosis based on ribosomal DNA spacer region has been unsuccessful in identifying *S. scabiei* to the host level [20] and immunological diagnosis is also very challenging as different varieties produce immunologically identical proteins [21]. Recently, mitochondrial cytochrome oxidase 1 (COX1) based marker has shown promising results in varieties level confirmation of *S. scabiei* infestation. No information on genetic characterization or molecular confirmation in *S. scabiei* is available from India. The present study seems to be the first report on the genetic characterization of *S. scabiei* from India.

Sarcoptic mange is a very common ectoparasite disease in pigs [22,23]. It affects growth rate and reproductive performance in pigs and increases piglet mortality [24–26] leading to significant economic loss to the swine industry. A rare probability of zoonotic potential of *S. scabiei* var suis cannot be ignored [27]; it may be transmitted to affect a variety of different animal host species as well as pig handlers [28,29] causing severe itching. As pigs in semi-intensive systems in developing and under-developed countries are generally reared by tribes who keep a close association with the animals, there is a chance of transmission to humans, especially children and immunocompromised adults [4]. Moreover, in developed countries, pigs have been kept as pets which can transmit the disease to humans. Sarcoptic mange is a very common problem in the pig rearing regions of India, especially Northeast and Eastern parts of India [30]. In the Andaman and Nicobar Islands, Nicobari pigs are reared by Nicobarese tribal people and close association between tribal people and pigs is observed.

S. scabiei modulates host immunity, inflammatory, and complement reactions in order to be established to the host skin [31–33]. *S. scabiei* infestation triggers multiple reactions including allergic reactions, inflammation, innate immune reactions, and activation of immune components in the skin [34]. The salivary solutions of burrowing mites contain different bioactive compounds with potential to influence host physiological functions [35]. Mites are reported to influence the cytokine and chemokine secretion from keratinocytes and dermal fibroblasts [31,33] and disturb the balance between Th1 and Th2 immune responses [36]. In addition, *S. scabiei* has been reported to disturb the antioxidant defense system in mammalian hosts [37,38]. However, the studies were mostly in vitro using skin equivalents [31]. Little is known about the host parasite interaction in *S. scabiei* infestation at the systemic level. Therefore, the study aims at genetic characterization of *S. scabiei* isolated from Nicobari pigs and a better understanding on host parasite interactions with special emphasis on serum biochemical parameters, lipid profile, oxidant/antioxidant balance, stress parameters, and immune responses.

2. Materials and Methods

2.1. Ethics Approval

This research has been approved by the Institute Animal Ethics Committee (IAEC) of ICAR-Central Island Agricultural Research Institute (ICAR-CIARI), Port Blair, Andaman and Nicobar Islands, India on 12 January 2020 and the ethical approved project identification code is 'ICAR-CIARI/AS/AICRP-Pig/IAEC/4960 dated 12 January 2020'. All the methods were performed in accordance with the relevant national guidelines and regulations.

2.2. Study Area and the Animals

The study was conducted on Nicobari pigs maintained at the institute pig farm of ICAR-CIARI (11.6060° N, 92.7058° E). Sarcoptic mange infestation in pigs was identified during a routine visit to the farm. Fifteen *S. scabiei*-infested animals (5–6 months old, 8 male and 7 female) participated in the present study. Ten healthy animals (5–6 months old, 5 male and 5 female) were treated as control. Absence of *S. scabiei* infestation in control animals was confirmed by dermatological and microscopic examination. All the animals were examined for presence of fleas, lice, ticks, or any other ectoparasite, and animals negative for these were only considered in the study. Coproscopic examination for light and heavy eggs [12] confirmed that the animals were free from gastrointestinal parasites and the animals were serologically negative for classical swine fever which is endemic to the Andaman and Nicobar Islands. Both control and infested animals were maintained in 12-h light-dark cycle on concrete floor in separate pens to avoid chance of transmission of sarcoptic mange and fed with a commercial diet containing 18% crude protein, 2% crude fat, 6% crude fiber, 4% acid insoluble ash and 10% moisture. The animals were offered water ad libitum.

2.3. Collection and Processing of Clinical Samples for Microscopic Analysis

Skin scrapings were collected from the infested Nicobari pigs using surgical blades. Mineral oil was applied on the surgical blades and scrapings were taken from the crusted area of skin till little blood appeared. Collected material was boiled in 10% sodium hydroxide (NaOH) (*w/v*) solution to dissolve the keratinized tissue and was centrifuged at 500× *g* for five minutes and the sediment was examined under 100× magnification in a light microscope (Trinocular Compound Microscope, Quasmo,

Ambala, Haryana, India 180129/53268). Photographs of different stages of mite and eggs present were taken at 400× magnification.

2.4. DNA Extraction

DNA from skin scrapings was isolated by a commercial kit (DNeasy Blood and Tissue kit, Cat. No. 69504, Qiagen, Hilden, Germany). Briefly, 25 mg of skin scrapping was ground in lysis buffer using a sterile pestle and motor. For cell lysis, proteinase K was added to the ground scrapings and it was incubated at 56 °C in a water bath overnight. The completely lysed samples were used for genomic DNA extraction as per protocol recommended by the manufacturer. The DNA samples were kept at −80 °C until further use.

2.5. Amplification and Sequencing of Cytochrome C Oxidase Subunit 1 (COX1) and Voltage Sensitive Sodium Channel (VSSC) Gene Segments

For molecular identification and characterization of the mite, two gene segments namely COX1 and VSCC were chosen [34]. COX1 and VSCC gene fragments were amplified using primers described earlier by Erster et al. [34]. The PCR was performed in Thermocycler (Eppendorf, Hamburg, Germany) with the cycling conditions mentioned previously [34]. The amplicons were purified using a commercial kit (MinElute PCR Purification Kit, Cat. No. 28004, Qiagen, Hilden, Germany) as per manufacturer's protocol and sequence information was generated by dideoxy fingerprinting.

2.6. Sequence Analysis

Representative COX1 and VSSC sequences of *S. scabiei* from different hosts distributed in different geographical regions were retrieved from GenBank (www.ncbi.nlm.nih.gov) and a summary of the information has been depicted in Table S1. Alignment of the sequences was done by ClustalW [39] in MEGAX [40]. Phylogenetic tree was constructed in MEGAX using Maximum Likelihood method and Tamura–Nei model [41]. The reliability of the tree was judged by 1000 bootstrap replications. For phylogenetic tree construction, we trimmed extra nucleotides from our sequences and GenBank retrieved sequences to make a homogeneous length of 193 bp for COX1 and 436 bp for VSSC. Evolutionary relationship among different sequences was deduced by median-joining networks constructed in Network v 10 with default settings [42].

2.7. Estimation of Biochemical Parameters

Commercially available kits (TBL, Solan, India) were used for estimation of serum biochemical parameters such as total protein, albumin, globulin, glucose, aspartate aminotransferase (AST), alanine aminotransferase (ALT), alkaline phosphatase (ALP), lactate dehydrogenase (LDH) and creatine kinase (CK). The parameters were assessed in an automated clinical chemistry analyzer (Transasia Biomedical Limited, Mumbai, India). The reference values of the parameters are presented in Table S2.

2.8. Estimation of Oxidative Stress Parameters

To investigate the oxidant/antioxidant balance in serum following *S. scabiei* infestation, the markers of oxidative stress such as serum level of total nitric oxide (NO), total serum antioxidant activity, malonyldialdehyde (MDA) concentration, as well as the levels of antioxidant enzymes; superoxide dismutase (SOD), glutathione-S-transferase (GSH), catalase were determined and compared to control animals.

NO level was evaluated by a commercial kit (EZAssay^{TM} Nitric Oxide Estimation kit, HiMedia Laboratories Pvt. Ltd., Nashik, India) according to the manufacturer's instructions. The principle of the assay is the reduction of NO_3 which was then converted to a blue-colored azo compound. The absorbance was recorded at 630 nm.

Antioxidant activity in serum was estimated by a colorimetric kit (EZAssay^{TM} Antioxidant Activity Estimation kit, HiMedia Laboratories Pvt. Ltd., Nashik, India) following the protocol recommended by the manufacturer.

Level of lipid peroxidation was measured by measurement of malonyldialdehyde (MDA). Levels of MDA in serum were measured by a commercial colorimetric based assay kit (EZAssay^{TM} TBARS estimation kit for lipid peroxidation, HiMedia Laboratories Pvt. Ltd., Nashik, India).

Serum levels of catalase, superoxide dismutase, glutathione-S-transferase were measured by ELISA based methodology using catalase assay kit, superoxide dismutase assay kit, and glutathione assay kit (Cayman chemicals, Ann Arbor, MI, USA) as per manufacturer's protocol.

2.9. Estimation of HSPs in Serum

Serum levels of four heat shock proteins (HSP20, HSP40, HSP70, and HSP90) in infested and control animals were measured by ELISA-based methodology on biotin double antibody sandwich technology using commercial kits (Porcine HSP20, HSP40, SHP70 and HSP90 ELISA kit, Arsh Biotech Pvt. Ltd., Life Technologies, Delhi, India). Briefly, serum samples were added to the wells pre-coated with the respective HSP monoclonal antibodies. This was followed by the addition of respective anti-HSP antibodies labeled with biotin-streptavidin-HRP complex. Washing was done to remove the unbound proteins. This was followed by the addition of substrates and measurement of absorbance at 450 nm using a microplate reader (SpectraMax Plus, Molecular Devices, San Jose, CA, USA).

2.10. Measurement of Cortisol in Serum

Cortisol level in serum was measured by ELISA, based on biotin double antibody sandwich technology using a commercial kit (Porcine Cortisol ELISA kit, Arsh Biotech Pvt. Ltd., Life Technologies, Delhi, India) as per the protocol of the manufacturer. Briefly, serum samples were added to cortisol monoclonal antibodies pre-coated wells. This was followed by the addition of anti-cortisol antibodies labeled with biotin-streptavidin-HRP complex. Washing was done to remove the unbound proteins. This was followed by the addition of substrates and measurement of absorbance at 450 nm using a microplate reader (SpectraMax Plus, Molecular Devices, San Jose, CA, USA). The reference value of cortisol is presented in Table S2.

2.11. Lipid Profile Analysis

Serum levels of total cholesterol (TC), high-density lipoprotein cholesterol (HDLc), low-density lipoprotein cholesterol (LDLc), and triglycerides (TG) were evaluated. Levels of TC, HDLc, and TG were determined by enzymatic methods using commercially available kits (Jeev Diagnostics Pvt. Ltd., Chennai, India; Pathozyme Diagnostics, Kholapur, India; and Spinreact, S.A., Spain respectively). The concentration of LDL was deduced by the following formula: LDL = TC − HDL − (TG ÷ 5) [43]. Moreover, we calculated cardiac risk factor (CRF) and atherogenic index (AI) using the following formulas: CRF =TC/HDL [44] and AI = (TC − HDL)/HDL [45]. The reference values of the parameters are presented in Table S2.

2.12. Estimation of Serum Immune Parameters

The concentrations of total IgG, IgA, and IgM in serum were determined using quantitative ELISA kits (Arsh Biotech Pvt. Ltd., Delhi, India) according to the manufacturer's instructions.

Serum levels of cytokines (IL-2, IL-4, IL6, IL-8, IL-12, IL-1β, IFN-γ) in the serum of *S. scabiei*-infested and control animals were measured by sandwich ELISA methodology using commercial kits from Life Technologies, New Delhi, India.

2.13. Estimation of Apoptotic Markers

Serum concentrations of three apoptotic markers (Caspase-3, Caspase-7, and Casepase-8) were estimated by Nori® porcine ELISA kits (Life Technologies Pvt. Ltd., Delhi, India) according to the manufacturer's manual.

2.14. Histopathology

Skin scrapings from infested animals were collected and fixed in formalin solution (10%) at room temperature and a routine process was adopted for histopathology. In brief, samples were passed through ascending grades of alcohol (70–100%) for dehydration. The dehydrated samples were then cleared in benzene, transferred to melted paraffin (60 °C) for impregnation, and paraffin blocks were finally prepared in metal molds. Sections were prepared by using microtome (Leica, Wetzlar, Germany; Catalogue No. RM2235) at 5μM Slides were prepared as per standard protocol and were stained with hematoxylin and eosin.

2.15. Statistical Analysis

Data were presented as mean ± standard deviation and were normally distributed. Differences among control and infested groups were calculated by *t*-test using GraphPad Prism software (San Diego, CA, USA) (http://www.graphpad.com). The mean values with a significance of $p < 0.05$ were considered to be statistically significant.

3. Results

3.1. Clinical Signs of the Infested Animals

Clinical examination showed intense pruritus associated with hyperkeratosis and crusts in the skin. Skin of the animals was thickened and wrinkled in appearance which was a characteristic feature of crusted scabies/Norwegian scabies. Alopecia in the affected area was observed (Figure 1a).

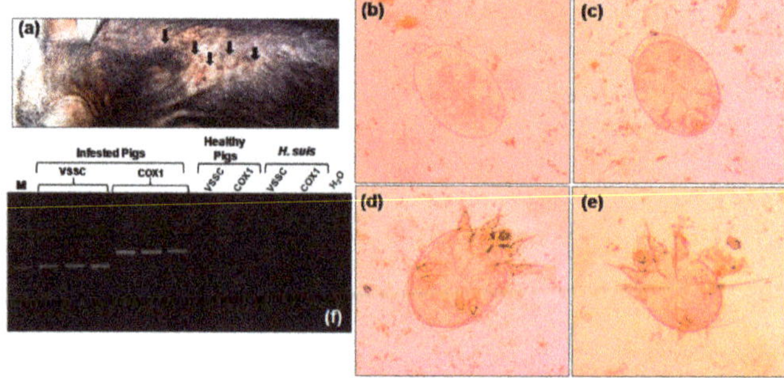

Figure 1. *Sarcoptes scabiei* infestation in Nicobari pig. (**a**) An infested animal with thickened and wrinkled skin and hair loss (black arrows), (**b**) mite egg (400× magnification), (**c**) mite egg containing larvae (400× magnification), (**d**) mite nymph (400× magnification), (**e**) larval stage of mite (400× magnification), (**f**) amplification of cytochrome C oxidase subunit 1 (COX1) and voltage sensitive sodium channel (VSSC) gene segments. Infested animals were positive for both of the gene fragments whereas control animals were negative. Primer specificity was verified using the lice (*Haematopinus suis*) DNA as negative control.

3.2. Microscopic Examination

On the basis of clinical symptoms, skin scrapings from infested animals were examined under microscope for the presence of mites. Microscopic examination confirmed the presence of eggs, eggs with six-legged larvae, and nymphal stage of *S. scabiei* (Figure 1b–e). The nymphal stages had short legs. The third and fourth pair of legs never projected beyond the body. Morphologically the mites were indistinguishable from *Sarcoptes* [12].

3.3. Molecular Identification and Characterization of S. scabiei

For molecular confirmation, one mitochondrial gene (COX1) and one nuclear gene (VSSC) were amplified using *Sarcoptes* specific primers and amplicons of both the genes were detected in infested pigs whereas in control pig, no amplicons were observed (Figure 1f). Specificity of the PCR was verified by using pig lice (*Haematopinus suis*) DNA as a negative control for both the genes (Figure 1f).

For differential diagnosis and molecular characterization, we generated partial sequence information of COX1 (2 samples) and VSSC (one sample) and the sequences were deposited in GenBank and obtained accession number MN986997-MN986998 for COX1 and MN986999 for VSSC. The sequences of COX1 and VSSC were 686 bp and 498 bp in length respectively. For phylogenetic analysis, we retrieved representative *S. scabiei* COX1 and VSSC sequences from different hosts distributed in different geographical regions. COX1 based phylogenetic tree (Figure 2a) indicated three distinct clades; two for human isolates (Clade A and B) and one for animal isolates (Clade C). In Clade A, two clearly defined clusters were detected; one includes human isolates from Hong Kong, South Korea, China and the other includes human isolates from France and Australia. Clade B contains human isolates from Panama. On the contrary, no host or region-specific clusters were observed in animal clade (Clade C). From the phylogenetic tree, it was evident that the Andaman isolates were phylogenetically close to pig isolate of Israel. A median-joining network was constructed to understand the evolutionary relationship of *S. scabiei* isolates. Two human-specific clades and one animal-specific clade were detected (Figure 2b).

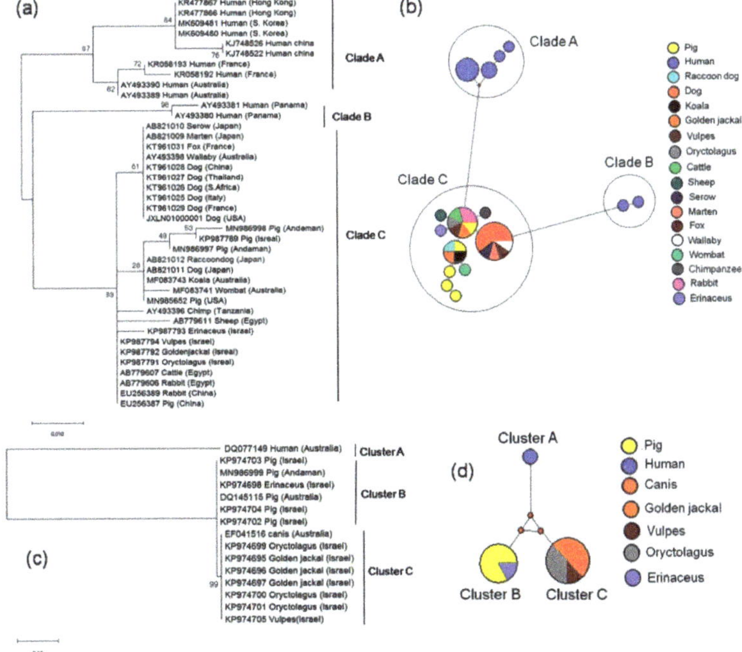

Figure 2. Evolutionary relationship of different isolates of *S. scabiei*. (**a**) COX1 based phylogenetic tree, (**b**) COX1 based network profile, (**c**) VSSC based phylogenetic tree, (**d**) VSSC-based network profile. Phylogenetic tree was constructed based on maximum likelihood method using Tamura–Nei model [41] implemented in MEGAX following 1000 bootstrap replications. Network was drawn in Network 10 with default settings [42].

Phylogenetic tree based on VSSC sequences showed three host-specific distinct clusters, one for human isolate (Cluster A) and the other two for animal isolates (Figure 2c). Within animal isolates, pig and erinaceus isolates formed one cluster (Cluster B) whereas oryctolagus, golden jackal, vulpes, and canis isolates formed another cluster (Cluster C). Pig isolates of Andaman belonged to Cluster B with pig isolates from Israel, Australia, and erinaceus isolate from Israel. The VSSC based network tree is presented in Figure 2d and three clearly defined clusters were depicted.

3.4. Serum Biochemical Parameters in Crusted Scabies

A significant decrease in plasma, total protein, albumin, and glucose, and significant increase in globulin, ASL, ALT, CK, LDL, and ALP in infested group as compared to those of control group was observed (Figure 3). Moreover, the ration of albumin and globulin (A:G) in infested group (0.6359 ± 0.048) was decreased than that of control group (1.121 ± 0.066).

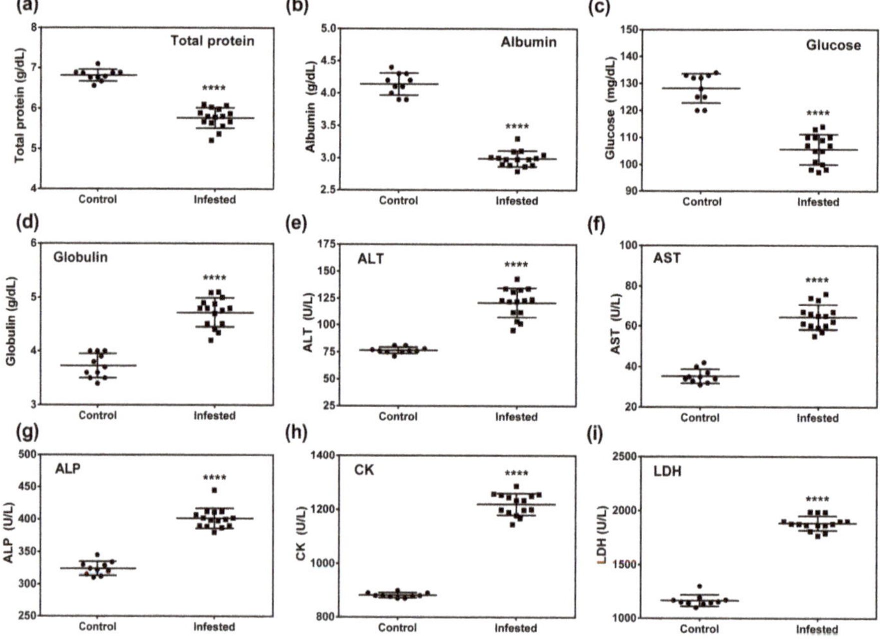

Figure 3. Effect of crusted scabieson serum biochemical parameters in a porcine model. (**a**) Serum total protein, (**b**) serum albumin, (**c**) serum glucose, (**d**) serum globulin, (**e**) alanine aminotransferase (ALT), (**f**) aspartate aminotransferase (AST), (**g**) alkaline phosphatase (ALP), (**h**) creatine kinase (CK), (**i**) lactate dehydrogenase (LDH). Data in scatter plot are shown as mean ± SD. *t*-test was performed to find out significant difference between the two groups. **** denotes $p \leq 0.0001$.

3.5. Crusted Scabies Leads to Dyslipidemia in Pigs

The results of the lipid profile analysis are presented in Figure 4. In the infested group, an atherogenic dyslipidemic profile indicated by elevated levels of total cholesterol, low-density lipoprotein cholesterol (LDL-c) and triglycerides (TG), and a decreased level of high-density lipoprotein cholesterol (HDLc) as compared to control group was observed. This was further supported by higher CRF and AI values in infested animals than those of control animals.

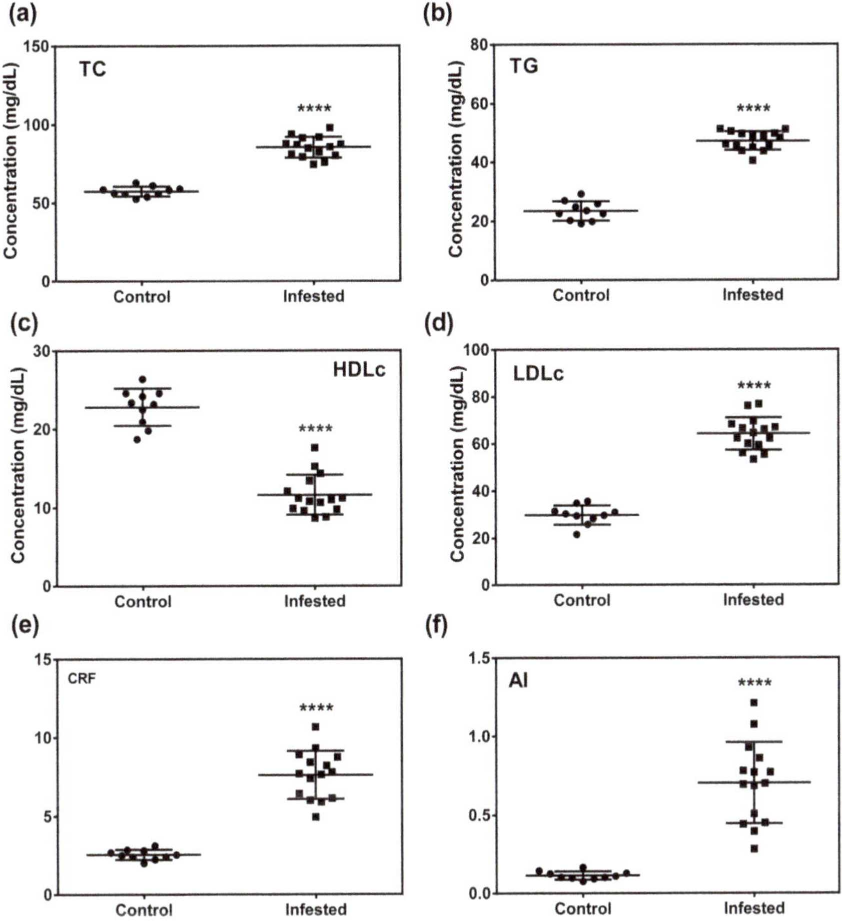

Figure 4. Effect of crusted scabies on serum lipid profile in a porcine model. (**a**) Total cholesterol (TC) concentration, (**b**) triglycerides (TG) concentration, (**c**) high-density lipoprotein cholesterol (HDLc) concentration, (**d**) low-density lipoprotein cholesterol (LDLc) concentration, (**e**) cardiac risk factor (CRF), (**f**) atherogenic index (AI). Data in scatter plot are shown as mean ± SD. *t*-test was performed to find out significant difference between the two groups. **** denotes $p \leq 0.0001$.

3.6. Crusted Scabies Induces Oxidative Stress

To investigate the effect of *S. scabiei* infestation on oxidant/antioxidant balance in serum, the markers of oxidative stress such as total antioxidant activity (T-AOC), total nitric oxide (TNO) concentration, levels of lipid peroxides, and malonyldialdehyde (MDA) as well as the levels of antioxidant enzymes; superoxide dismutase (SOD), glutathione-S-transferase (GSH) and catalase were determined and compared to control animals (Figure 5). The infested group showed significantly higher serum total nitric oxide (TNO) concentration than the control group. Total antioxidant activity (T-AOC) in the infested group was significantly lower than that of control group. Activity of catalase, SOD, and concentration of GSH in infested animals decreased significantly as compared to those of control animals. On the other hand, MDA concentration in the infested group was found significantly higher than its corresponding value in the control group (Figure 5).

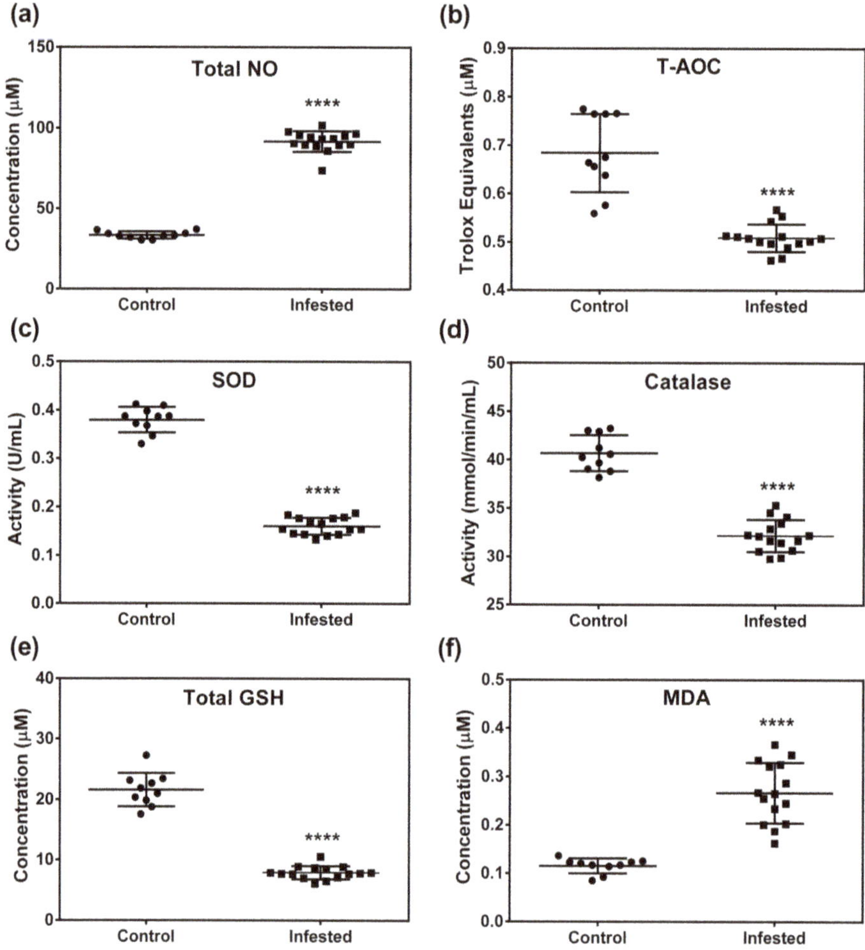

Figure 5. Effect of crusted scabies on antioxidant profiles and oxidative stress indicator in a porcine model. (**a**) Total nitric oxide (NO) concentration, (**b**) total antioxidant capacity (T-AOC), (**c**) superoxide dismutase (SOD) activity, (**d**) catalase activity, (**e**) glutathione-S-transferase (GSH) concentration, (**f**) malonyldialdehyde (MDA) concentration. Data in scatter plot are shown as mean ± SD. *t*-test was performed to find out significant difference between the two groups. **** denotes $p \leq 0.0001$.

3.7. Crusted Scabies Is Associated with Stress

To investigate whether *S. scabiei* infestation leads to stress in host animals, level of stress biomarkers such as serum cortisol concentration and serum levels of four heat shock proteins (HSP20, HSP40, HSP70, and HSP90) in control and infested animals were analyzed. Serum cortisol concentration in infested group was found significantly higher than that of control group (Figure 6).

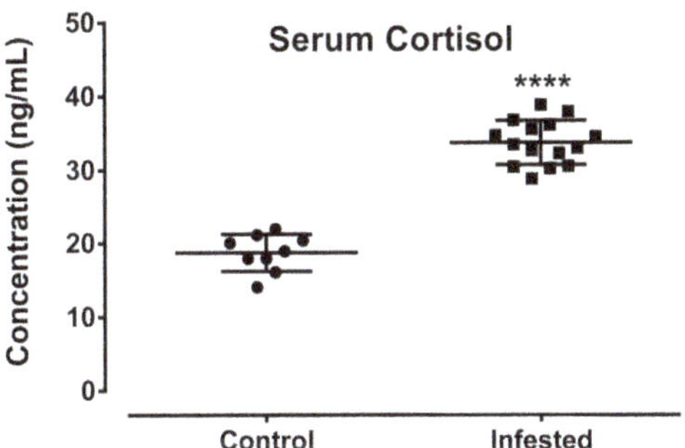

Figure 6. Effect of crusted scabieson serum cortisol concentration in a porcine model. Data in scatter plot are shown as mean ± SD. *t*-test was performed to find out significant difference between the two groups. **** denotes $p \leq 0.0001$.

HSP20, HSP70, and HSP90 were up-regulated in infested animals compared with the control animals. On the other hand, no significant change in HSP40 level between control and infested animals was detected (Figure 7).

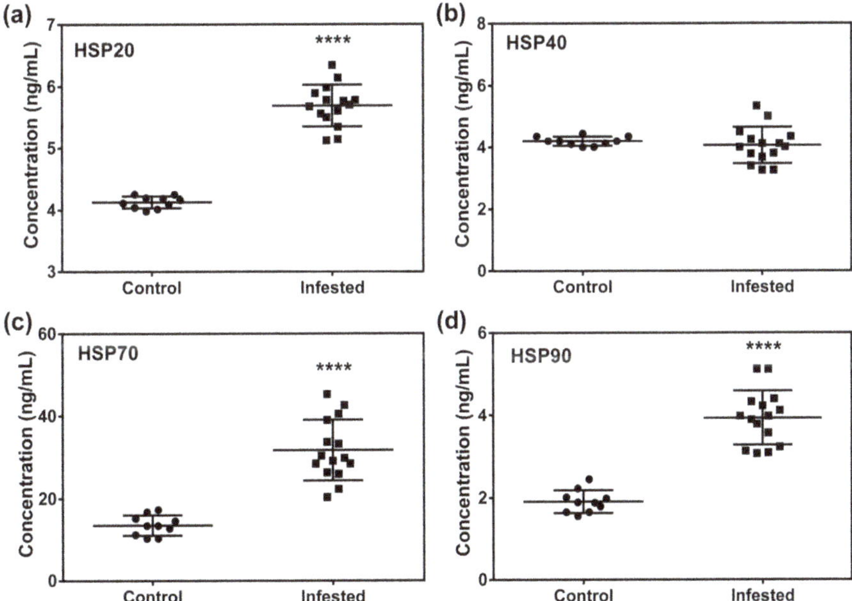

Figure 7. Effect of crusted scabies on serum heat shock proteins (HSPs) concentration in a porcine model. (**a**) HSP20, (**b**) HSP40, (**c**) HSP70, and (**d**) HSP90.Data in scatter plot are shown as mean ± SD. *t*-test was performed to find out significant difference between the two groups. **** denotes $p \leq 0.0001$.

3.8. Crusted Scabies Up-Regulates Serum Levels of Apoptotic Markers

Serum concentrations of all three apoptotic markers (Caspase 3, 7, and 8) in infested and control animals were recorded. Up-regulation of all three apoptotic markers was detected in infested group as compared to control group (Figure 8).

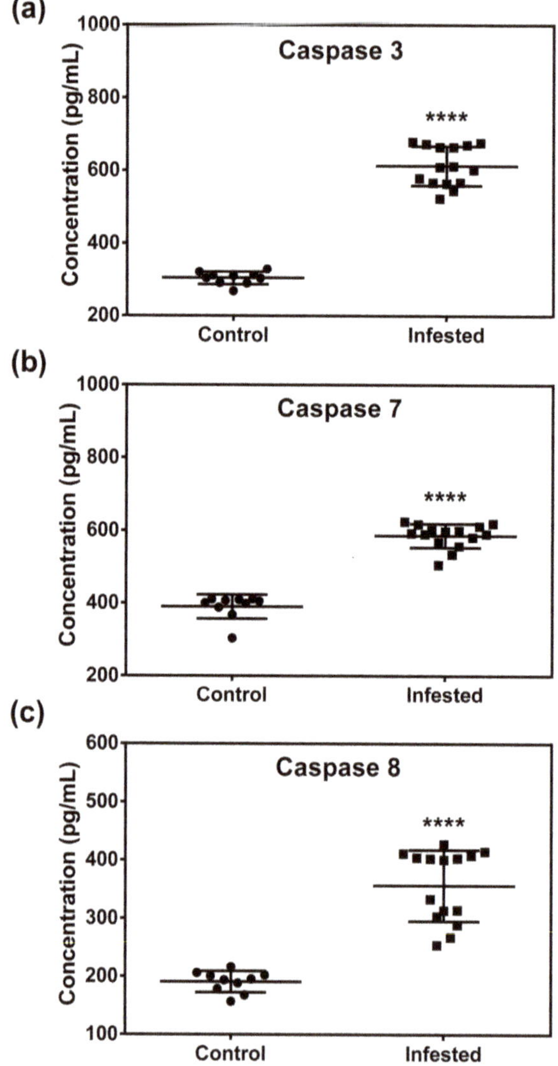

Figure 8. Effect of crusted scabies on serum levels of apoptotic markers in a porcine model. (**a**) Caspase 3, (**b**) caspase 7, (**c**) caspase 8. Data in scatter plot are shown as mean ± SD. *t*-test was performed to find out significant difference between the two groups. **** denotes $p \leq 0.0001$.

3.9. Crusted Scabies Alters Immune Response

Effect of *S. scabiei* infestation on the concentration of immunoglobulins and the level of cytokines were evaluated. Significantly higher concentrations of serum IgA, IgG, and IgM were detected in the infested group as compared to those of the control group (Figure 9a–c). Up-regulation of pro-inflammatory cytokines (IL-2, IL-6, IL-12, IL-1β, and IFN-γ) and anti-inflammatory cytokine IL-4 were detected in the infested group than the control group (Figure 9d–j). On the other hand, no significant difference in IL-8 concentration between the two groups was detected (Figure 9g).

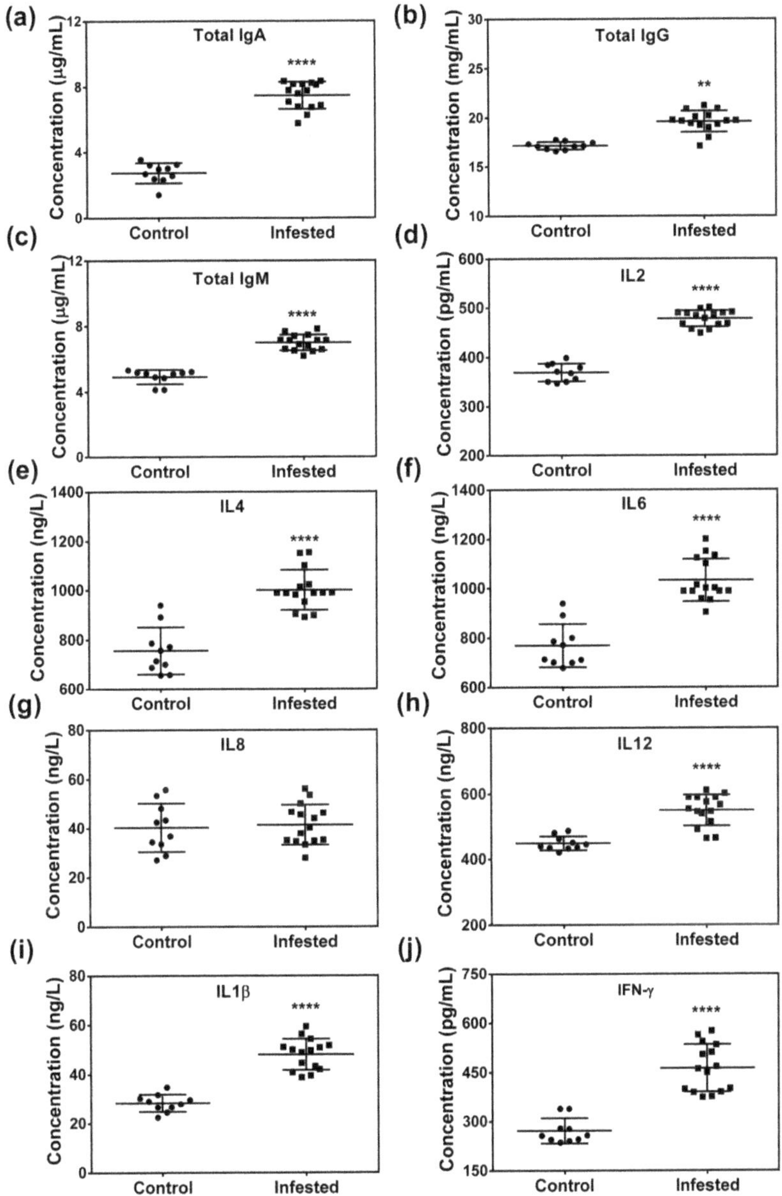

Figure 9. Effect of crusted scabieson total serum immunoglobulins and cytokine response in a porcine model. (**a**) Total IgA, (**b**) total IgG, (**c**) total IgM, (**d**) IL-2, (**e**) IL-4, (**f**) IL-6, (**g**) IL-8, (**h**) IL-12, (**i**) IL-1β, and (**j**) IFN-γ. Data in scatter plot are shown as mean ± SD. *t*-test was performed to find out significant difference between the two groups. ** denotes $p \leq 0.01$, **** denotes $p \leq 0.0001$.

3.10. Histopathology

Parakeratotic hyperkeratosis characterized by thickening of the stratum corneum and the presence of nuclei was observed. Mite with its exoskeleton and remnants were detected in the tunnel of stratum corneum of the epidermis and appeared as a cleft in the upper epidermis. Epidermis exhibited acanthosis and spongiosis associated with dense eosinophilic dermal infiltrate. Superficial perivascular or diffuse infiltrate of lymphocytes and histiocytes, accompanied by neutrophils and eosinophils were evident in the dermis. There was psoriasiform hyperplasia characterized by epidermal projections into the dermis for interdigitaion with dermal papillae. Excessive production of squames suggested crust formation (Figure 10).

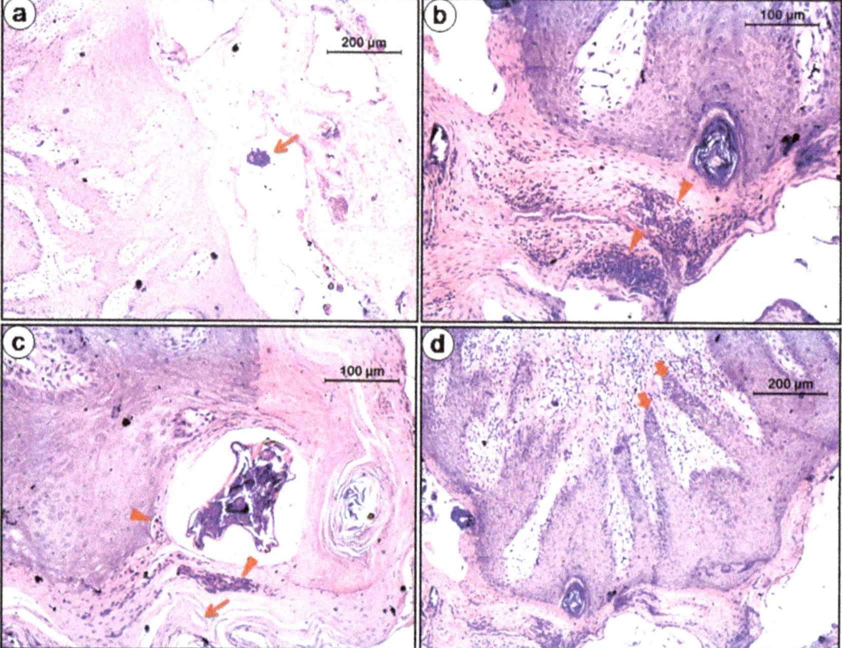

Figure 10. Histopathology of infested skin (**a**) Mite (arrow) with its exoskeleton and remnants prevailed in the tunnel of stratum cornium of epidermis (original magnification 100×, scale bar: 200 µm), (**b,c**) Parakeratotic hyperkeratosis characterized by upsurge in the thickness of the stratum corneum and excessive production of squames (arrow) followed by infiltration (arrow head) into epidermis suggested crust formation. There was infiltration of cells especially eosinophils (arrow head) into epidermis and dermal vascular area causing dermatitis (original magnification 200×, scale bar: 100 µm), (**d**) Psoriasiform hyperplasia: Note the epidermal projections (block arrow) into the dermis for interdigitaion with dermal papillae (original magnification 100×, scale bar: 200 µm).

4. Discussion

Scabies is considered a disease of resource-poor communities of underdeveloped and developing countries and a neglected tropical disease [46]. Wide host range and presence of different varieties necessitate differential identification for epidemiological study and for designing eradication strategy. *S. scabiei* infestation in multiple hosts including humans, companion animals, farm animals, lab animals, and wild animals throughout the globe points toward pathogen dispersal and spillover [47,48]. Bidirectional interactions among human, animals, pathogen, and environment might be the underlying reason behind its multi-host adaptation [49]. As morphological and serological tests are not much successful at variety level identification, molecular markers have emerged as alternative differential diagnostic tools. Mitochondrial cytochrome C oxidase subunit I gene (COX1) is a well-established universal marker for species identification due to its high variability [50,51] and is being used extensively for DNA bar-coding of medically important parasites [52]. Erster et al. [34] used nuclear marker

VSSC for genetic characterization of *S. scabiei* from multiple hosts in Israel. In the present study, we used COX1 and VSSC markers for molecular characterization of the *S. scabiei* isolated from pig host. COX1 based phylogenetic tree (Figure 2a) indicated that isolates under animal clade (Clade C) were phylogenetically closer to Clade B (human isolates from Panama) than Clade A (human isolates from Hong Kong, South Korea, China, France, and Australia). In general, human isolates of *S. scabiei* are heterogeneous in nature, and proximity of Clade B and Clade C may be due to low gene flow between mites of those groups [53]. COX1 and VSSC based phylogenetic study could not distinguish host or geographical specific isolates belonging to animal origin. Absence of host-specific cluster among animal isolates argues against the hypothesis of delineating *S. scabiei* as per host origin and provides support to the hypothesis that *S. scabiei* from different hosts and different geographical location arose from a single speciation event [16,54,55]. Evolution of different varieties in different hosts might be due to mitochondrial capture or selective sweep phenomenon which is considered as the driving force behind uniparental inheritance [56]. This hypothesis has been proposed as the adaptation of *Sarcoptes* in different marsupials in Australia [48]. Moreover, parasites with low independent dispersal capacities face strong population bottlenecks; under such situations they lower their genetic diversity to maintain their adaptive potential [57,58]. This might explain the low genetic diversity in *S. scabiei* among animal isolates.

S. scabiei is a contagious disease with a broad host range of more than 100 species and its effects across species are generally conserved [59]. Therefore, it is reasonable to assume that understanding the changes in one host may have significance in other hosts [60]. In recent years, pigs have gained popularity as a biomedical model in translational research due to its anatomical and physiological similarity with humans [61,62]. Close resemblance between human and pig in clinical manifestation and disease progression of *S. scabiei* [63,64] underscore the importance of porcine model in understanding host-parasite interactions of scabies/mange. The present study showed marked changes in serum biochemical parameters, lipid profiles, stress parameters, oxidant/antioxidant balance, and immune parameters following crusted scabies in a porcine model.

Increased ALT, AST, ALP, LDH, CK, and decreased albumin and glucose levels in infested animals indicate organ damage, especially compromised liver function [65]. Oxidative stress has been associated with cell necrosis and apoptosis through activation of several cell signaling pathways such as JNK, mitogen-activated protein kinase (MAPK) leading to cellular and organ damage [66,67]. It is a well-established fact that oxidative stress causes liver injury [68] as the liver is a major target organ of ROS [69]. Moreover, free radicals, especially ROS catalyze the oxidation of LDL to generate oxidized LDL and oxidized LDL also promotes cellular apoptosis in different organs [67,70]. In the present study, crusted scabies induced oxidative stress in pigs which might have detrimental effects on organs including the liver. Oxidative stress-induced organ damage might explain elevated levels of AST, ALT, and alkaline phosphatase. Alteration in the biochemical parameters was also reported in sarcoptic mange affected goats and dogs [37,38,71]. As the liver is the major organ for albumin synthesis and glucose generation through gluconeogenesis and glucogenolysis [72], impaired liver function may attribute low levels of glucose and albumin in infested animals. In addition, high ROS has been reported to impair insulin signaling [73,74] causing an increase in insulin secretion; which might be another reason for low glucose concentration in infested animals. On the contrary, up-regulation in globulin production in infested animals might be due to the production of antibodies against *S. scabiei* antigens [38]. Low glucose and albumin and high globulin were also reported in Sarcoptes-infested goats [37] and dogs [38].

Abnormal blood lipid metabolites including elevated TC, LDLc, and TG were detected in infested animals than control animals. Imbalance in lipid metabolites is considered to be a predisposing cause of atherosclerosis and cardiovascular disorders [75]. In the present study, atherogenic dyslipidemia associated with high CRF and AI values in infested animals is an indicator of their susceptibility towards cardiovascular diseases. Imbalance in lipid metabolism is associated with many dermatological disorders especially chronic inflammatory skin diseases such as psoriasis, litchen planus, granuloma annulare, and histiocytosis [76]. Chronic inflammation in animals with crusted scabies may explain the abnormal lipid profiles. Moreover, oxidative stress which was also observed in the present study induced production of TNF-Alpha by Kupffer cells, which might induce inflammation [68]. Increased levels of pro-inflammatory cytokines following infestation might be the underlying cause behind dyslipidemia [76]. Increased serum levels of pro-inflammatory cytokines such as IL-2, IL-6, IL-12, IL-1β, and IFN-γ in infested animals were detected in the present study. Pro-inflammatory cytokines such as IL-6, IFN-γ inhibit the activity of lipoprotein lipase (LPL) [77] which hampers the clearance of VLDL and LDL cholesterol [78,79] leading to their increased concentration in blood. Moreover, crusted scabies induced oxidative stress in animals, and under oxidative stress, ROS induced activation of atherogenic genes via NF-kappa B [67] may further explain the atherogenic dyslipidemia in infested animals. In addition, alopecia due to infestation results in disruption of pelt-environment interface which in turn hampers thermoregulation as excessive heat loss to the environment occurs through the skin. This creates a negative energy balance to the host [60]. Mobilization of fat from fat stores to meet the energy deficiency may induce imbalance in lipid profile. An imbalance in fatty acid composition in adipose tissue with up-regulated omega-6 acids and down-regulated omega-3 acids in *S. scabiei* infested wombat host was reported [60]. The results of the study indicate that crusted scabies may shift lipid profile towards atherogenic dyslipidemia with high susceptibility towards cardiovascular disorder.

We determined the serum concentration of cortisol which is considered a biomarker of stress. The study found a significant increase in serum cortisol concentration in infested animals than control animals which indicated that infestation triggered stress response. Hypothalamic-pituitary-adrenal (HPA) axis, an important hormonal regulator of stress, is thought to be involved in the production of cortisol hormone [80]. In stressful conditions, activation of HPA axis followed by elevated plasma cortisol level is an adaptive response of the host [81]. HSPs are constitutively produced in normal physiological conditions whereas stress of any kind including environmental stress, viral infection, glucose deprivation, exposure of toxins, and oxidative stress induce over-expression of HSPs or cell-stress proteins [82]. HSPs are a classic example of proteins with 'moonlighting functions' [83] as they are involved in a number of physiological functions. Besides acting as chaperone and protein folding, involvement of HSPs in immune function such as antigen processing and presentation, expression of innate receptors and stimulation of innate immunity is well documented [84]. Moreover, HSPs, especially HSP70 and 27 play vital roles in stress-induced DNA repair by modulating DNA-repair enzymes, thus maintaining genome integrity [84]. Stress up-regulates the intracellular production of HSPs and they are released into the circulation following a non-classical secretion pathway [85]. In the present study, crusted scabies induced cellular and oxidative stress which in turn might have stimulated over-expression of HSPs. Thus, HSP orchestrated activation of heat shock response is an adaptation mechanism under stress [86]. Oxidative stress shifts the intracellular environment towards the oxidative state which in turn disturbs the protein native conformation and proteins lose their folded structure [86]. HSPs act as sensors to cellular redox changes both in prokaryotes and in eukaryotes; then give signals for activation of the heat stress response [86]. HSP70 and HSP90 act as the primary sensors of protein misfolding. Moreover, HSP27 has been reported to act as an antioxidant by preventing oxidation of glutathione under oxidative stress [87]. HSP70 and HSP90 act as anti-apoptotic proteins; they exert their role by binding to apoptosis protease activity factor 1 (Apaf-1) blocking downstream cascades of apoptosis pathway [88,89]. In the present study, the infested animals under oxidative stress might have induced cellular apoptosis. This is supported by the up-regulation of serum levels of apoptotic markers in infested animals. The up-regulation of HSPs in infested animals might be an adaptation mechanism to prevent cellular apoptosis.

Oxidative stress has been reported to be involved in the pathogenesis of several diseases including parasitic infections [90]. Cellular metabolism produces reactive oxygen species (ROS) and reactive nitrogen species (RNS) [91]. At a low to moderate level, they are involved in several physiological functions including cell signaling as well as immune functions [92]. On the contrary, at a high level, they cause damage to molecules including lipids, proteins, lipoproteins, and nucleic acids starting a chain reaction of free radical formation leading to a condition known as oxidative stress [93,94]. Therefore, tight regulation in production of ROS and RNS and delicate balance between beneficial and detrimental effects is vital for cellular homeostasis. Under oxidative stress, the antioxidant defense system elicits several antioxidant enzymes to fight against free radicals and maintain homeostasis. Superoxide dismutase, catalase, and glutathione peroxidase act as first-line defense antioxidants to suppress or prevent the formation of reactive species in cells [95]. SOD catalyzed the conversion of O_2^- into H_2O_2 which is further reduced to H_2O and O_2 by either catalase or glutathione peroxidase [96]. In the present study, it was observed that crusted scabies caused an increase in the production of serum total NO in pigs. Though we did not measure ROS, oxidative stress induction in infested animals was evident from the reduction of total antioxidant capacity, activity of catalase, SOD, and concentration of GSH as well as an increase in MDA concentration as compared to control animals. The down-regulation of T-AOC and antioxidant enzymes indicate that they have been over-consumed to tackle oxidative stress [97]. The results of our study are in agreement with the report of Beighet al. [98], in which decreased catalase and GSH in dogs suffering from dermatophytosis was reported. A decrease in antioxidant enzyme activities was also reported in buffaloes infested by sarcoptic mange [99] and in sheep, which was suffering from psoroptic mange [100]. Increased MDA concentration in infested animals indicated oxidative injury. Cell membranes with a rich store of polyunsaturated fatty acids are excellent targets for free radical attacks [101] and oxidation of lipid molecules induce production of MDA [102]. So, MDA, the end product of lipid peroxidation, is considered as a hallmark of oxidative damage [103]. In the present study, increased MDA concentration in infested animals might be associated with cellular damage. An increase in skin MDA in *S. scabiei* infested dogs were reported by Nwufoh et al. [104].

S. scabiei infestation is associated with modulation of various parameters of host immune responses; humoral, adaptive, and inflammatory immune responses [31,33]. IgA, IgG, and IgM are the major immunoglobulins associated with humoral immunity. IgM is produced at the early stage of antibody response against a foreign antigen. IgG, being the most abundant antibody in blood and extracellular fluid, is involved in systemic immune response. IgA is generally associated with mucosal immunity and primarily found in secretions [105]. In the present study, up-regulation of total IgA, IgG, and IgM in infested pigs than the control group was observed. Elevated levels of immuniglobulins in infested animals indicated that mite had elicited antibody-mediated immune response against mite antigens although it was not possible to know whether or not the antibody response was *S. scabiei* specific. Up-regulation of total IgG might be a consequence of secondary bacterial infection which is very common in scabies infestation. Increased IgG in *S. scabiei* infested rabbits, dogs [21,106], and human [107] was reported. Elevated levels of circulatory IgA were reported in human crusted scabies [107].

We further assessed the systemic inflammatory response of the animals by determining serum levels of pro-inflammatory cytokines including IL-1β, IL-2, IL-6, IL-12, and IFN-γ and anti-inflammatory cytokine IL-4. Cytokines play key roles in the host immune and inflammatory responses as well as in the maintenance of tissue integrity [108]. The present study recorded dysregulation of cytokine balance which might be due to modulation of immunity by *S. scabiei* released antigens. Th1 and Th2 are the two major subpopulations of T helper cells and parasitic infections are associated with either Th1 or Th2 polarized immune response [109]. IL-2, TNF-β, and IFN-γ are Th1 cells specific cytokines whereas Th2 cells release IL-4, IL-5, IL-10, and IL-13 as signature cytokines [110]. Naïve helper T cells (Th0) can be differentiated to either Th1 or Th2 based on distinct activation pathways; cytokines such as IFN-γ andIL-12 activates Th0 to Th1 cells, otherwise, IL-10 and IL-4 are crucial stimulatory cytokines for Th2 cell polarization [111]. Intricate balance among different cytokines is crucial for homeostasis of mammalian species and imbalance may pose threat to health [112]. In the present study, *S. scabiei* infestation was associated with up-regulation of IL-2, IFN-γ, IL-1β, and IL-4 indicating both Th1 and Th2 response. The most probable cause of the mixed Th1/Th2 response might be dysregulation of cytokine balance due to modulation of immunity by *S. scabiei* released antigens or changes of cytokine levels after a chronic infection. Th1 immune response is generally associated with protection in most infectious diseases whereas Th2 immune response elicits high titers of antibody production and cell-mediated inflammation [113]. Moreover, allergic inflammatory diseases are dominated by Th2 immune response [6]. Induction of mixed Th1/Th2 immune response in the current study indicates that the immune response to mange infestation is complex and local immune response analysis is necessary for a better understanding of the disease. Th2 response had been described as a predominant immune response in crusted scabies and other allergic inflammatory disorders [6]. Bioactive compounds in arthropod saliva trigger different immune response to host. Low molecular weight salivary components cannot act as antigen but can bind to skin proteins as haptens and stimulate Th1 response. Some salivary antigens may cause basophil hypersensitivity (Th1 response) by binding to epidermal langerhans cells. Salivary antigens also may trigger Th2 response in association with IgE production and type I hypersensitivity [107].

The histopathological findings in the study mimic the classical narrative of sarcoptic mange in mammalian species [3]. Antigenic materials including excretion and secretions of mites trigger hypersensitivity reactions in the skin [114] which might be the reason behind eosinophilia observed in the infested animals. Our observations are consistent with those in vulpes [115], racoon dog [116], wombats [117] and Iberian ibex [118].

5. Conclusions

This communication is the first report on genetic characterization of *S. scabiei* from India. Mitochondrial (COX1) or nuclear (VSSC) based markers could not distinguish *S. scabiei* at a variety level especially for animal isolates which suggests that delineating varieties based on host origin is not warranted. The study contributes to the rich pool of knowledge on the consequences of crusted scabies on host physiology. We could establish a new connection showing that mange infestation results in an atherogenic dyslipidemia in the host.

Supplementary Materials: The following are available online at http://www.mdpi.com/2076-2615/10/12/2312/s1, Table S1: COX1 and VSSC sequences used in the study, Table S2: Reference values of different biochemical parameters in pigs.

Author Contributions: Conceptualization, A.K.D. and S.S.; methodology, A.K.D., S.S., P.P., S.M., S.K.R., G.S. and K.M.; software, A.K.D. and S.B.; validation, D.M. and A.K.; formal analysis, A.K.D. and A.K.B.; investigation, A.K.T., D.B., G.S., and A.K.D.; resources, A.K.D., D.B. and A.K.; data curation, A.K.D.; writing—original draft preparation, A.K.D.; writing—review and editing, D.B.; visualization, D.B.; supervision, S.B.; project administration, A.K.D.; funding acquisition, A.K.D. All authors have read and agreed to the published version of the manuscript.

Funding: This research was funded by a Grant from All India Coordinated Research Project on Pig (AICRP on Pig), Indian Council of Agricultural Research, New Delhi, India with Grant number AICRP-Pig/ICAR-CIARI.

Acknowledgments: The authors acknowledge the Director, ICAR-Central Island Agricultural Research Institute (CIARI), Port Blair, Andaman and Nicobar Islands, India for providing all the necessary facilities to carry out the present study.

Conflicts of Interest: The authors declare no conflict of interest. The funders had no role in the design of the study; in the collection, analyses, or interpretation of data; in the writing of the manuscript, or in the decision to publish the results.

References

1. Heukelbach, J.; Wilcke, T.; Winter, B.; Feldmeier, H. Epidemiology and morbidity of scabies and *pediculosis capitis* in resource-poor communities in Brazil. *Br. J. Dermatol.* **2005**, *153*, 150–156. [CrossRef] [PubMed]
2. Alasaad, S.; Rossi, L.; Heukelbach, J.; Pérez, J.M.; Hamarsheh, O.; Otiende, M.; Zhu, X.Q. The neglected navigating web of the incomprehensibly emerging and re-emerging *Sarcoptes* mite. *Infect. Genet. Evol.* **2013**, *17*, 253–259. [CrossRef] [PubMed]

3. Bornstein, S.; Morner, T.; Samuel, W.M. Sarcoptesscabiei and sarcoptic mange. In *Parasitic Diseases of Wild Mammals*, 2nd ed.; Samuel, W.M., Pybus, M.J., Kocan, A.A., Eds.; Manson Publishing: London, UK, 2001; pp. 107–119.

4. Walton, S.F.; Currie, B.J. Problems in diagnosing scabies, a global disease in human and animal populations. *Clin. Microbiol. Rev.* **2007**, *20*, 268–279. [CrossRef] [PubMed]

5. Dagleish, M.P.; Ali, Q.; Powell, R.K.; Butz, D.; Woodford, M.H. Fatal *Sarcoptesscabiei* infection of blue sheep (*Pseudoisnayaur*) in Pakistan. *J. Wildl. Dis.* **2007**, *43*, 512–517. [CrossRef] [PubMed]

6. Bhat, S.A.; Mounsey, K.E.; Liu, X.; Walton, S.F. Host immune responses to the itch mite, *Sarcoptesscabiei*, in humans. *Parasites Vectors* **2017**, *10*, 385. [CrossRef] [PubMed]

7. Hengge, U.R.; Currie, B.J.; Jäger, G.; Lupi, O.; Schwartz, R.A. Scabies: A ubiquitous neglected skin disease. *Lancet Infect. Dis.* **2006**, *6*, 769–779. [CrossRef]

8. Terry, B.C.; Kanjah, F.; Sahr, F.; Kortequee, S.; Dukulay, I.; Gbakima, A.A. *Sarcoptesscabiei* infestation among children in a displacement camp in Sierra Leone. *Public Health* **2001**, *115*, 208–211. [CrossRef]

9. Carapetis, J.R.; Connors, C.; Yarmirr, D.; Krause, V.; Currie, B.J. Success of a scabies control program in an Australian aboriginal community. *Pediatr. Infect. Dis. J.* **1997**, *16*, 494–499. [CrossRef]

10. Andrews, R.M.; McCarthy, J.; Carapetis, J.R.; Currie, B.J. Skin disorders, including pyoderma, scabies, and tinea infections. *Pediatr. Clin. N. Am.* **2009**, *56*, 1421–1440. [CrossRef]

11. Currie, B.J.; McCarthy, J.S. Permethrin and ivermectin for scabies. *N. Engl. J. Med.* **2010**, *362*, 717–725. [CrossRef]

12. Soulsby, E.J.L. *Helminths, Arthropods and Protozoa of Domesticated Animals*, 7th ed.; Bailliere Tindall: London, UK, 1982.

13. Walton, S.F.; Dougall, A.; Pizzutto, S.; Holt, D.; Taplin, D.; Arlian, L.G.; Morgan, M.; Currie, B.J.; Kemp, D.J. Genetic epidemiology of *Sarcoptesscabiei* (Acari: Sarcoptidae) in northern Australia. *Int. J. Parasitol.* **2004**, *34*, 839–849. [CrossRef] [PubMed]

14. Daszak, P.; Cunningham, A.A.; Hyatt, A.D. Emerging infectious diseases of wildlife—Threats to biodiversity and human health. *Science* **2000**, *287*, 443–449. [CrossRef] [PubMed]

15. Alasaad, S.; Oleaga, Á.; Casais, R.; Rossi, L.; Min, A.M.; Soriguer, R.C.; Gortázar, C. Temporal stability in the genetic structure of *Sarcoptesscabiei* under the host-taxon law: Empirical evidences from wildlife-derived *Sarcoptes* mite in Asturias, Spain. *Parasites Vectors* **2011**, *4*, 151. [CrossRef] [PubMed]

16. Berrilli, F.; D'Amelio, S.; Rossi, L. Ribosomal and mitochondrial DNA sequence variation in *Sarcoptes* mites from different hosts and geographical regions. *Parasitol. Res.* **2002**, *88*, 772–777. [CrossRef] [PubMed]

17. Menzano, A.; Rambozzi, L.; Rossi, L. Outbreak of scabies in human beings, acquired from chamois (*Rupicaprarupicapra*). *Vet. Rec.* **2004**, *155*, 568. [CrossRef]

18. Wallgren, P.; Bornstein, S. The spread of porcine sarcoptic mange during the fattening period revealed by development of antibodies to *Sarcoptesscabiei*. *Vet. Parasitol.* **1997**, *73*, 315–324. [CrossRef]

19. Fain, A. Epidemiological problems of scabies. *Int. J. Dermatol.* **1978**, *17*, 20–30. [CrossRef]

20. Alasaad, S.; Soglia, D.; Spalenza, V.; Maione, S.; Soriguer, R.C.; Pérez, J.M.; Rasero, R.; Degiorgis, M.P.; Nimmervoll, H.; Zhu, X.Q.; et al. Is ITS-2 rDNA suitable marker for genetic characterization of *Sarcoptes* mites from different wild animals in different geographic areas? *Vet. Parasitol.* **2009**, *159*, 181–185. [CrossRef]

21. Arlian, L.G.; Morgan, M.S.; Arends, J.J. Immunologic cross-reactivity among various strains of *Sarcoptesscabiei*. *J. Parasitol.* **1996**, *82*, 66–72. [CrossRef]

22. Garcia, R.; Piche, C.; Davies, P.; Gross, S. Prevalence of sarcoptic mange mites and dermatitis in slaughter pigs in North America and Western Europe. In Proceedings of the 13th Conference of the International Pig Veterinary Society, Bangkok, Thailand, 26–30 June 1994; p. 250. Available online: https://agris.fao.org/agris-search/search.do?recordID=TH2000001208 (accessed on 5 October 2020).

23. Jensen, J.C.; Nielsen, L.H.; Arnason, T.; Cracknell, V. Elimination of mange mites *Sarcoptesscabiei* var. suis from two naturally infested Danish sow herds using a single injection regime with doramectin. *Acta Vet. Scand.* **2002**, *43*, 75–84. [CrossRef]

24. Davies, P.R. Sarcoptic mange and production performance of swine: A review of the literature and studies of associations between mite infestation, growth rate and measures of mange severity in growing pigs. *Vet. Parasitol.* **1995**, *60*, 249–264. [CrossRef]

25. Arends, J.J.; Stanislaw, C.M.; Gerdon, D. Effects of sarcoptic mange on lactating swine and growing pigs. *J. Anim. Sci.* **1990**, *68*, 1495–1499. [CrossRef] [PubMed]

26. Henken, A.M.; Verstegen, M.W.A.; Van Der Hel, W.; Boon, J.H.A. Pilot study of parasite worry and restlessness caused by sarcoptic mange in swine. In Proceedings of the 10th Conference of the International Pig Veterinary Society, Rio de Janerio, Brazil, 14–17 August 1988.

27. Grahofer, A.; Bannoehr, J.; Nathues, H.; Roosje, P. *Sarcoptes* infestation in two miniature pigs with zoonotic transmission—A case report. *BMC Vet. Res.* **2018**, *14*, 91. [CrossRef] [PubMed]

28. Sokolova, T.V.; Lange, A.B. Parazito-khoziainnaiaspetsifichnost' chesotochnogozudnia *Sarcoptesscabiei* (Acariformes: Sarcoptidae) chelovekaizhivotnykh (obzorliteratury) [The parasite-host specificity of the itch mite *Sarcoptesscabiei* (Acariformes: Sarcoptidae) in man and animals (a review of the literature)]. *Parazitologiia* **1992**, *26*, 97–104.

29. Chakrabarti, A. Pig handler's itch. *Int. J. Dermatol.* **1990**, *29*, 205–206. [CrossRef]

30. Laha, R. Sarcoptic mange infestation in pigs: An overview. *J. Parasit. Dis.* **2015**, *39*, 596–603. [CrossRef]

31. Morgan, M.S.; Arlian, L.G.; Markey, M.P. *Sarcoptesscabiei* mites modulate gene expression in human skin equivalents. *PLoS ONE* **2013**, *8*, e71143. [CrossRef]

32. Mullins, J.S.; Arlian, L.G.; Morgan, M.S. Extracts of *Sarcoptesscabiei* De Geer down-modulate secretion of IL-8 by skin keratinocytes and fibroblasts and of GM-CSF by fibroblasts in the presence of proinflammatory cytokines. *J. Med. Entomol.* **2009**, *46*, 845–851. [CrossRef]

33. Cote, N.M.; Jaworski, D.C.; Wasala, N.B.; Morgan, M.S.; Arlian, L.G. Identification and expression of macrophage migration inhibitory factor in *Sarcoptesscabiei*. *Exp. Parasitol.* **2013**, *135*, 175–181. [CrossRef]

34. Erster, O.; Roth, A.; Pozzi, P.S.; Bouznach, A.; Shkap, V. First detection of *Sarcoptesscabiei* from domesticated pig (*Sus scrofa*) and genetic characterization of *S. scabiei* from pet, farm and wild hosts in Israel. *Exp. Appl. Acarol.* **2015**, *66*, 605–612. [CrossRef]

35. Morgan, M.S.; Arlian, L.G. Response of human skin equivalents to *Sarcoptesscabiei*. *J. Med. Entomol.* **2010**, *47*, 877–883. [CrossRef] [PubMed]

36. Lalli, P.N.; Morgan, M.S.; Arlian, L.G. Skewed Th1/Th2 immune response to *Sarcoptesscabiei*. *J. Parasitol.* **2004**, *90*, 711–714. [CrossRef] [PubMed]

37. De, U.K.; Dey, S. Evaluation of organ function and oxidant/antioxidant status in goats with sarcoptic mange. *Trop. Anim. Health Prod.* **2010**, *42*, 1663–1668. [CrossRef] [PubMed]

38. Beigh, S.A.; Soodan, J.S.; Bhat, A.M. Sarcoptic mange in dogs: Its effect on liver, oxidative stress, trace minerals and vitamins. *Vet. Parasitol.* **2016**, *227*, 30–34. [CrossRef]

39. Thompson, J.D.; Higgins, D.G.; Gibson, T.J. CLUSTAL W: Improving the sensitivity of progressive multiple sequence alignment through sequence weighting, position-specific gap penalties and weight matrix choice. *Nucleic Acids Res.* **1994**, *22*, 4673–4680. [CrossRef] [PubMed]

40. Kumar, S.; Stecher, G.; Li, M.; Knyaz, C.; Tamura, K. MEGA X: Molecular Evolutionary Genetics Analysis across Computing Platforms. *Mol. Biol. Evol.* **2018**, *35*, 1547–1549. [CrossRef]

41. Tamura, K.; Nei, M. Estimation of the number of nucleotide substitutions in the control region of mitochondrial DNA in humans and chimpanzees. *Mol. Biol. Evol.* **1993**, *10*, 512–526. [CrossRef]

42. Bandelt, H.J.; Forster, P.; Röhl, A. Median-joining networks for inferring intraspecific phylogenies. *Mol. Biol. Evol.* **1999**, *16*, 37–48. [CrossRef]

43. Yu, I.T.; Ju, C.C.; Lin, J.; Wu, H.L.; Yen, H.T. Effects of probiotics and selenium combination on the immune and blood cholesterol concentration of pigs. *J. Anim. Feed Sci.* **2004**, *13*, 625–634. [CrossRef]

44. Marbut, M.M.; Majeed, B.M.; Rahim, S.M.; Yuusif, M.Y. Estimation of malondialdehyde as oxidative factor and glutathione as early detectors of hypertensive pregnant women. *Tikrit Med. J.* **2009**, *15*, 63–69.

45. Campbell, J.M.; Crenshaw, J.D.; Polo, J. The biological stress of early weaned piglets. *J. Anim. Sci. Biotechnol.* **2013**, *4*, 19. [CrossRef] [PubMed]

46. Steer, A. Scabies Joins the List of WHO Neglected Tropical Diseases. Lancet Global Health Blog. Global Health Blogs. 2014. Available online: http://globalhealth.thelancet.com/2014/07/07/scabies-joins-list-who-neglected-tropical-diseases (accessed on 3 October 2020).

47. Mounsey, K.E.; McCarthy, J.S.; Walton, S.F. Scratching the itch: New tools to advance understanding of scabies. *Trends Parasitol.* **2013**, *29*, 35–42. [CrossRef] [PubMed]

48. Fraser, T.A.; Shao, R.; Fountain-Jones, N.M.; Charleston, M.; Martin, A.; Whiteley, P.; Holme, R.; Carver, S.; Polkinghorne, A. Mitochondrial genome sequencing reveals potential origins of the scabies mite *Sarcoptesscabiei* infesting two iconic Australian marsupials. *BMC Evol. Biol.* **2017**, *17*, 233. [CrossRef]

49. Alexander, K.A.; Carlson, C.J.; Lewis, B.L.; Getz, W.M.; Marathe, M.V.; Eubank, S.G.; Sanderson, C.E.; Blackburn, J.K. The Ecology of pathogen spillover and disease emergence at the human-wildlife-environment interface. In *The Connections between Ecology and Infectious Disease*; Advances in Environmental Microbiology; Hurst, C., Ed.; Springer: Cham, Switzerland, 2018; Volume 5, pp. 267–298.

50. Mandal, S.D.; Chhakchhuak, L.; Gurusubramanian, G.; Kumar, N.S. Mitochondrial markers for identification and phylogenetic studies in insects—A Review. *DNA Barcodes* **2014**, *2*, 1–9. [CrossRef]

51. Pentinsaari, M.; Salmela, H.; Mutanen, M.; Roslin, T. Molecular evolution of a widely-adopted taxonomic marker (COI) across the animal tree of life. *Sci. Rep.* **2016**, *6*, 35275. [CrossRef] [PubMed]

52. Ondrejicka, D.A.; Locke, S.A.; Morey, K.; Borisenko, A.V.; Hanner, R.H. Status and prospects of DNA barcoding in medically important parasites and vectors. *Trends Parasitol.* **2014**, *30*, 582–591. [CrossRef]

53. Andriantsoanirina, V.; Ariey, F.; Izri, A.; Bernigaud, C.; Fang, F.; Charrel, R.; Foulet, F.; Botterel, F.; Guillot, J.; Chosidow, O.; et al. *Sarcoptesscabiei* mites in humans are distributed into three genetically distinct clades. *Clin. Microbiol. Infect.* **2015**, *21*, 1107–1114. [CrossRef]

54. Zahler, M.; Essig, A.; Gothe, R.; Rinder, H. Molecular analyses suggest monospecificity of the genus *Sarcoptes* (Acari: Sarcoptidae). *Int. J. Parasitol.* **1999**, *29*, 759–766. [CrossRef]

55. Gu, X.B.; Yang, G.Y. A study on the genetic relationship of mites in the genus *Sarcoptes* (Acari: Sarcoptidae) in China. *Int. J. Acarol.* **2008**, *34*, 183–190. [CrossRef]

56. Christie, J.R.; Beekman, M. Selective sweeps of mitochondrial DNA can drive the evolution of uniparental inheritance. *Evolution* **2017**, *71*, 2090–2099. [CrossRef]

57. Gandon, S. Local adaptation and the geometry of host–parasite coevolution. *Ecol. Lett.* **2002**, *5*, 246–256. [CrossRef]

58. Appelgren, A.S.C.; Saladin, V.; Richner, H.; Doligez, B.; McCoy, K.D. Gene flow and adaptive potential in a generalist ectoparasite. *BMC Evol. Biol.* **2018**, *18*, 99. [CrossRef] [PubMed]

59. Arlian, L.G.; Morgan, M.S. A review of *Sarcoptesscabiei*: Past, present and future. *Parasites Vectors* **2017**, *10*, 297. [CrossRef] [PubMed]

60. Martin, A.M.; Fraser, T.A.; Lesku, J.A.; Simpson, K.; Roberts, G.L.; Garvey, J.; Polkinghorne, A.; Burridge, C.P.; Carver, S. The cascading pathogenic consequences of *Sarcoptesscabiei* infection that manifest in host disease. *R. Soc. Open Sci.* **2018**, *5*, 180018. [CrossRef] [PubMed]

61. Prather, R.S.; Walters, E.M.; Wells, K.D. Swine in Biomedical Research 2014. *Lab Anim.* **2014**, *44*, 9. [CrossRef]

62. Monticello, T.M.; Haschek, W.M. Swine in translational research and drug development. *Toxicol. Pathol.* **2016**, *44*, 297–298. [CrossRef]

63. Mounsey, K.; Ho, M.F.; Kelly, A.; Willis, C.; Pasay, C.; Kemp, D.J.; McCarthy, J.S.; Fischer, K. A tractable experimental model for study of human and animal scabies. *PLoS Negl. Trop. Dis.* **2010**, *4*, e756. [CrossRef]

64. Rampton, M.; Walton, S.F.; Holt, D.C.; Pasay, C.; Kelly, A.; Currie, B.J.; McCarthy, J.S.; Mounsey, K.E. Antibody responses to *Sarcoptesscabiei* apolipoprotein in a porcine model: Relevance to immunodiagnosis of recent infection. *PLoS ONE* **2013**, *8*, e65354. [CrossRef]

65. Giannini, E.G.; Testa, R.; Savarino, V. Liver enzyme alteration: A guide for clinicians. *CMAJ* **2005**, *172*, 367–379. [CrossRef]

66. Quinn, M.T.; Parthasarathy, S.; Fong, L.G.; Steinberg, D. Oxidatively modified low density lipoproteins: A potential role in recruitment and retention of monocyte/macrophages during atherogenesis. *Proc. Natl. Acad. Sci. USA* **1987**, *84*, 2995–2998. [CrossRef]

67. Ogura, S.; Shimosawa, T. Oxidative stress and organ damages. *Curr. Hypertens Rep.* **2014**, *16*, 452. [CrossRef] [PubMed]

68. Li, S.; Tan, H.Y.; Wang, N.; Zhang, Z.; Lao, L.; Wong, C.; Feng, Y. The role of oxidative stress and antioxidants in liver diseases. *Int. J. Mol. Sci.* **2015**, *16*, 26087–26124. [CrossRef] [PubMed]

69. Sánchez-Valle, V.; Chávez-Tapia, N.C.; Uribe, M.; Méndez-Sánchez, N. Role of oxidative stress and molecular changes in liver fibrosis: A review. *Curr. Med. Chem.* **2012**, *19*, 4850–4860. [CrossRef] [PubMed]

70. Ogura, S.; Kakino, A.; Sato, Y.; Fujita, Y.; Iwamoto, S.; Otsui, K.; Yoshimoto, R.; Sawamura, T. Lox-1: The multifunctional receptor underlying cardiovascular dysfunction. *Circ. J.* **2009**, *73*, 1993–1999. [CrossRef] [PubMed]

71. Sharma, S.K.; Soodan, J.S.; Sharma, N. Haemato-biochemical alterations in canine dermatitis. *Indian Vet. J.* **2011**, *88*, 56–58.

72. Han, H.S.; Kang, G.; Kim, J.S.; Choi, B.H.; Koo, S.H. Regulation of glucose metabolism from a liver-centric perspective. *Exp. Mol. Med.* **2016**, *48*, e218. [CrossRef] [PubMed]

73. Ogihara, T.; Asano, T.; Katagiri, H.; Sakoda, H.; Anai, M.; Shojima, N.; Ono, H.; Fujishiro, M.; Kushiyama, A.; Fukushima, Y.; et al. Oxidative stress induces insulin resistance by activating the nuclear factor-kappa B pathway and disrupting normal subcellular distribution of phosphatidylinositol 3-kinase. *Diabetologia* **2004**, *47*, 794–805. [CrossRef]

74. Fridlyand, L.E.; Philipson, L.H. Reactive species and early manifestation of insulin resistance in type 2 diabetes. *Diabetes Obes. Metab.* **2006**, *8*, 136–145. [CrossRef]

75. Misra, A.; Luthra, K.; Vikram, N.K. Dyslipidemia in Asian Indians: Determinants and significance. *J. Assoc. Physicians India* **2004**, *52*, 137–142.

76. Shenoy, C.; Shenoy, M.M.; Rao, G.K. Dyslipidemia in dermatological disorders. *N. Am. J. Med. Sci.* **2015**, *7*, 421–428. [CrossRef]

77. Hahn, B.H.; Grossman, J.; Chen, W.; McMahon, M. The pathogenesis of atherosclerosis in autoimmune rheumatic diseases: Roles of inflammation and dyslipidemia. *J. Autoimmun.* **2007**, *28*, 69–75. [CrossRef] [PubMed]

78. Kobashigawa, J.A.; Kasiske, B.L. Hyperlipidemia in solid organ transplantation. *Transplantation* **1997**, *63*, 331–338. [CrossRef] [PubMed]

79. Derfler, K.; Hayde, M.; Heinz, G.; Hirschl, M.M.; Steger, G.; Hauser, A.C.; Balcke, P.; Widhalm, K. Decreased postheparin lipolytic activity in renal transplant recipients with cyclosporin A. *Kidney Int.* **1991**, *40*, 720–727. [CrossRef] [PubMed]

80. Stephens, M.A.; Wand, G. Stress and the HPA axis: Role of glucocorticoids in alcohol dependence. *Alcohol Res.* **2012**, *34*, 468–483.

81. Li, L.A.; Yang, J.J.; Li, Y.; Lv, L.; Xie, J.J.; Du, G.M.; Jin, T.M.; Qin, S.Y.; Jiao, X.L. Effect of weaning age on cortisol release in piglets. *Genet. Mol. Res.* **2016**, *15*, 1–10. [CrossRef]

82. Pockley, A.G.; Henderson, B. Extracellular cell stress (heat shock) proteins-immune responses and disease: An overview. *Philos. Trans. R. Soc. Lond. B Biol. Sci.* **2018**, *373*, 20160522. [CrossRef]

83. Campbell, R.M.; Scanes, C.G. Endocrine peptides 'moonlighting' as immune modulators: Roles for somatostatin and GH-releasing factor. *J. Endocrinol.* **1995**, *147*, 383–396. [CrossRef]

84. Bolhassani, A.; Agi, E. Heat shock proteins in infection. *Clin. Chim. Acta* **2019**, *498*, 90–100. [CrossRef]

85. Pockley, A.G.; Bulmer, J.; Hanks, B.M.; Wright, B.H. Identification of human heat shock protein 60 (Hsp60) and anti-Hsp60 antibodies in the peripheral circulation of normal individuals. *Cell Stress Chaperones* **1999**, *4*, 29–35. [CrossRef]

86. Kalmar, B.; Greensmith, L. Induction of heat shock proteins for protection against oxidative stress. *Adv. Drug Deliv. Rev.* **2009**, *61*, 310–318. [CrossRef]

87. Arrigo, A.P. The cellular "networking" of mammalian Hsp27 and its functions in the control of protein folding, redox state and apoptosis. *Adv. Exp. Med. Biol.* **2007**, *594*, 14–26. [CrossRef] [PubMed]

88. Yuan, J.; Murrell, G.A.; Trickett, A.; Wang, M.X. Involvement of cytochrome c release and caspase-3 activation in the oxidative stress induced apoptosis in human tendon fibroblasts. *Biochim. Biophys. Acta* **2003**, *1641*, 35–41. [CrossRef]

89. Annunziato, L.; Amoroso, S.; Pannaccione, A.; Cataldi, M.; Pignataro, G.; D'Alessio, A.; Sirabella, R.; Secondo, A.; Sibaud, L.; Di Renzo, G.F. Apoptosis induced in neuronal cells by oxidative stress: Role played by caspases and intracellular calcium ions. *Toxicol. Lett.* **2003**, *139*, 125–133. [CrossRef]

90. Ellah, M.R.A. Involvement of free radicals in parasitic infestations. *J. Appl. Anim. Res.* **2013**, *41*, 69–76. [CrossRef]

91. Birben, E.; Sahiner, U.M.; Sackesen, C.; Erzurum, S.; Kalayci, O. Oxidative stress and antioxidant defense. *World Allergy Organ.* **2012**, *5*, 9–19. [CrossRef]

92. Valko, M.; Leibfritz, D.; Moncol, J.; Cronin, M.T.D.; Mazur, M.; Telser, J. Free radicals and antioxidants in normal physiological functions and human disease. *Int. J. Biochem. Cell Biol.* **2007**, *39*, 44–84. [CrossRef]

93. Nordberg, J.; Arnér, E.S.J. Reactive oxygen species, antioxidants, and the mammalian thioredoxin system. *Free Radic. Biol. Med.* **2001**, *31*, 1287–1312. [CrossRef]

94. Phaniendra, A.; Jestadi, D.B.; Periyasamy, L. Free Radicals: Properties, sources, targets, and their implication in various diseases. *Indian J. Clin. Biochem.* **2015**, *30*, 11–26. [CrossRef]

95. Kurutas, E.B. The importance of antioxidants which play the role in cellular response against oxidative/nitrosative stress: Current state. *Nutr. J.* **2016**, *15*, 71. [CrossRef]

96. Griendling, K.K.; FitzGerald, G.A. Oxidative stress and cardiovascular injury: Part I: Basic mechanisms and in vivo monitoring of ROS. *Circulation* **2003**, *108*, 1912–1916. [CrossRef]

97. Beigh, S.A.; Soodan, J.S.; Nazki, S.; Khan, A.M. Oxidative stress, hematobiochemical parameters, trace elements and vitamins in dogs with zinc responsive dermatosis. *VeterinarskiArhiv* **2014**, *84*, 591–600.

98. Beigh, S.A.; Soodan, J.S.; Singh, R.; Khan, A.M.; Dar, M.A. Evaluation of trace elements, oxidant/antioxidative status, vitamin C and β-carotene in dogs with dermatophytosis. *Mycoses* **2014**, *57*, 358–365. [CrossRef]

99. Dimri, U.; Sharma, M.C.; Swarup, D.; Ranjan, R.; Kataria, M. Alterations in hepatic lipid peroxides and antioxidant profile in Indian water buffaloes suffering from sarcoptic mange. *Res. Vet. Sci.* **2008**, *85*, 101105. [CrossRef] [PubMed]

100. Dimri, U.; Sharma, M.C.; Yamdagni, A.; Ranjan, R.; Zama, M.M.S. Psoroptic mange infestation increases oxidative stress and decreases antioxidant status in sheep. *Vet. Parasitol.* **2010**, *68*, 318322. [CrossRef] [PubMed]

101. Brand, A.; Bauer, N.G.; Hallott, A.; Goldbaum, O.; Ghebremeskel, K.; Reifen, R.; Richter-Landsberg, C. Membrane lipid modification by polyunsaturated fatty acids sensitizes oligodendroglial OLN-93 cells against oxidative stress and promotes up-regulation of heme oxygenase-1 (HSP32). *J. Neurochem.* **2010**, *113*, 465–476. [CrossRef] [PubMed]

102. Ardestani, A.; Yazdanparast, R. Antioxidant and free radical scavenging potential of *Achillea santolina* extracts. *Food Chem.* **2007**, *104*, 21–29. [CrossRef]

103. DelRio, D.; Stewart, A.J.; Pellegrini, N. A review of recent studies on malondialdehyde as toxic molecule and biological marker of oxidative stress. *Nutr. Metab. Cardiovasc. Dis.* **2005**, *15*, 316–328. [CrossRef]

104. Nwufoh, O.C.; Sadiq, N.A.; Emikpe, B.O.; Omobowale, T. Assessment of cutaneous and serum oxidative stress changes in dogs infested with *Sarcoptesscabiei* var. *Canis*. *Acta Vet. Eurasia* **2020**, *46*, 1–6. [CrossRef]

105. Luo, Q.; Cui, H.; Peng, X.; Fang, J.; Zuo, Z.; Deng, J.; Liu, J.; Deng, Y. Intestinal IgA+ cell numbers as well as IgA, IgG, and IgM contents correlate with mucosal humoral immunity of broilers during supplementation with high fluorine in the diets. *Biol. Trace Elem. Res.* **2013**, *154*, 62–72. [CrossRef]

106. Morgan, M.S.; Arlian, L.G. Serum antibody profiles of *Sarcoptesscabiei* infested or immunized rabbits. *Folia Parasitol.* **1994**, *41*, 223–227.

107. Roberts, L.J.; Huffam, S.E.; Walton, S.F.; Currie, B.J. Crusted scabies: Clinical and immunological findings in seventy-eight patients and a review of the literature. *J. Inf. Secur.* **2005**, *50*, 375–381. [CrossRef] [PubMed]

108. Hu, C.H.; Xiao, K.; Luan, Z.S.; Song, J. Early weaning increases intestinal permeability, alters expression of cytokine and tight junction proteins, and activates mitogen-activated protein kinases in pigs. *J. Anim. Sci.* **2013**, *91*, 1094–1101. [CrossRef]

109. Jankovic, D.; Sher, A.; Yap, G. Th1/Th2 effector choice in parasitic infection: Decision making by committee. *Curr. Opin. Immunol.* **2001**, *13*, 403–409. [CrossRef]

110. Romagnani, S. T-cell subsets (Th1 versus Th2). *Ann. Allergy Asthma Immunol.* **2000**, *85*, 9–21. [CrossRef]

111. Kaiko, G.E.; Horvat, J.C.; Beagley, K.W.; Hansbro, P.M. Immunological decision-making: How does the immune system decide to mount a helper T-cell response? *Immunology* **2008**, *123*, 326–338. [CrossRef] [PubMed]

112. Cicchese, J.M.; Evans, S.; Hult, C.; Joslyn, L.R.; Wessler, T.; Millar, J.A.; Marino, S.; Cilfone, N.A.; Mattila, J.T.; Linderman, J.J.; et al. Dynamic balance of pro- and anti-inflammatory signals controls disease and limits pathology. *Immunol. Rev.* **2018**, *285*, 147–167. [CrossRef]

113. Spellberg, B.; Edwards, J.E., Jr. Type 1/Type 2 immunity in infectious diseases. *Clin. Infect. Dis.* **2001**, *32*, 76–102. [CrossRef]

114. Pence, D.B.; Ueckermann, E. Sarcoptic mange in wildlife. *Rev. Sci. Tech. Off. Int. Epizoot.* **2002**, *21*, 385–398. [CrossRef]

115. Nimmervoll, H.; Hoby, S.; Robert, N.; Lommano, E.; Welle, M.; Ryser-Degiorgis, M.P. Pathology of sarcoptic mange in red foxes (*Vulpes vulpes*): Macroscopic and histologic characterization of three disease stages. *J. Wildl. Dis.* **2013**, *49*, 91–102. [CrossRef]

116. Ninomiya, H.; Ogata, M. Sarcoptic mange in free-ranging raccoon dogs (*Nyctereutesprocyonoides*) in Japan. *Vet. Dermatol.* **2005**, *16*, 177–182. [CrossRef]

117. Ryser-Degiorgis, M.P.; Ryser, A.; Bacciarini, L.N.; Angst, C.; Gottstein, B.; Janovsky, M.; Breitenmoser, U. Notoedric and sarcoptic mange in free-ranging lynx from Switzerland. *J. Wildl. Dis.* **2002**, *38*, 228–232. [CrossRef] [PubMed]

118. Espinosa, J.; Ráez-Bravo, A.; López-Olvera, J.R.; Pérez, J.M.; Lavín, S.; Tvarijonaviciute, A.; Cano-Manue, F.J.; Fandos, P.; Soriguer, R.C.; Granados, J.E.; et al. Histopathology, microbiology and the inflammatory process associated with *Sarcoptesscabiei* infection in the Iberian ibex, *Capra pyrenaica*. *Parasites Vectors* **2017**, *10*, 596. [CrossRef] [PubMed]

Publisher's Note: MDPI stays neutral with regard to jurisdictional claims in published maps and institutional affiliations.

Article

Changes in the Percentages of B- and T-Lymphocytes and Antibody Titres in Laying Hens Infested with *Dermanyssus gallinae*—A Preliminary Study

Sylwia Koziatek-Sadłowska and Rajmund Sokół *

Department of Parasitology and Invasive Diseases, Faculty of Veterinary Medicine,
University of Warmia and Mazury in Olsztyn, ul. Oczapowskiego 13, 10-719 Olsztyn, Poland;
sylwia.koziatek@uwm.edu.pl
* Correspondence: rajmund.sokol@uwm.edu.pl; Tel.: +48-89-523-35-63

Received: 28 April 2020; Accepted: 3 June 2020; Published: 5 June 2020

Simple Summary: *Dermanyssus gallinae,* a hematophagous ectoparasite, adversely affects the health status of laying hens, leading to reduced egg production and significant economic losses in commercial poultry farms. The aim of this study was to determine the effect of *D. gallinae* on the immunological parameters (three lymphocyte subpopulations) and post-vaccination antibody titres (against most important avian pathogens) in layer hens during the egg production cycle. A total of 80 blood samples were collected at four time-points (B1–B4) from 10 Hy-Line Brown hens naturally infested with *D. gallinae*, which were randomly selected from a commercial layer farm. The infestation was monitored and treated twice with Biobeck PA 910 (active ingredient AI – silicon dioxide). The samples were collected before and after each treatment. The percentages of subpopulations of B cells and helper (Th) and cytotoxic (Tc) T cells were determined by flow cytometry. Antibody titres were determined by the immunoenzymatic method. The percentage of Th cells and post-vaccination anti-infectious bronchitis virus (anti-IBV) and anti-Newcastle disease virus (anti-NDV) antibodies decreased significantly at the second infestation peak when the number of parasites was twice higher than at the first infestation peak. There were non-significant correlations between the number of mites and antibody titres. These findings suggested that *D. gallinae* might inhibit humoral immune responses since the percentages of B cells and Th cells were negatively correlated with the number of mites. The percentage of Tc cells was positively correlated with the number of mites, which indicated that *D. gallinae* could stimulate cellular immune responses in infested laying hens. However, further research is needed to determine whether *D. gallinae* suppresses the production of vaccine-induced antibodies.

Abstract: (1) Background: *Dermanyssus gallinae,* a hematophagous ectoparasite, adversely affects the health status of laying hens, leading to reduced egg production and significant economic losses in commercial farms. The aim of this study was to determine the effect of *D. gallinae* on the development of post-vaccination immune responses in layer hens. (2) Methods: A total of 80 blood samples were collected at four time-points (B1–B4) from 10 Hy-Line Brown hens, randomly selected from a commercial layer farm. The flock was naturally infested with *D. gallinae* and treated twice with Biobeck PA 910 (AI silicon dioxide). The samples were collected before and after each treatment. The percentages of IgM+ B cells, CD3+/CD4+ T cells and CD3+/CD8a+ T cells were determined by flow cytometry; the titres of antibodies against avian encephalomyelitis, infectious bronchitis virus, Newcastle disease virus, *Ornithobacterium rhinotracheale*, reticuloendotheliosis virus and avian reovirus were determined by the immunoenzymatic method. (3) Results: The percentage of Th cells and post-vaccination anti-IBV and anti-NDV antibodies decreased significantly at the second infestation peak when the number of parasites was twice higher than at the first infestation peak. Non-significant negative correlations were found between the number of mites and the percentage of B cells (R = −0.845, $p > 0.05$) and between the number of mites and the percentage of Th cells (R = −0.522,

$p > 0.05$), and a significant positive correlation was noted between the number of mites and the percentage of Tc cells ($R = -0.982$, $p < 0.05$). There were non-significant correlations between the number of mites and antibody titres. (4) Conclusion: The present findings suggested that *D. gallinae* might inhibit immune responses since the percentages of B cells and Th cells were negatively correlated with the number of mites. The percentage of Tc cells was positively correlated with the number of mites, which indicated that *D. gallinae* could stimulate cellular immune responses in infested laying hens. However, further research is needed to determine whether *D. gallinae* suppresses the production of vaccine-induced antibodies.

Keywords: *D. gallinae*; hematophagous ectoparasite; poultry red mite; antibody titre; lymphocyte subpopulation

1. Introduction

Dermanyssus gallinae (De Geer, 1778) (*D. gallinae*), commonly known as poultry red mite (prm), is a temporary ectoparasite, feeding on the blood of wild and domestic birds [1–4]. It is currently the most serious mite pest in commercial caged layer hen farms [5–9]. Farm conditions favour mass occurrence and uncontrolled growth of the parasite. Such conditions include high, constant temperature, high relative air humidity, constant access to the host body and long production cycle (80 weeks) [8,10,11]. The parasite settles in numerous gaps in the poultry house construction and equipment where it goes through the development cycle of five life stages (egg, larva, protonymph, deutonymph and adults). It needs the host's blood for growth in the last three stages [1,12]. They feed briefly and return to their refugia [13,14]. This way of life makes it poorly accessible to acaricides [8,10,15]. Moreover, there have been increasing numbers of reports on the prm resistance to acaricides [16–18]. The implementation of enriched cages (in line with the EU Directive 1999.74/EC) has probably led to a significant and uncontrolled proliferation of *D. gallinae* [8,19]. Under heavy infestation, the welfare of the laying hens significantly decreases. Upon puncturing the skin of the host, the mite introduces toxic saliva that can cause itching and irritation, which leads to distress and, consequently, poor feed conversion, reduced egg production and increased bird mortality [10,20,21].

Birds in commercial flocks of laying hens are given preventive vaccination until the 20th week of their lives. Not only are they important from the economic perspective, but they ensure consumer safety. Vaccination effectiveness may be affected by feed, genetic, environmental, chemical and biological (bacteria and viruses) factors. According to the literature, hens, which are chronically exposed to *D. gallinae*, do not produce a strong immune response against this parasite. A non-significant correlation between serum IgY level and red mite population levels has been observed [22]. Another study has revealed that *D. gallinae* feeding stimulates Th1 and pro-inflammatory cytokines/chemokines, initially followed by their subsequent down-regulation [23]. There is also a theory that *D. gallinae* might adopt a feeding strategy of minimal host interference, while *D. gallinae* could determine host immune status by nymphal/larval survival rates [24]. Our recent studies revealed that *D. gallinae* infestation caused somatic and psychogenic stress in layer hens, which led to lower humoral immunity decrease of γ-globulin in the blood of hens infested by *D. gallinae* [25,26]. Therefore, the question may be formulated, whether the pressure (feeding) of *D. gallinae* may have an immunosuppressive effect and affect the development of post-vaccination immune responses in laying hens. There has been no scientific research on the subject. The research on *D. gallinae*, in an ectoparasite–host immunological relationship, mainly concerns vaccines against this parasite.

Therefore, the aim of this study was to determine the percentages of subpopulations of B IgM+ cells, T cells CD3+/CD4+ (Th) and T cells CD3+/CD8a+ (Tc) by flow cytometry and the titres of antibodies against AE (avian encephalomyelitis), IBV (infectious bronchitis virus), NDV (Newcastle disease virus), ORT (*Ornithobacterium rhinotracheale*), REV (reticuloendotheliosis virus) and REO (avian reovirus) using the immunoenzymatic method with ELISA tests (IDEXX, USA) in the peripheral blood of laying hens infested with *D. gallinae*, controlled with Biobeck PA 910 (active ingredient (AI) silicon dioxide).

2. Materials and Methods

2.1. Birds

Blood samples for analyses were collected from 10 randomly selected laying hens from an industrial laying hen farm naturally infested with *D. gallinae*. A flock of 56,400 hens of the Hy-Line Brown line was kept in battery cages for 54 weeks in conditions complying with the valid standards. Hens were fed with high-quality feed and watered *ad libitum* from automatic drinkers. The air temperature in the hen house was 20–22 °C, and the relative humidity (RH) was 40%–60%. The birds were given prophylactic vaccinations as per the standard programme: *Salmonella enteritidis*—6–10 days, and 70–74 days, *Staphylococcus gallinarum*—42–46 days and 16 weeks, *Mycoplasma gallisepticum*, *Mycoplasma synoviae*—56–60 days, infectious bronchitis virus (IBV) and Newcastle disease virus (NDV)—35–39 days and 13 weeks, Gumboro disease (IBD)—22–26 days and 30–34 days, avian encephalomyelitis (AE) and egg drop syndrome (EDS′76)—12–13 weeks. Day or week indicates the age range of the hen while vaccinated.

All procedures were conducted according to the prevailing national legislation on the use of animals in research. The experiment was carried out in compliance with the recommendations of the Local Ethics Committee for Animal Experimentation in Olsztyn (no. 59/2006).

2.2. Monitoring of the Dermanyssus gallinae Infestation

The infestation of *D. gallinae* was monitored in 2-week intervals. The parasites were caught with a system of traps fixed to the cages under the conveyor belt for egg collection, four at each of fixed selected places, on the second and the fourth row of cages [27]. The traps from each site were collected separately to a twist-type jar (0.9 L) and sent to the laboratory. The jars with their contents were cooled down for 30 min at a temperature of −20 °C to immobilise the parasites and count them. The deposit was poured onto a 2 × 2 cm chequered Petri dish, and the parasite life stages; (a) adults, (b) larvae + nymphs, (c) eggs were counted with a binocular magnifying glass (Olympus, Type SZ, magn. 40×, Tokyo, Japan). The infestation size was presented as the average number of each life stage (Figure 1).

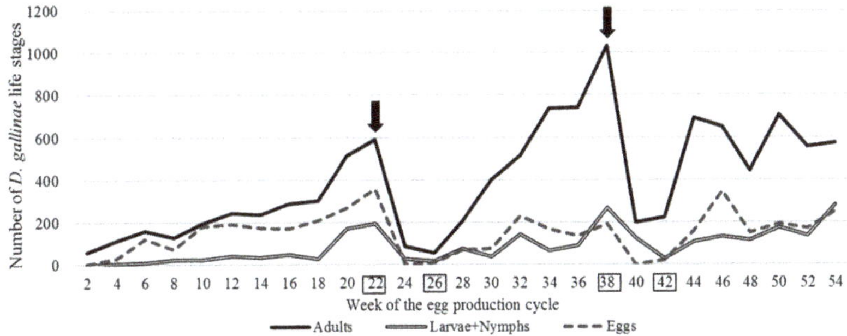

Figure 1. The size of the *Dermanyssus gallinae* population and its life stages following two applications of the acaricide product Biobeck PA 910 (AI silicon dioxide, Biologische Preparate, Brilon, Germany) during a 54-week production cycle. (Mite treatment MTI and MTII dates are marked by arrows, blood sampling B1–4 dates are marked by squares).

2.3. Controlling of the Dermanyssus gallinae Infestation

The infestation of *D. gallinae* was controlled with Biobeck PA 910 (AI silicon dioxide, Biologische Preparate, Brilon, Germany), which was applied in accordance with the manufacturer's guidelines at 1–2 g/hen by spraying onto the cages and hens with a device imparting static electricity to it. Two dates for mite treatment were established based on the results of ongoing monitoring of the number of different life stages. It was performed in the 22nd and 38th week of the production cycle. The first mite treatment (MTI) was performed after a rapid increase in the number of *D. gallinae* was observed—a nearly two-fold increase in the number of adults and eggs and a four-fold increase in the number of larvae + nymphs compared to week 16 (Figure 1). The second mite treatment (MTII) was performed after a two-fold increase, in the number of adults and larvae + nymphs, was recorded compared to week 32 (the number of eggs was similar).

2.4. Collection of the Blood Samples

Blood samples for cytometric and serological tests were collected at four time-points during the egg production cycle, from the wing vein of 10 randomly selected hens for each analysis. Blood samples were collected in the 22nd—B1 (before MTI), 26th—B2 (after MTI), 38th—B3 (before MTII), 42nd—B4 (after MTII) week of the production cycle. Hens were 40, 44, 56, 60 weeks old, respectively, in B1–B4. A total of 80 blood samples were tested (40 for each analysis). Hens were marked, and the blood was taken from the same hens each time.

2.5. Flow Cytometry Analysis

Blood samples in the amount of 1.5 mL were collected into test tubes with an anticoagulant EDTA-K2 (Sarstedt). Subsequently, blood was diluted 1:1 with PBS (phosphate-buffered saline) with 1% fetal calf serum (FCS). The blood samples were layered on 3 mL of Histopaque-1077 gradient (Sigma-Aldrich, Darmstadt, Germany) and centrifuged in 15 mL FALCON tubes at 400× g at the temperature of 25 °C for 30 min. After centrifuging, the layer of mononuclear cells was carefully collected into sterile tubes and washed twice with PBS with an addition of 1% FCS and subsequently suspended in 1 mL PBS. The cells were then counted by the chamber method. A total of 10^6 cells were collected from the suspension and transferred to cytometric tubes. Two microlitres of antibodies targeted at surface domains of the CD3 receptor (Mouse AntiChicken CD3-FITC clone CT-3, SouthernBiotech, Birmingham, AL, USA), CD4 receptor (Mouse AntiChicken CD4-FITC clone CT-4, SouthernBiotech, Birmingham, AL, USA) and CD8 receptor (Mouse AntiChicken CD8a-Cy5 clone 3-298, SouthernBiotech, Birmingham, AL, USA), T cells and immunoglobulin BCR (IgM$^+$) of B cells (Mouse AntiChickenIgM-SPRD clone M-1, SouthernBiotech, Birmingham, AL, USA) were added to each tube. The samples were incubated on ice for 30 min in darkness. Subsequently, the cells were washed twice in PBS, centrifuged at 250× g for 7 min at 4 °C, and the resulting cell pellets were suspended in 0.5 mL of fixing buffer (CellFIX, Becton Dickinson, Franklin Lakes, NJ, USA) and refrigerated. The samples were tested after 18 h with a flow cytometer FACSCanto II (Becton Dickinson, Franklin Lakes, NJ, USA). Immunophenotypic analysis of the lymphocytes was performed with a specialist software FlowJo (Tree Star, Inc., Ashland, OR, USA).

2.6. Serological Analysis (ELISA)

Blood samples were taken to sterile tubes and refrigerated for 1 h to separate the serum from morphotic elements. The blood was then centrifuged for 20 min at 4 °C at 1000× g. Portions of 1.5 mL of serum were transferred to microcentrifuge tubes. The titres of antibodies against AE (avian encephalomyelitis), IBV (infectious bronchitis virus), NDV (Newcastle disease virus), ORT (*Ornithobacterium rhinotracheale*), REO (avian reovirus) and REV (reticuloendotheliosis virus) were determined in accordance with the manufacturer's guidelines for commercial tests ELISA Pro Flok HEV Ab (Zoetis, Parsippany, NJ, USA). The absorbance was read out with a BioTek ELx800

(Winooski, VT, USA) plate reader. The data analysis and calculations were performed in the xCheck IDEXX (Westbrook, ME, USA) software environment.

2.7. Statistical Analysis

The results were analysed statistically by calculating mean values, standard error (SE) and coefficient of variation (CV%) for antibody titres. Differences between B1–B4 for lymphocyte subpopulations and antibody titres were analysed using repeated-measures ANOVA test or MANOVA test if the assumption of sphericity was violated (statistically significant Mauchly test). Tukey's post hoc test was used. The statistical significance was declared at $p < 0.05$. Pearson correlation coefficient test was used to determine the relationships between mite numbers and lymphocyte subpopulations and antibody titres.

3. Results

3.1. Development of the Dermanyssus gallinae Infestation

The mean number of different life stages of *D. gallinae* (adults, larvae + nymphs and eggs) in consecutive weeks of the production cycle of laying hens; dates of mite treatment (MTI and MTII); dates of blood sampling for assays (B1–B4) are shown in Figure 1. Two peaks in mite population (PI and PII, in the study called 'infestation peak') during the 54-week production cycle induced by mite treatment were recorded. The growth of the *D. gallinae* population was moderate until week 18. Subsequently, the growth rate increased. A two-fold increase in the number of adults, larvae + nymphs and eggs of *D. gallinae* was recorded in week 22 of the production cycle (592 adults, 194 larvae + nymphs and 357 eggs) compared to week 18. Mite treatment I (MTI) was performed, as a result of which the number of life stages of *D. gallinae* was reduced significantly (found: 56 adult forms, 16 larvae + nymphs and 10 eggs) in week 24 of the production cycle compared to week 22. The infestation remained at a low level for 2 weeks. Subsequently, it started to grow rapidly, starting with week 26. The reinfestation peak (PII) was observed in week 38 of the production cycle (found: 1034 adult forms, 268 larvae + nymphs and 191 eggs). The second mite treatment (MTII) was performed, which resulted in a statistically significant reduction in the number of individual life stages of *D. gallinae* in week 40 of the production cycle (found: 199 females, 124 larvae-nymphs and 1 egg). The effect persisted for the next 2 weeks. After that, the population started to regenerate, but statistically non-significant fluctuations of the numbers of individual life stages were observed. Adults in a number of 574, 282 larvae-nymphs and 251 eggs were observed in the 54th (final) week.

Each mite treatment procedure resulted in a reduction in the number of individual life stages for 2 weeks. After that time, the population re-grew, and the infestation development rate after the second mite treatment procedure was 2.4 times higher than the infestation development rate compared to MTI.

3.2. Changes in the Percentages of Lymphocytes Subpopulations

The percentages and changes in the populations of Th cells, Tc cells and B cells in laying hens infested with *D. gallinae* controlled with Biobeck PA 910 (AI silicon dioxide, Biologische Preparate, Brilon, Germany) are shown in Figure 2. The Th cells accounted for the highest portion of all the lymphocytes, followed by B cells and Tc cells in each assay (B1–4). There were 68.52% of Th cells of total lymphocytes at the infestation peak I (B1). The *D. gallinae* population decreased after Biobeck PA 910 (AI silicon dioxide, Biologische Preparate, Brilon, Germany) was applied (B2), and the percentage of the lymphocytes decreased non-significantly (61.25%), and the decrease was continued significantly at the infestation peak II (B3) (50.34%) compared to B1 ($p = 0.0002$) and B2 ($p = 0.018$). Subsequently, the percentage increased to 60.6% ($p = 0.028$) after another mite treatment compared to B3. The analysis of correlation revealed a non-significant negative correlation between mite number and B cells (R = -0.845, $p > 0.05$), a non-significant negative correlation between mite number and Th cells (R = −0.522, $p > 0.05$) and a significant positive correlation between mite number and Tc cells (R = −0.982, $p < 0.05$).

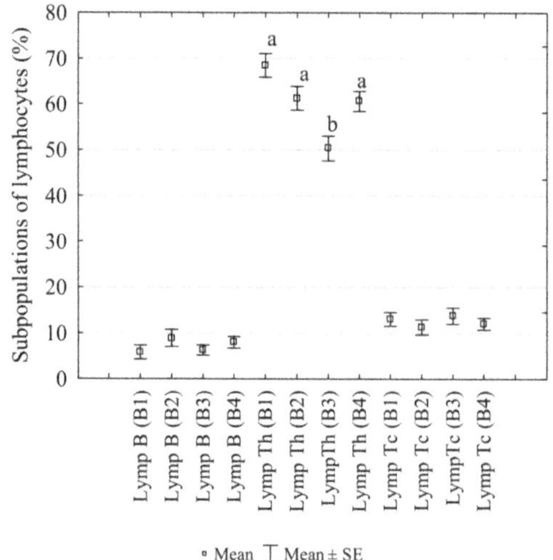

Figure 2. Changes in the percentages of B IgM+ cells (Lymp B), T CD3+CD4+ cells (Lymp Th) and T CD3+CD8a+ cells (Lymp Tc) in the peripheral blood of laying hens infested with *Dermanyssus gallinae*, controlled with Biobeck PA 910 (AI silicon dioxide, Biologische Preparate, Brilon, Germany). (B1–B4: blood sampling dates; a,b: different letters indicate statistical differences between groups).

3.3. The Changes in the Titres of Antibodies

The titres of antibodies against AE, IBV, NDV, REV, REO and ORT in laying hens infested with *D. gallinae*, controlled with Biobeck PA 910 (AI silicon dioxide, Biologische Preparate, Brilon, Germany), are shown in Figure 3. The study found a significant decrease in the titre of post-vaccination antibodies against IBV and NDV at the second infestation peak (B3) ($p = 0.014$ and $p = 0.001$, respectively). The titres of post-vaccination antibodies against AE did not change significantly during the study. However, a non-significant negative correlation was observed between mite numbers and AE titre (R = −0.564, $p > 0.05$). Antibodies against REO, REV and ORT were found despite the hens were not immunised earlier against those pathogens. The titres of antibodies against REO in subsequent weeks of the study were 7472, 8163, 2246, 1556. The titres of antibodies against REV were 121, 14, 19, 1 The titres of antibodies against ORT increased during the production cycle and were 402, 1904, 2921, 2150.

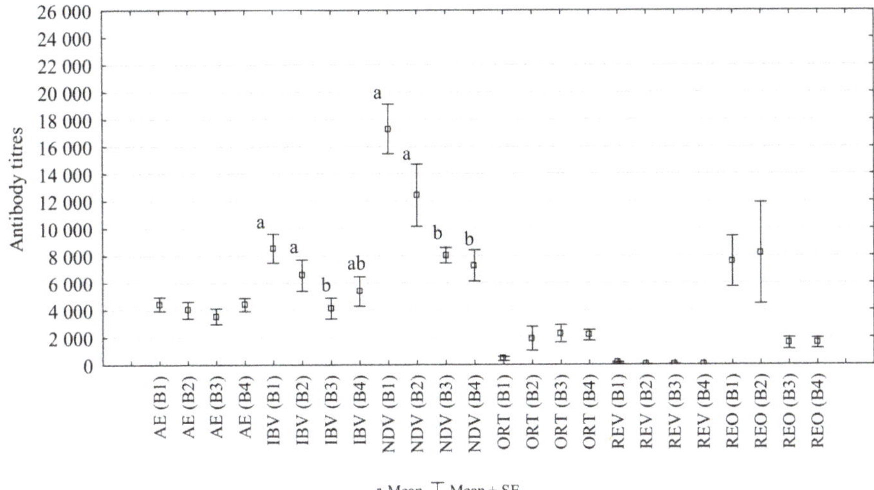

Figure 3. Changes in the titres of antibodies against AE (avian encephalomyelitis), IBV (infectious bronchitis virus), NDV (Newcastle disease virus), ORT (*Ornithobacterium rhinotracheale*), REV (reticuloendotheliosis virus) and REO (avian reovirus) in the peripheral blood of laying hens infested with *D. gallinae*, controlled with Biobeck. (B1–B4: blood sampling dates; a, ab, b: different letters indicate statistical differences between groups).

4. Discussion

This paper was the first report on the effect of *D. gallinae* infestation on a post-vaccination immune response in laying hens during the production cycle. Due to difficulties with access to the hen house free from the infestation of *D. gallinae*, we could not include a control group. Nevertheless, blood samples for cytometric and serological analyses were collected at the infestation peaks (on the day of mite treatment) and after the infestation subsided was recorded. Thus, we could observe the changes in the studied parameters under high and low mite pressure, which showed how *D. gallinae* might affect these parameters.

In the first infestation peak, no significant change in the percentages of B, Th or Tc cells or antibodies was observed. In the re-infestation of *D. gallinae* (infestation peak II), a significant decrease in the Th cells and IBV and NDV post-vaccination antibodies was observed.

Helper lymphocytes (Th) take part in the development of humoral response. They facilitate activation, proliferation and differentiation of B and Tc cells and have a CD4+ receptor on their surface. They also recognise antigens after they bind to class II of major histocompatibility complex (MHC) cells [28,29]. Our study showed a significant decrease in the percentage of these lymphocytes at the infestation peak II and a subsequent increase when the infestation subsided. These findings suggested that *D. gallinae* could inhibit the development of humoral immune response in infested laying hens.

Cytotoxic lymphocytes (Tc) play an important role in acquiring post-vaccine immunity. They are also called antigen-specific since they have a CD8+ receptor on their surface. A CD3+ receptor is present on the surface of most mature T cells and thymocytes. They are capable of destroying cells infected by a virus, cancer cells and of recognising proteins and glycoproteins (antigens) [28]. The percentage of these lymphocytes, as observed in this study, did not change significantly during the experiment. However, a strong positive correlation between mite numbers and those cells was reported. The percentage of these lymphocytes was seen to decrease during the periods of low *D. gallinae* infestation. These findings suggested that the parasite could stimulate the cellular immune response in infested laying hens, but the infestation was not sufficiently high to produce a significant effect.

B cells play a crucial role in the humoral immune response. They differentiate into memory cells and/or plasmatic cells, which produce antibodies specific to protein antigens in a complex process in which antigen-specific CD4+ helper cells take part [29,30]. Like Tc cells, the percentage of these lymphocytes, observed in this study, did not change significantly during the experiment. Only an increasing trend was observed for these lymphocytes during the period of low *D. gallinae* infestation. Despite the absence of statistical significance, these findings might suggest that *D. gallinae* could inhibit the post-vaccination humoral immune response. This was also testified to by a decrease in the post-vaccination antibody titre during the period of high *D. gallinae* infestation (Figure 3).

The titres of post-vaccination antibodies of AE, IBV, NDV changed during the cycle in a similar manner (Figure 3). They decreased with an increasing infestation (B1–B3), and their level increased after mite treatment (B4). A similar trend was observed in antibodies against REO, against which the hens were not immunised. On the other hand, antibodies against ORT showed an increasing trend in the B1–B3. The presence of antibodies against ORT in the blood of the hens under study, despite the absence of vaccination, indicated the occurrence of field infection. There was no correlation between mite numbers and antibody titres. Thus, there was insufficient evidence to support the hypothesis that *D. gallinae* has a negative effect on the production of post-vaccination antibodies in layer hens. Further study, including a larger sample size and negative control group, is required.

5. Conclusions

D. gallinae is currently considered to be the most bothersome ectoparasite in the commercial laying hen farms. Vaccines protect hens against pathogens, which is important from an economic perspective and consumer safety. Development of vaccine-induced immune responses might be affected by many factors. The role of *D. gallinae* as a potential immunosuppressor has been unknown. The present study was the first report in this area. The present findings suggested that *D. gallinae* might inhibit immune responses since the percentages of B cells and Th cells were negatively correlated with the number of mites. The percentage of Tc cells was positively correlated with the number of mites, which indicated that *D. gallinae* could stimulate cellular immune responses in infested laying hens. However, further research is needed to determine whether *D. gallinae* suppresses the production of vaccine-induced antibodies.

Author Contributions: Conceptualization, S.K.-S. and R.S.; methodology, S.K.-S. and R.S.; validation, R.S.; formal analysis, S.K.-S.; investigation, S.K.-S. and R.S.; resources, R.S.; data curation, R.S.; writing—original draft preparation, S.K.-S.; writing—review and editing, R.S.; visualization, S.K.-S.; supervision, R.S.; project administration, R.S.; funding acquisition, S.K.-S. and R.S. All authors have read and agreed to the published version of the manuscript.

Funding: This research received no external funding

Acknowledgments: Publication financially co-supported by the Minister of Science and Higher Education in the range of the program entitled "Regional Initiative of Excellence" for the years 2019–2022, Project No. 010/RID/2018/19, amount of funding PLN 12,000,000 PLN.

Conflicts of Interest: The authors declare no conflict of interest. The funders had no role in the design of the study; in the collection, analyses, or interpretation of data; in the writing of the manuscript, or in the decision to publish the results.

Abbreviations

AE	avian encephalomyelitis
ANOVA	analysis of variance
B1–B4	blood collection dates correspond to the assay groups
BCR	B-cell receptor
CD	cluster of differentiation
CV%	coefficient of variation
D. gallinae	Dermanyssus gallinae

FCS	fetal calf serum
IBV	infectious bronchitis virus
IGM	immunoglobulin M
Lymp Tc	lymphocytes T cells CD3+/CD8a+ (Tc)
Lymp B	lymphocytes B IgM+ cells
Lymp Th	lymphocytes T cells CD3+/CD4+ (Th)
MANOVA	multivariate analysis of variance
MHC	major histocompatibility complex
MTI-II	mite tratment dates
NDV	Newcastle disease virus
ORT	Ornithobacterium rhinotracheale
PBS	phosphate-buffered saline
PI-II	the peak in the mite population, in this study, referred to as infestation peak
prm	poultry red mite
REO	avian reovirus
REV	reticuloendotheliosis virus
RH	relative humidity
SE	standard error

References

1. Wood, H.P. The chicken mite: Its life history and habits. *US Dep. Agric.* **1917**, *553*, 1–14.

2. Chmielewski, W. Roztocze zamieszkujące gniazda wróbla domowego (Passer domesticus L). *Wiad Parazytol.* **1982**, *28*, 5–107.

3. Romaniuk, K.; Owczarzak-Podziemska, I. Występowanie saprobiontycznych i pasożytniczych roztoczy w ściółce ferm indyków. *Med. Weter.* **2002**, *58*, 298–299.

4. Roy, L.; Chauve, C. Historical review of the genus *Dermanyssus* Duges, 1834 (Acari: Mesostigmata: Dermanyssidae). *Parasite* **2007**, *14*, 87–100. [CrossRef]

5. Hoglund, J.; Nordenfors, H.; Uggla, A. Prevalence of the poultry red mite, *Dermanyssus gallinae*, in different types of production systems for egg layers in Sweden. *Poult. Sci.* **1995**, *74*, 1793–1798. [CrossRef] [PubMed]

6. Fiddes, M.D.; Le Gresley, S.; Parsons, D.G.; Epe, C.; Coles, G.C.; Stafford, K.A. Prevalence of the poultry red mite (*Dermanyssus gallinae*) in England. *Vet. Rec.* **2005**, *157*, 233–235. [CrossRef]

7. Van Emous, R.A. Wage war against the red mite! *Poult. Int.* **2005**, *44*, 26–33.

8. Sparagano, O.A.E.; Pavlicevic, A.; Murano, T.; Camarda, A.; Sahibi, H.; Kilpinen, O.; Mul, M.; Van Emous, R.A.; Le Bouguin, S.; Hoel, K.; et al. Prevalence and key figures for the poultry red mite *Dermanyssus gallinae* infections in poultry farm systems. *Exp. Appl. Acarol.* **2009**, *48*, 3–10. [CrossRef]

9. Sparagano, O.A.E.; George, D.R.; Harrington, D.W.; Giangaspero, A. Significance and control of the poultry red mite, *Dermanyssus gallinae*. *Ann. Rev. Entomol.* **2014**, *59*, 447–466. [CrossRef]

10. Chauve, C. The poultry red mite *Dermanyssus gallinae* (Dr Geer, 1778): Current situation and future prospects for control. *Vet. Parasitol.* **1998**, *79*, 239–245. [CrossRef]

11. Maurer, V.; Baumgartner, J. Temperature influence on life table statistics of the chicken mite *Dermanyssus gallinae* (Acari: Dermanyssidae). *Exp. Appl. Acarol.* **1992**, *15*, 27–40. [CrossRef] [PubMed]

12. Moss, W.W. The mite genus *Dermanyssus*: A survey, with description of *Dermanyssus trochilinis*, n. sp., and a revised key to the species (Acari: Mesostigmata: Dermanyssidae). *J. Med. Entomol.* **1978**, *14*, 627–640. [CrossRef]

13. Entrekin, D.L.; Oliver, J.H., Jr. Aggregation of the chicken mite, *Dermanyssus gallinae* (Acari: Dermanyssidae). *J. Med. Entomol.* **1982**, *1*, 671–678. [CrossRef] [PubMed]

14. Koenraadt, C.J.M.; Dicke, M. The role of volatiles in aggregation and host-seeking of the haematophagous poultry red mite *Dermanyssus gallinae* (Acari: Dermanyssidae). *Exp. Appl. Acarol.* **2010**, *50*, 191–199. [CrossRef]

15. Roy, L.; Chauve, C.; Buronfosse, T. Contrasted ecological repartition of the northern fowl mite *Ornithonyssus sylviarum* (Mesostigmata: Macronyssidae) and the chicken red mite *Dermanyssus gallinae* (Mesostigmata: Dermanyssidae). *Acarologia* **2010**, *50*, 207–219. [CrossRef]

16. Abdel-Ghaffar, F.; Semmler, M.; Al-Rasheid, K.; Mehlhorn, H. In vitro efficacy of ByeMite® and Mite-Stop® on developmental stages of the red chicken mite *Dermanyssus gallinae*. *Parasitol. Res.* **2009**, *105*, 1469–1471. [CrossRef]

17. Beugnet, F.; Chauve, C.; Gauthey, M.; Beert, L. Resistance of the red poultry mite to pyrethroids in France. *Vet. Rec.* **1997**, *140*, 577–579. [CrossRef]

18. Mul, M.; Van Niekerk, T.; Chirico, J.; Maurer, V.; Kilpinen, O.; Sparagano, O.A.E.; Thind, B.; Zoons, J.; Moore, D.; Bell, B.; et al. Control methods for *Dermanyssus gallinae* in systems for laying hens: Results of an international seminar. *Worlds Poult. Sci. J.* **2009**, *65*, 589–599. [CrossRef]

19. Chirico, J.; Tauson, R. Traps containing acaricides for the control of *Dermanyssus gallinae*. *Vet. Parasitol.* **2002**, *110*, 109–116. [CrossRef]

20. Cosoroaba, I. Massive *Dermanyssus gallinae* invasion in battery-husbandry raised fowls. *Rev. Med. Vet.* **2001**, *152*, 89–96.

21. Cencek, T. Prevalence of *Dermanyssus gallinae* in poultry farms in Silesia Region in Poland. *Bull. Vet. Inst. Pulawy.* **2003**, *47*, 465–469.

22. Arkle, S.; Guy, J.H.; Sparagano, O.A.E. Immunological effects and productivity variation of red mite (*Dermanyssus gallinae*) on laying hens—Implications for egg production and quality. *Worlds Poul. Sci. J.* **2006**, *62*, 249–257. [CrossRef]

23. Harrington, D.W.; Robinson, K.; Guy, J.; Sparagano, O.A.E. Characterization of the immunological response to *Dermanyssus gallinae* infestation in domestic fowl. *Transbound. Emerg. Diss.* **2010**, *57*, 107–110. [CrossRef] [PubMed]

24. Harrington, D.W.; Robinson, K.; Sparagano, O.A.E. Immune responses of the domestic fowl to *Dermanyssus gallinae* under laboratory conditions. *Parasitol. Res.* **2010**, *106*, 1425–1434. [CrossRef]

25. Kowalski, A.; Sokół, R. Influence of *Dermanyssus gallinae* (poultry red mite) invasion on the plasma levels of corticosterone, catecholamines and proteins in layer hens. *Pol. J. Vet. Sci.* **2009**, *12*, 231–235.

26. Sokół, R.; Koziatek-Sadłowska, S.; Michalczyk, M. The influence of *Dermanyssus gallinae* and different lighting regimens on selected blood proteins, corticosterone levels and egg production in layer hens. *Vet. Res. Commun.* **2019**, *43*, 31–36. [CrossRef]

27. Sokół, R.; Romaniuk, K. Przebieg i dynamika inwazji *Dermanyssus gallinae* w fermie kur niosek. *Med. Weter* **2007**, *63*, 484–486.

28. Arstila, T.P.; Vainio, O.; Lassila, O. Central role of CD4+ T cells in avian immune response. *Poult. Sci.* **1994**, *73*, 1019–1026. [CrossRef]

29. Chen, X.; Jensen, P.E. The role of B lymphocytes as antigen-presenting cells. *Arch. Immunol. Ther. Exp.* **2008**, *56*, 77–83. [CrossRef]

30. Parker, D.C. T cell-dependent B cell activation. *Ann. Rev. Immunol.* **1993**, *11*, 331–360. [CrossRef]

animals

Article

Prevalence Rate and Molecular Characteristics of *Oestrus ovis* L. (Diptera, Oestridae) in Sheep and Goats from Riyadh, Saudi Arabia

Dina M. Metwally [1,2,*], Shurug A. Albasyouni [1], Ibrahim A.H. Barakat [1,3], Isra M. Al-Turaiki [4], Amal M. Almuhanna [1], Muhammad A. Bashir [5], Hanadi B. Baghdadi [6,7], Manal F. El-Khadragy [8] and Reem A. Alajmi [1]

1   Zoology Department, College of Science, King Saud University, P.O. Box 2455, Riyadh 11451, Saudi Arabia; shurugaalbasyouni@gmail.com (S.A.A.); ibrahimahb@yahoo.com (I.A.H.B.); aalmuhanna@KSU.EDU.SA (A.M.A.); ralajmi@ksu.edu.sa (R.A.A.)
2   Parasitology Department, College of Veterinary Medicine, Zagazig University, Zagazig 44519, Egypt
3   Cell Biology Department, National Research Center, 33 Bohouth St., Dokki, Giza 12622, Egypt
4   Department of Information Technology, College of Computer and Information Sciences, King Saud University, Riyadh 11451, Saudi Arabia; ialturaiki@ksu.edu.sa
5   Department of Plant Protection, Faculty of Agricultural Science Ghazi University, Dera Ghazi Khan 32200, Pakistan; abashir@gudgk.edu.pk
6   Biology Department, College of Science, Imam Abdulrahman Bin Faisal University, Dammam City 31441, Saudi Arabia; hbbaghdadi@iau.edu.sa
7   Basic and Applied Scientific Research Center, Imam Abdulrahman Bin Faisal University, Dammam City 31441, Saudi Arabia
8   Department of Biology, Faculty of Science, Princess Nourah Bint Abdelrahman University, Riyadh 11451, Saudi Arabia; mfelkhadragy@pnu.edu.sa
*   Correspondence: dhasanin@ksu.edu.sa

Citation: Metwally, D.M.; Albasyouni, S.A.; Barakat, I.A.H.; Al-Turaiki, I.M.; Almuhanna, A.M.; Bashir, M.A.; Baghdadi, H.B.; El-Khadragy, M.F.; Alajmi, R.A. Prevalence Rate and Molecular Characteristics of *Oestrus ovis* L. (Diptera, Oestridae) in Sheep and Goats from Riyadh, Saudi Arabia. *Animals* **2021**, *11*, 689. https://doi.org/10.3390/ani11030689

Academic Editor: Robert W. Li

Received: 28 January 2021
Accepted: 26 February 2021
Published: 4 March 2021

Publisher's Note: MDPI stays neutral with regard to jurisdictional claims in published maps and institutional affiliations.

Simple Summary: In the current study, we investigated the prevalence rate of *Oestrus ovis* (*O. ovis*) larvae in heads of sheep and goats obtained from abattoirs in Riyadh, Saudi Arabia. It was found that season did not have any significant effect on the characteristics of the larvae that affect sheep, except for the length of larva, for which those collected in spring differed from those collected in winter and autumn. In goats, the L3 instar number was significantly higher during summer. *Oestrus ovis* larvae were detected in sheep and goats by molecular analysis and they were most closely related to *O. ovis* larvae recovered from GenBank.

Abstract: Heads of sheep (*n* = 600) and goats (*n* = 800) slaughtered at Al-Aziziah Abattoir in Riyadh, Saudi Arabia, were inspected for the presence of *O. ovis* larvae (L). Heads were split along the longitudinal axes, and larvae (L1, L2, and L3) were gathered. The infestation rate was significantly higher in goats (44.5%; 356/800) than that in sheep (22.3%; 134/600). Out of the 151 collected larvae from sheep, 0% were L1, 1.3% were L2, and 98.7% were L3. Out of the total of 468 larvae from goats, 0% were L1, 1.2% were L2, and 98.8% were L3. The infestation rate was significantly higher in males than that in females. Myiasis-causing larvae collected from Riyadh, Saudi Arabia, were authenticated as *O. ovis*, according to morphological characteristics. Polymerase chain reaction (PCR) amplification of a partial fragment (600 bp) of the mitochondrial cytochrome c oxidase subunit I (*mtCOI*) gene further confirmed the species. Phylogenetic analysis based on the partial *mtCOI* gene sequence demonstrated that 23 unique sequences showed high similarity based on nucleotide pairs of *O. ovis* accessions retrieved from GenBank.

Keywords: myiasis; prevalence; *Oestrus ovis*; *mtCOI*; sheep; goats; Saudi Arabia

1. Introduction

The sheep gadfly *O. ovis* Linnaeus 1761 (Diptera, Oestridae) is one cause of myiasis in sheep and goats. Sneezing and nasal discharge are the most prominent clinical symptoms of infested animals [1]. Myiasis lowers the health of animals and causes major economic losses to the livestock industry due to abortion, reduced milk production, loss of weight gain, and fertility [2]. *Oestrus ovis* larvae persist in cranial cavities of sheep [3]. The incidence of *O. ovis*-related myiasis in the Jazan area was previously reported to be as high as 53.54% [1]. Clinical appearance, morphological characterizations of the larvae (e.g., slits of the posterior spiracles found on the posterior spiracle plates), and occasional identification of the adult fly are among the many approaches leading to the diagnosis of myiasis [4]. Molecular methods facilitate in the diagnosis and detection of a wide variety of species, including flies that cause myiasis [5,6]. Molecular approaches can be efficiently used to identify all larval stages [7]. Multiple genetic markers of ribosomal DNA (16S rRNA, 28S rRNA) and mitochondrial DNA (cytochrome oxidase genes (*COI*, *COII*, and 12S mtDNA)) have been used [8]. A widely used method for the molecular identification of myiasis-causing larvae belonging to the Oestridae family is the sequencing of the mitochondrial cytochrome c oxidase subunit I (*mtCOI*) gene, which contributes to an essential database for molecular identification of organisms [9,10]. Phylogenetic relationships can be assessed by polymerase chain reaction (PCR) and PCR restriction fragment length polymorphism (PCR-RFLP) assays [7]. For some species belonging to Calliphoridae, Sarcophagidae, and Oestridae, the mitochondrial gene subunit has been isolated, proving effective for molecular phylogenetic studies [4,11–16]. *Oestrus* spp. larvae identification can be complicated, as there are considerable morphological similarities within the species. The objectives of the present study were to investigate the seasonal prevalence of *O. ovis* among slaughtered sheep and goats in Riyadh, Saudi Arabia, and to use the sequencing of the *mtCOI* gene to identify the phylogenies of *O. ovis* larvae. One genetic marker, *COI*, was used to identify larvae collected from domestic sheep and goats to accurately identify *Oestrus* parasites [17].

2. Materials and Methods

2.1. Larvae Collection and Morphological Identification

This study was conducted between March 2019 and January 2020 in the city of Riyadh, Saudi Arabia. Riyadh City is in the center of the Kingdom of Saudi Arabia. It extends between latitude: 24°38′ North and longitude:46°43′ East. It has a very dry and arid climate; summer temperatures rise very dramatically during the day but fall at night, ranging between 35 °C and 43 °C. In the winter, the temperature drops dramatically and may reach 0 °C. Riyadh City is located 600 m above sea level, with several valleys and rich sand dunes. A total of 600 sheep and 800 goat heads from the Al-Aziziah Abattoir were analyzed for *O. ovis* larvae. The sex and age (<1 year is young and >1 year old is adult) of slaughtered animals were registered. At the investigation site, the nasal and frontal sinuses were longitudinally divided into the dorsoventral planes using a sharp handsaw. *Oestrus ovis* present in the mucous membrane of the nasal septa and nasal segments were collected with forceps and preserved in 70% ethanol for molecular studies and morphometric identification. Larvae were identified as the first, second, and third larval stages [18,19]. The percentage of each larval instar was calculated according to the equation (Larval stage percent = number of each larval stage/Total number of larvae × 100). In each infested head, the larvae were counted with the aid of identification keys [20]. The severity of the infestation was calculated as the number of larvae/number of animals infested.

2.2. DNA Extraction and Polymerase Chain Reaction (PCR)

DNA from each individual larva (about 0.025 gm from anterior section of the larva) was extracted using a DNeasy Blood & Tissue Kit (250) (Cat No./ID: 69506, Qiagen, Hilden, Germany) following the manufacturers' protocol. *mtCOI* gene was amplified using the primer pairs described in Table 1 [21] and a thermal cycler (Veriti® 96-Well Thermal Cycler, Model 9902, Biosystem). This procedure was performed in a mixture (20 μL) consisting of master mix (5×, 4 μL), RNase-free water (12 μL), and DNA template (2 μL). The PCR program consisted of a denaturation step of 94 °C for 2 min, followed by

40 cycles of denaturation for 30 s at 94 °C. PCR products were analyzed by 1.5% agarose gel electrophoresis.

Table 1. Set of primers used for the amplification of mitochondrial cytochrome c oxidase subunit I (*mtCOI*) gene using extracted DNA of *O. ovis* larvae.

Gene	Primers	Sequences	References
mtCOI	FFCOI	5′-GGAGCATTAATYGGRGAYG-3′	[21]
	RHCO	5′-TAAACTTCAGGGTGACCAAAAATCA-3′	

2.3. Nucleotide Sequences of the mtCOI Gene

A total of 40 sequences (Table S1), which were coded according to properties of their host, were analyzed in this study. The sequences were processed using Geneious Prime Build 2020-04-07 08:42 [22]. Prior to processing, all sequences were truncated using the error probability method with a limit of 0.05 on both sides. Related sequences were retrieved from GenBank after performing a BLAST [23] search. Multiple sequence alignments were then generated using CLUSTAL Omega [24] implemented in Geneious software. A phylogenetic tree was created using the neighbor-joining method [25] with the Tamura_Nei model and using 10,000 replicates. All samples were deposited to GenBank, and accession numbers have been provided.

2.4. Statistical Analysis

All data concerning the effect of different seasons were analyzed using one-way analysis of variance (ANOVA), followed by Duncan's test to compare means. Statistical analysis was conducted using SPSS software program version 24, and all results were expressed as means $\pm$ SEM. Differences were considered significant at $p \leq 0.05$.

3. Results

3.1. Prevalence of Infestation over the Course of the Study in Sheep and Goats

This study, which was conducted on 600 sheep and 800 goats (Table 2), showed that infestation rate with larvae was much higher in goats (44.5%; 356/800) than that in sheep (22.3%; 134/600). This also applied to the total number of larvae, where the total number of larvae in 356 head of goats was 486, while it was 151 in 134 head of sheep.

Table 2. Prevalence of *Oestrus* larvae in sheep and goats over the course of the study.

Species	No. of Head Examined	Infestation Rate (%)	Mean of Infestation	No. of Larvae (%)			Total Larvae (%)
				L1	L2	L3	
Sheep	600	134 (22.3)	1.78 ± 0.02 [a]	0.0 (0.0)	2.0 (1.5)	149.0 (111.2)	151.0 (112.7)
Goat	800	356 (44.5)	1.56 ± 0.02 [b]	0.0 (0.0)	6.0 (1.7)	482.0 (135.4)	486 (136.5)

Mean values with different superscript in column differed significantly at $p \leq 0.05$.

Results of statistical analysis concerning infestation rate showed that the ovine species is significantly more susceptible to infection with larvae than caprine species: the infestation mean in sheep was 1.78, and in goats was 1.56 at $p \leq 0.05$, as a significant level.

The findings in Table 3 show that, in both species (goats and sheep), there was a disparity in the percentage of infestation between males and females, where the rate of infestation in males was higher than that in females (32.66% (98/300) and 65.50% (262/400) in sheep and goats, respectively). Additionally, the incidence of infestation was higher in female goats (32.50 percent; 94/400) than in female sheep (12.0 percent; 36/300). The seasonal effect on the characteristics of larvae specifically infesting goats (Figure 1) showed that there was a significant difference in the total number of larvae and L3 instar.

The mean values of total larvae and L3 instar were significantly higher in the summer (1.94 ± 0.12 and 1.91 ± 0.12, respectively) than in other seasons.

Table 3. Sex differences on the prevalence of *Oestrus* larvae in sheep and goats.

Species	Sex	No. of Examined Animals	No. of Infested (%)
Sheep	Male	300	98 (32.66)
	Female	300	36 (12)
Goat	Male	400	262 (65.5)
	Female	400	94 (32.5)
Total		1400	490 (35.00)

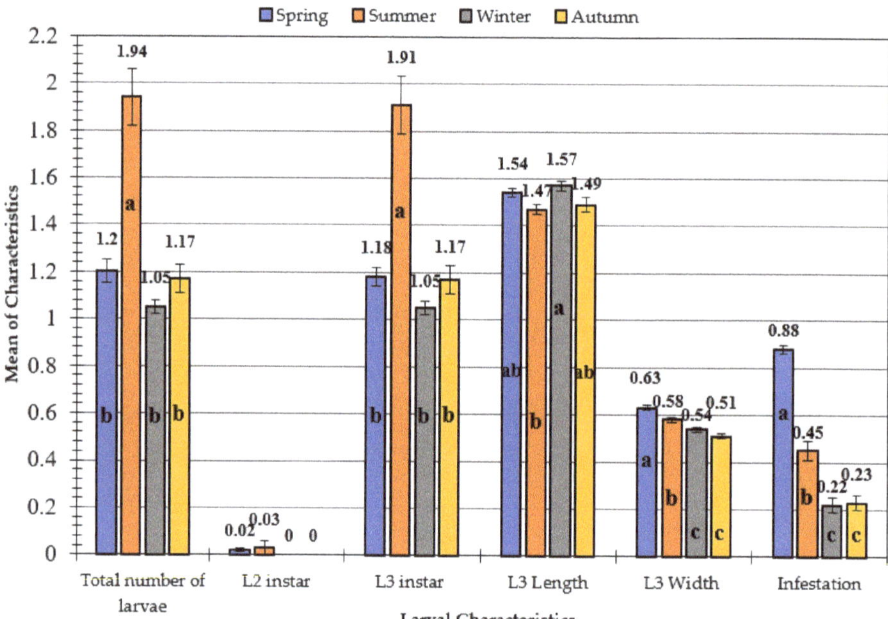

Figure 1. Effect of season on larval characteristics in infested goats. Mean values with different letters (a, b, c) are significantly different at $p \leq 0.05$ between seasons.

In addition, it was found that there was no significant difference between the means of total number of larvae and L3 in the spring, winter, and autumn. Concerning the larvae width and infestation, the trend was the same in the two characteristics. No significant difference between the winter and autumn was observed, while the significant differences in width and infestation appeared between spring and summer. As for the larvae length characteristic, it was found that only the winter and summer seasons varied significantly, while there were no significant variations between the spring and autumn seasons. The results of the impact of season on the studied characteristics differed only in goats. It was found that seasons of the year did not have any significant effect on the characteristics of the larvae that affect sheep, in general, except for the length of larva, in which those collected in spring differed from those collected in winter and autumn (Figure 2).

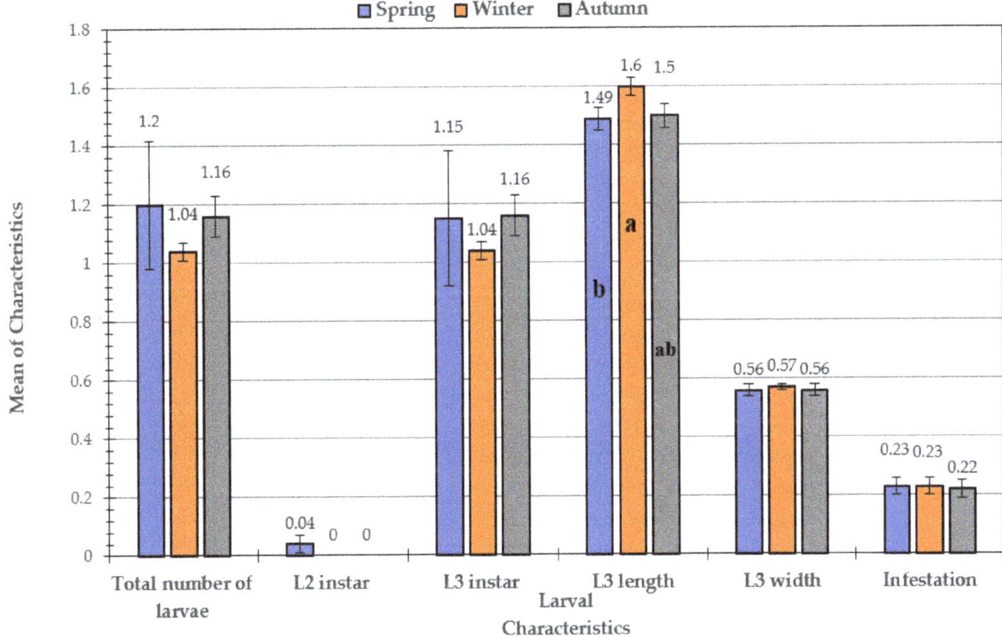

Figure 2. Effect of seasons on larval characteristics in infested sheep. Mean values with different letters (a and b) are significantly different at $p \leq 0.05$ between seasons.

3.2. Morphological Examination

The identification of mature *O. ovis* larvae was based on morphological characteristic features of the larval stages. Second instar larvae (L2) were found only in young male goats in spring and summer; the average length and width were 0.84 ± 0.08 cm and 0.26 ± 0.05 cm, respectively, in spring, while the average length and width were 0.90 ± 0.10 cm and 0.30 ± 0.00 cm, respectively, in the summer. These L2 larvae are white in color, and the dorsal side shows only a few weak denticles on the second segment (Figure 3A). On the ventral side, the segments have rows of peculiarly shaped currycomb-like spines (Figure 3B). Antennary lobes are less separated giving to pseudocephalon triangular shape. The buccal funnel is well structured. The hooks of the first segment are less robust and more curved (Figure 3C). Furthermore, the median part of the postanal bulge is spinulose (Figure 3D). Ventrally the segments are provided with rows of peculiarly shaped currycomb-like spines, the thoracic and abdominal segments on the ventral side, show single-ended, caudally projected spines (Figure 3E). The posterior peritremes are more or less circular (Figure 3F), with the channel indicated by a distinct suture (Figure 3G,H).

Third instar larvae (L3) are yellow in color when young (Figure 4A), changing to a light brown later in the mature stage, and show broad transverse blackish bands on the dorsal side (Figure 4B). The second segment on the dorsal side presents with a variable number of small denticles; the following segments are bare but have a rough, leatherlike skin pattern, which is distinct only on the darkened parts. On the ventral side, the segments bear rows of strong spines (Figure 4C), which are irregularly placed on the third segment but are fairly regular on the following ones. The number of rows varies from two to five (Figure 4D). The postanal bulge shows fewer spines, while the preanal bulge is bare (Figure 4E). The posterior peritremes are circular, with a central button and lacking a distinct suture and the channel appears in the posterior peritremes (Figure 4F).

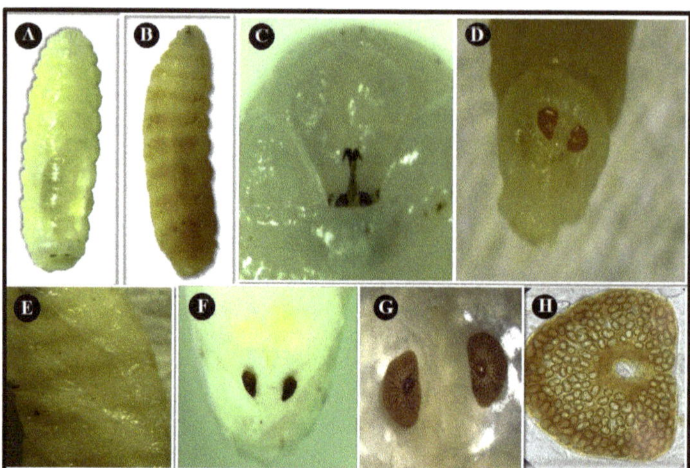

Figure 3. Second instar larva (L2). (**A**) Dorsal view; (**B**) ventral view; (**C**) the front end containing the cephaloskeleton ; (**D**) the median part of the postanal bulge is spinulose; (**E**) ventral view shows spines on anterior segment ; (**F**) posterior spiracle; (**G**) stigmal plates and channel appear with clear distinct suture; (**H**) posterior spiracle (10×).

Figure 4. Third instar larva (L3). (**A**) Yellow color; (**B**) dorsal view shows broad transverse blackish bands; (**C**) ventral view shows spines on segments; (**D**) magnified view of the spines on the ventral segments; (**E**) postanal bulge shows fewer spines; (**F**) posterior spiracle is in a D-shape with central button but no distinct suture.

3.3. PCR Amplification and Nucleotide Sequence Analysis of the Partial mtCOI Gene

PCR amplification was performed using a set of primers (FFCOI and RHCO) to amplify a 600-bp fragment of the *mtCOI* gene from *O. ovis* larvae. A total of 23/40 sequences were analyzed in this study (sequences with HQ% (percent high quality) score less than 40 were discarded). The pairwise identity of sheep sequences was 97.9%, and that of goat sequences was 99.2%. Overall, the 23 sequences shared 98.1% pairwise identity and had 35.6% GC content. The results of BLAST search showed that all the sequences had high similarity to the *O. ovis* "sheep botfly" in GenBank. Figure 5 shows the phylogenetic tree indicating respective evolutionary relationships. MT787554 (*Chrysomya bezziana*) is used as an out-group.

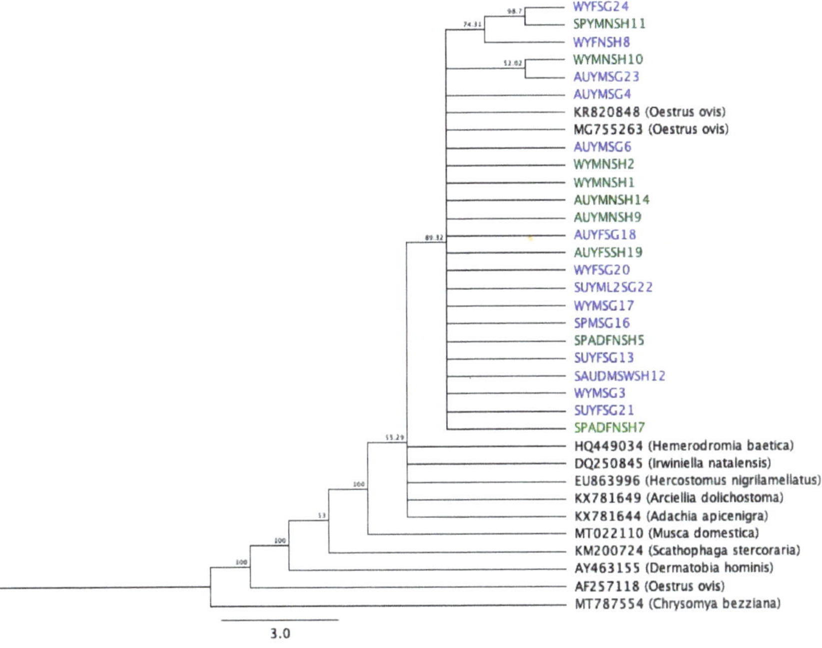

Figure 5. Genetic relationships of the *mtCOI* gene collected samples with other species retrieved from GenBank. The tree is generated using the neighbor-joining method, implemented in the Geneious software, with the Tamura_Nei model.

4. Discussion

Out of the examined 600 sheep and 800 goat heads, 22.3% and 44.5%, respectively, were infested with *O. ovis* larvae. These percentages were high compared with previous studies in Saudi Arabia [26], which recorded 5.9% in Riyadh Region and 0.299% from adult *O. ovis* spp. in Jeddah [27], but similar to the 53.5% (257/480) reported in the Jazan Region [1]. Our results were also similar to those reported in Greece (43.2%) [28], Turkey (22.52%) [29], and Libya (42.33%) [30]. Our results were higher than those reported by other researchers around the world, such as by [31] in Egypt (8.67%), but less than those in Spain (71.1%) [32] and Italy (91.0%) [33]. These differences may be a result of differences in geographical area, animal breed, and climate. The infestation rate was higher in male sheep and goats than females (16.35% and 32.75%, respectively, compared to 6% and 11.75%; $p \leq 0.05$). This was consistent with a report investigating sex differences in infestation in Libya [30].

In the central region of Saudi Arabia, the spring months are the best time of the year, with low temperature and humidity [26]. The disappearance of fly activity in most of the year is probably due to the hot, dry, and windy weather. The 27 °C was optimum temperature for adults emerge from pupa, while constant temperatures below 16 °C and above 32 °C were fatal [34].

Our results did not demonstrate any significant seasonal prevalence of infestation in sheep, suggesting that time of the year does not have any significant effect. This is in disagreement with a previous study [35], which mentioned that the peaks of infestation in Turkey were in the summer and spring, while the lowest infestation rates were in autumn. Another study in Egypt [36] concluded that heavier infestations were recorded in autumn (17.91%), while the lowest seasonal peak was in winter (7.85%). In Iraq, the highest recorded range was in the autumn, which was 76.92%, and the lowest range was in winter (30.76%) [37]. The reason for these differences may be changes in environmental conditions.

The present study recorded that 0% were L1, 1.5% were L2, and 111.2% were L3 of the total of 151 larvae collected from sheep, and 0% were L1, 1.7% were L2, and 135.4% were L3 of the of total 468 larvae collected from goats. The L1 result was lower than that recorded in previous studies [30,35] that reported that L1 larvae prevailed at 90.5% of the overall weight, and another study [37] that reported 13.14% L1 larvae. Our results were in agreement with one study [38] that reported a higher level of larvae in stage L3 (63.4%) than that in L2 (26.6%) and L1 (10%). It is possible that L1 larvae might have passed unseen due to their small size, that they were present in hidden places such as turbinates and ethmoid bones, or that numerous L1 were demolished in the nasal holes during the hypo-biotic period [39]. The morphological characters of L2 and L3 were in accordance with those stated in the identification key [20].

PCR analysis was performed using set of primers (FFCOI and RHCO) and successfully amplified 606bp fragment of *mtCOI* gene from *O.ovis* larvae. The sequence of amplified fragment showed genome arrangement typical of *O. ovis* sequence in Genbank. PCR amplification of *mtCOI* gene from eighteen species of Oestridae causing myiasis resulted in 686 bp [12]. Rooted phylogenetic trees elucidated phylogenetic relationships between *O.ovis* larvae amplicons and other members of Family: Oestridae published in Genbank.

The results of BLAST search showed that all the sequences had high similarity to the *O. ovis* "sheep botfly" identified in Brazil (accession number, KR820703) [40], with pairwise identity between 94.8% and 99.5% and coverage more than 86%. The sequences are also similar to the *O. ovis* sheep botfly identified in Turkey (accession number, MT124626) [41], with pairwise identity between 94.2% and 98.70% and coverage greater than 86.18%.

5. Conclusions

In this study, the results indicated that the infestation rate was significantly higher in sheep than goat, male than female, and spring season than other seasons of the year.

Supplementary Materials: The following are available online at https://www.mdpi.com/2076-2615/11/3/689/s1, Table S1: Accession numbers of samples deposited to GenBank.

Author Contributions: All the authors contributed equally to this work. All authors have read and agreed to the published version of the manuscript.

Funding: This research was funded by Deanship of Scientific Research at Princess Nourah Bint Abdulrahman University through the Fast-Track research funding program.

Institutional Review Board Statement: Ethical review and approval were waived for this study, due to collection of samples from slaughtered animals. The study did not involve alive animals.

Data Availability Statement: All relevant data are within the paper.

Conflicts of Interest: The authors declare no conflict of interest. The funders had no role in the design of the study; in the collection, analyses, or interpretation of data; in the writing of the manuscript, or in the decision to publish the results.

References

1. Hanan, B.A. Seasonal prevalence of *Oestrus ovis* L. (Diptera: Oestridae) larvae in infested sheep in Jazan Region. Saudi Arabia. *J. Parasitol. Vector Biol.* **2013**, *5*, 66–71.
2. Traversa, D.; Otranto, D. A new approach for the diagnosis of myiasis of animals: The example of horse nasal myiasis. *Vet. Parasitol.* **2006**, *141*, 186–190. [CrossRef]
3. Hidalgo, A.; Palma, H.; Oberg, C.; Fonseca-Salamanca, F. *Oestrus ovis* infestation of grazing sheep during summer in southern Chile. *Pesqui. Veterinária Bras.* **2015**, *35*, 497–500. [CrossRef]
4. Ogo, N.I.; Onovoh, E.; Okubanjo, O.O.; Galindo, R.C.; De La Lastra, J.-M.P.; De La Fuente, J.; De La Lastra, J.M.P. Molecular identification of *Cordylobia anthropophaga Blanchard* (Diptera: Calliphoridae) larvae collected from dogs (*Canis familiaris*) in Jos South, Plateau State, Nigeria. *Onderstepoort J. Vet. Res.* **2012**, *79*, 4. [CrossRef]
5. Otranto, D.; Traversa, D. Molecular evidence indicating that *Przhevalskiana silenus*, *P. aegagri* and *P. crossii* (Diptera, Oestridae) are one species. *Acta Parasitol.* **2004**, *49*, 173–176.
6. Hall, M.; Adams, Z.; Wyatt, N.; Testa, J.; Edge, W.; Nikolausz, M.; Farkas, R.; Ready, P. Morphological and mitochondrial DNA characters for identification and phylogenetic analysis of the myiasis causing flesh fly *Wohlfahrtia magnifica* and its relatives, with a description of *Wohlfahrtia monegrosensis* sp. n. Wyatt & Hall. *Med. Vet. Entomol.* **2009**, *23*, 59–71.
7. Ames, C.; Turner, B.; Daniel, B. The use of mitochondrial cytochrome oxidase I gene (COI) to differentiate two UK blowfly species—*Calliphora vicina* and *Calliphora vomitoria*. *Forensic Sci. Int.* **2006**, *164*, 179–182. [CrossRef]
8. Otranto, D.; Stevens, J.R. Molecular approaches to the study of myiasis-causing larvae. *Int. J. Parasitol.* **2002**, *32*, 1345–1360. [CrossRef]
9. Otranto, D.; Puccini, V. Cytochrome oxidase I (COI) gene of some obligate myiasis causing larvae (Diptera: Oestridae): Which perspective for phylogenetic studies? Preliminary data. *Proc. Cost Action* **2000**, *833*, 195–201.
10. Otranto, D.; Colwell, D.D.; Mililillo, P.; Di Marco, V.; Paradies, P.; Napoli, C.; Giannetto, S. Report in Europe of nasal myiasis by *Rhinoestrus* spp. in horses and donkeys: Seasonal patterns and taxonomical considerations. *Vet. Parasitol.* **2004**, *122*, 79–88. [CrossRef] [PubMed]
11. Otranto, D.; Colwell, D.D.; Traversa, D.; Stevens, J.R. Species identification of *Hypoderma* affecting domestic and wild ruminants by morphological and molecular characterization. *Med. Vet. Entomol.* **2003**, *17*, 316–325. [CrossRef]
12. Otranto, D.; Traversa, D.; Guida, B.; Tarsitano, E.; Fiorente, P.; Stevens, J.R. Molecular characterization of the mitochondrial cytochrome oxidase I gene of Oestridae species causing obligate myiasis. *Med. Vet. Entomol.* **2003**, *17*, 307–315. [CrossRef]
13. Otranto, D.; Mililillo, P.; Traversa, D.; Colwell, D.D. Morphological variability and genetic identity in *Rhinoestrus* spp. causing horse nasal myiasis. *Med. Vet. Entomol.* **2005**, *19*, 96–100. [CrossRef]
14. Li, X.; Cai, J.F.; Guo, Y.D.; Wu, K.L.; Wang, J.F.; Liu, Q.L.; Wang, X.H.; Chang, Y.F.; Yang, L.; Lan, L.M.; et al. The availability of 16S rRNA for the identification of forensically important flies (Diptera: Muscidae) in China. *Trop. Biomed.* **2010**, *27*, 155–166. [PubMed]
15. Xinghua, W.; Jufeng, C.; Yadong, G.; Yufeng, C.; Kunlu, W.; Qinlai, L.; Jifeng, C.; Yunfeng, C.; Jiangfeng, W.; Yang, L.; et al. The availability of 16SrDNA gene for identifying forensically important blowflies in China. *Rom. J. Leg. Med.* **2010**, *18*, 43–50. [CrossRef]
16. Hendawy, S.H.; Allam, N.A.; Kandil, O.M.; Zayed, A.A.; Desouky AR, A.; Mahmoud, A. Partial COI and 16S rRNA genes sequences of *Cephalopina titillator* mitochondrial dna: Evidence for variation in evolutionary rates within myiasis causing species. *Indian J. Anim. Res.* **2012**, *52*, 105–110.
17. Moreno, V.; Romero-Fernández, I.; Marchal, J.A.; Beltran, M.; Granados, J.E.; Habela, M.A.; Tamadon, A.; Rakhshandehroo, E.; Sarasa, M.; Pérez, J.M.; et al. Molecular characterization of bot flies, *Oestrus* spp., (Diptera, Oestridae), from domestic and wild Bovidae hosts. *Vet. Parasitol.* **2015**, *212*, 473–477. [CrossRef]
18. Ferrar, P. *A Guide to the Breeding Habits and Immature Stages of Diptera Cyclorrhapha*; EJ Brill/Scandinavian Science Press: Copenhagen, Denmark, 1987; Volume 8, pp. 1–2.
19. Smith, K.G. *An Introduction to the Immature Stages of British Flies: Diptera Larvae, with Notes on Eggs, Puparia and Pupae*; Royal Entomological Society: London, UK, 1989; Volume 10, p. 14.
20. Zumpt, F. Myiasis in man and animals in the Old World. A textbook for physicians, veterinarians and zoologists. In *Myiasis in Man and Animals in the Old World: A Textbook for Physicians, Veterinarians and Zoologists*; Butterworth & Co. (Publishers) Ltd.: London, UK, 1965.
21. Bosly, A.H. Molecular identification studies on *Oestrus ovis* L. (Diptera: Oestridae) larvae infested sheep in jazan region, Saudi Arabia. *Indian J. Anim. Res.* **2018**, *52*, 105–110.
22. Drummond, A.J.; Ashton, B.; Buxton, S.; Cheung, M.; Cooper, A.; Duran, C.; Field, M.; Heled, J.; Kearse, M.; Markowitz, S.; et al. Geneious v5.3. Biomatters Ltd: Auckland, New Zealand. 2010. Available online: http://www.geneious.com (accessed on 28 January 2021).
23. Altschul, S.F.; Gish, W.; Miller, W.; Myers, E.W.; Lipman, D.J. Basic local alignment search tool. *J. Mol. Biol.* **1990**, *215*, 403–410. [CrossRef]
24. Sievers, F.; Wilm, A.; Dineen, D.; Gibson, T.J.; Karplus, K.; Li, W.; Lopez, R.; McWilliam, H.; Remmert, M.; Söding, J.; et al. Fast, scalable generation of high-quality protein multiple sequence alignments using Clustal Omega. *Mol. Syst. Biol.* **2011**, *7*, 539. [CrossRef] [PubMed]

25. Saitou, N.; Nei, M. The neighbor-joining method: A new method for reconstructing phylogenetic trees. *Mol. Biol. Evol.* **1987**, *4*, 406–425. [CrossRef]
26. Alahmed, A.M. Seasonal infestation of *Oestrus ovis* larvae in sheep heads in central region of Saudi Arabia. *J. Egypt. Soc. Parasitol.* **2000**, *30*, 895–901.
27. Alikhan, M.; Al Ghamdi, K.; Al Zahrani, F.S.; Khater, E.I.; Allam, A.M. Prevalence and Salient Morphological Features of Myiasis Causing Dipteran Flies in Jeddah, Saudi Arabia. *Biosci. Biotechnol. Res. Asia* **2018**, *15*, 101–109. [CrossRef]
28. Papadopoulos, E.; Chaligiannis, I.; Morgan, E.R. Epidemiology of *Oestrus ovis* L. (Diptera: Oestridae) larvae in sheep and goats in Greece. *Small Rumin. Res.* **2010**, *89*, 51–56. [CrossRef]
29. Karatepe, B.; Karatepe, M.; Güler, S. Epidemiology of *Oestrus ovis* L. infestation in sheep in Nigde province, Turkey. *Rev. Médecine Vétérinaire* **2014**, *165*, 225–230.
30. Mohsen, M.N.; Raham, S.E.; Gasim, M.H. *Oestrus ovis* larval infestation among sheep and goats of green moun-tain areas in Libya. *J. Adv. Vet. Anim. Res.* **2015**, *2*, 382–387.
31. Amin, A.R.; Morsy, T.A.; Shoukry, A.; Mazyad, S.A. Oestrid head maggots in slaughtered sheep in Cairo abattoir. *J. Egypt. Soc. Parasitol.* **1997**, *27*, 855–861. [PubMed]
32. Alcaide, M.; Reina, D.; Sánchez, J.; Frontera, E.; Navarrete, I. Seasonal variations in the larval burden distribution of *Oestrus ovis* in sheep in the southwest of Spain. *Vet. Parasitol.* **2003**, *118*, 235–241. [CrossRef] [PubMed]
33. Caracappa, S.; Rilli, S.; Zanghi, P.; Di Marco, V.; Dorchies, P. Epidemiology of ovine oestrosis (*Oestrus ovis* Linné 1761, Diptera: Oestridæ) in Sicily. *Vet. Parasitol.* **2000**, *92*, 233–237. [CrossRef]
34. Rogers, C.E.; Knapp, F.W. Bionomics of the Sheep Bot Fly, *Oestrus Ovis*. *Environ. Entomol.* **1973**, *2*, 11–23. [CrossRef]
35. Arslan, M.O.; Kara, M.; Gicik, Y. Epidemiology of Oestrus ovis infestations in sheep in Kars province of north-eastern Turkey. *Trop. Anim. Health Prod.* **2009**, *41*, 299–305. [CrossRef] [PubMed]
36. Ramadan, M.Y.; Khater, H.F.; Omer, S.F.; Rahman, A.A. Epidemiology of *Oestrus ovis* infesting Egyptian sheep. In Proceedings of the XX International Congress of Mediterranean Federation of Health and Production of Ruminants, Assiut, Egypt, 19–22 February 2013; pp. 19–22.
37. Mohammed, R.G.; Josef, S.S.; Abed, K.J. Prevalence of *Oestrus ovis* Larvae in Slaughtered Sheep of Misan City, Iraq. *Syst. Rev. Pharm.* **2020**, *11*, 652–655.
38. Yilma, J.; Dorchies, P. Epidemiology of *Oestrus ovis* in southwest France. *Vet. Parasitol.* **1991**, *40*, 315–323. [CrossRef]
39. Bart, A.G.; Minár, J. Probability description of regulation on the level of population and individual in the host-parasite system using *Oestrus ovis* (Diptera: Oestridae) as an example. *Folia Parasitol.* **1992**, *39*, 75–83.
40. Marinho, M.A.T.; Wolff, M.; Ramos-Pastrana, Y.; de Azeredo-Espin, A.M.L.; Amorim, D.D.S. The first phylo-genetic study of Mesembrinellidae (Diptera: Oestroidea) based on molecular data: Clades and congruence with morpholog-ical characters. *Cladistics* **2017**, *33*, 134–152. [CrossRef]
41. Karademir, G.K.; Usluğ, S.; Okur, M.; Inci, A.; Yıldırım, A. Molecular Characterization and Phylogenetic Analyses of *Oestrus ovis* Larvae Causing Human Naso-pharyngeal Myiasis Based on CO1 Barcode Sequences. *Turk. J. Parasitol.* **2020**, *44*, 43–47. [CrossRef] [PubMed]

Review

Helminth Parasites among Rodents in the Middle East Countries: A Systematic Review and Meta-Analysis

Md Mazharul Islam [1,2,*], Elmoubashar Farag [3,*], Mohammad Mahmudul Hassan [4], Devendra Bansal [3], Salah Al Awaidy [5], Abdinasir Abubakar [6], Hamad Al-Rumaihi [3] and Zilungile Mkhize-Kwitshana [7,8]

[1] Department of Animal Resources, Ministry of Municipality and Environment, Doha, P.O. Box 35081, Qatar
[2] School of Laboratory Medicine and Medical Sciences, College of Health Sciences, University of KwaZulu Natal, Durban 4000, South Africa
[3] Ministry of Public Health, Doha, P.O. Box 42, Qatar; dbansal@moph.gov.qa (D.B.); halromaihi@moph.gov.qa (H.A.-R.)
[4] Faculty of Veterinary Medicine, Chattogram Veterinary and Animal Sciences University, Khulshi, Chattogram 4225, Bangladesh; miladhasan@yahoo.com
[5] Ministry of Health, P.O. Box 393, PC 113 Muscat, Oman; salah.awaidy@gmail.com
[6] Infectious Hazard Preparedness (IHP) Unit, WHO Health Emergencies Department (WHE), World Health Organization, Regional Office for the Eastern Mediterranean, Cairo 11371, Egypt; abubakara@who.int
[7] School of Life Sciences, College of Agriculture, Engineering & Science, University of KwaZulu Natal, Durban 40000, South Africa; mkhizekwitshanaz@ukzn.ac.za
[8] South African Medical Research Council, Cape Town 7505, South Africa
* Correspondence: walidbdvet@gmail.com (M.M.I.); eabdfarag@moph.gov.qa (E.F.); Tel.: +974-66064382 (M.M.I.); +974-55543816 or +974-44070396 (E.F.)

Received: 3 October 2020; Accepted: 30 October 2020; Published: 9 December 2020

Simple Summary: The review was conducted to establish an overview of rodent helminths in the Middle East as well as their public health importance. Following a systematic search, 65 field research were identified, studied, and analyzed. The overall prevalence of cestodes, nematodes, and trematodes were 24.88%, 32.71%, and 10.17%, respectively. The review detected 21 species of cestodes, 56 nematodes, and 23 trematodes, from which 22 have zoonotic importance. *Capillaria hepatica*, *Hymenolepis diminuta*, *Hymenolepis nana*, and *Cysticercus fasciolaris* were the most frequent and widespread zoonotic helminths. The review identified that there is an information gap on rodent helminths at the humans-animal interface level in this region. Therefore, the public health importance of rodent-borne helminth parasites is not fully recognized. Countrywide detailed studies on rodent helminths, along with the impact on public health, should be conducted in this region.

Abstract: Rodents can be a source of zoonotic helminths in the Middle East and also in other parts of the world. The current systematic review aimed to provide baseline data on rodent helminths to recognize the threats of helminth parasites on public health in the Middle East region. Following a systematic search on PubMed, Scopus, and Web of Science, a total of 65 research studies on rodent cestodes, nematodes, and trematodes, which were conducted in the countries of the Middle East, were analyzed. The study identified 44 rodent species from which *Mus musculus*, *Rattus norvegicus*, and *Rattus rattus* were most common (63%) and recognized as the primary rodent hosts for helminth infestation in this region. Cestodes were the most frequently reported ($n = 50$), followed by nematodes (49), and trematodes (14). The random effect meta-analysis showed that the pooled prevalence of cestode (57.66%, 95% CI: 34.63–80.70, $I^2\% = 85.6$, $p < 0.001$) was higher in Saudi Arabia, followed by nematode (56.24%, 95% CI: 11.40–101.1, $I^2\% = 96.7$, $p < 0.001$) in Turkey, and trematode (15.83%, 95% CI: 6.25–25.1, $I^2\% = 98.5$, $p < 0.001$) in Egypt. According to the overall prevalence estimates of individual studies, nematodes were higher (32.71%, 95% CI: 24.89–40.54, $I^2\% = 98.6$, $p < 0.001$) followed by cestodes (24.88%, 95% CI: 19.99–29.77, $I^2\% = 94.9$, $p < 0.001$) and trematodes (10.17%,

95% CI: 6.7–13.65, $I^2\%$ = 98.3, p < 0.001) in the rodents of the Middle East countries. The review detected 22 species of helminths, which have zoonotic importance. The most frequent helminths were *Capillaria hepatica*, *Hymenolepis diminuta*, *Hymenolepis nana*, and *Cysticercus fasciolaris*. There was no report of rodent-helminths from Bahrain, Jordan, Lebanon, Oman, United Arab Emirates, and Yemen. Furthermore, there is an information gap on rodent helminths at the humans-animal interface level in Middle East countries. Through the One Health approach and countrywide detailed studies on rodent-related helminths along with their impact on public health, the rodent control program should be conducted in this region.

Keywords: rodent; helminth; cestode; trematode; nematode; Middle East; meta-analysis

---

1. Introduction

Helminths are among the most diverse and geographically widespread groups of parasites that infect both humans and animals [1]. Although they are from different phyla or class (nematode, cestode, and trematode), the mode of transmission, infection, and pathogenesis, as well as host immune-responsiveness of these pathogens, follows a typical pattern. Approximately one-third of the world population is infected with one or more types of helminths. From amongst 300,000 species of helminths that typically infect vertebrates, 287 of them infect humans, from which 95% are either zoonoses or have evolved from animal parasites [2]. About 100 of the zoonotic helminths cause asymptomatic infection or mild symptoms in humans, while only a small percentage of them cause severe or even fatal infections [1]. In resource-poor countries, livestock is a source of food, production, income source, and deposit of wealth. Parasite infections in animals indirectly affect human health through financial hardship and malnutrition. Based on the burden of death, sickness, and treatment cost for both humans and animals for helminth infestation, the zoonotic parasites' socioeconomic burden was presented as high or low socioeconomic impact [3].

Rodents are significant sources of parasitic zoonosis in humans, serving as reservoirs and vectors of at least 70 zoonotic diseases, of which 16 are helminth parasites [4]. Consumption of uncooked/improperly cooked food contaminated with the infective larvae, eggs, or metacercariae is the primary source of humans infestation with helminth parasites [5,6]. When pilfering humans food, rodents pass stool or urine that contaminates said food, leading to transmission of zoonotic helminths from rodents to humans [7].

The Middle East is an intercontinental region with a total population of over 411 million [8] in 17 sovereign countries, including Bahrain, Cyprus, Egypt, Iran, Iraq, Israel, Jordan, Kuwait, Lebanon, Oman, Palestine, Qatar, Kingdom of Saudi Arabia (KSA), Syria, Turkey, United Arab Emirates (UAE), and Yemen. The majority of people in this region live in poverty [9], with the highest percentage being specific to Yemen, Syria, Egypt, Palestine, and Iraq [10,11]. Cultural diversity, weak economic policy, poor governance, rapid population growth, low educational structure, gender discrimination, underdeveloped infrastructure, and war and conflict have turned the region into a hot spot for many emerging and re-emerging diseases, including rodent-borne parasitic infections [9,12,13]. In the past, rodent-borne infections have led to multiple instances of a fatal epidemic, in part due to a lack of relevant information available on the subject, which makes it difficult to maintain public health sustainability [14,15].

Helminth infestations are mostly neglected diseases [16]. Therefore, the complete picture of zoonotic helminths is not well known in the Middle East area. Despite several studies being done on helminths in this region, no systematic review or meta-analysis was performed on rodent helminths, including zoonotic importance in the Middle East region. Our objective is to summarize baseline information on rodent helminths in this region using evidence-based records of the helminths detected in rodents in the Middle Eastern countries. The review also identifies the rodent helminths with public health importance in this region.

2. Materials and Methods

We followed PRISMA (Preferred Reporting Items for Systematic Reviews and Meta-Analysis) guidelines [17] to conduct the systematic review using a four-step approach: database search, evaluating relevant articles, data extraction, and summarizing. One author conducted the data search. Two authors were involved in critical evaluation and data extraction from the selected articles, while one author managed the compilation of said data. Afterward, two authors arranged the data and conducted the meta-analysis (Figure 1, Supplementary Table S1).

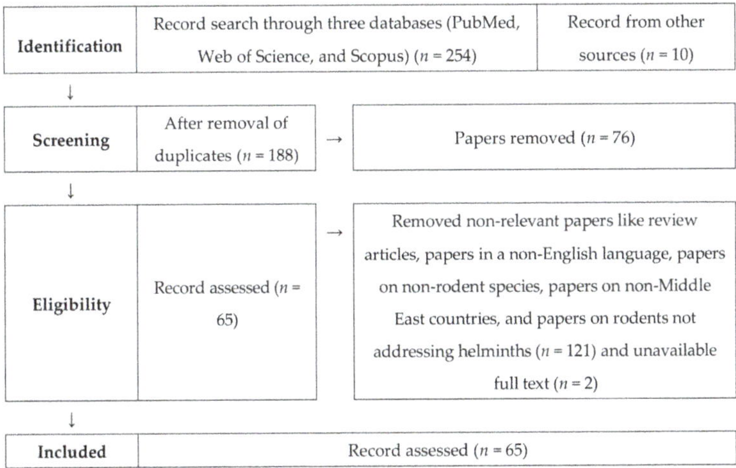

Figure 1. Systematic review PRISMA flow diagram describing the selection of published articles on rodent helminths in the Middle East and inclusion/exclusion process used in the study.

2.1. Search Strategy

A literature search on rodent helminth parasites in the Middle East was performed on 17 June 2019 through PubMed, Web of Science, and Scopus (Figure 1). The search included all the original research articles containing field evidence of helminth parasites (trematode, nematode, cestode) among rodents in the Middle East countries. The search did not have any date range of publication. The keywords included (Rodent OR Rat OR Jird OR Gerbil OR Vole OR Mouse OR Hamster OR Porcupine OR Squirrel OR Jerboa) AND (Endoparasite OR Helminth OR Cestode OR Trematode OR Nematode) AND (17 Middle East country names individually). Screening on the search was conducted as [Title/Abstract] in PubMed, [TITLE-ABS-KEY] in Scopus, and [Topic] in Web of Science.

2.2. Search of Relevant Articles

The data search results were processed using EndNote X9 (clarivate analytics, Philadelphia, PA, USA), which was also used to identify and exclude duplicate studies. Then we proceeded to peruse through the titles and abstracts to find the relevant articles. However, articles that were ambiguous regarding their relevance by their title and abstract were subjected to full-text analysis. Only documents published in English were considered for the review [18–82].

2.3. Data Extraction and Summarizing

Evidence-based field reports give a clear picture of any pathogen's availability, diversity, and dynamics in a locality [83,84]. We considered only the field reports containing rodent helminths for data abstraction. The extracted data included several variables such as country and location of sampling, season, year of sampling, rodent information (rodent species, sex, total rodent count, and the number of infected), helminth species and type, and possible associating factors of rodent infestation with helminth (Supplementary Table S2). The zoonotic rodent-borne helminths in this region were identified from the list of rodent helminths from this review with the support of published articles.

2.4. Data Analysis

The aggregated data was transcribed and stored in a Microsoft Excel spreadsheet, and then the data was forwarded to STATA/IC-13.0 (Stata Corp, 4905 Lakeway Drive, College Station, TX 77845, USA) for statistical analysis. Crude prevalence estimation was performed by dividing the total number of helminth-positive rodents with the total number of rodents sampled and expressed as a percentage. The crude estimate of prevalence was used throughout, the 95% confidence interval (CI), and the p-value were calculated on different types of helminths among the countries. Study variations among the studies were evaluated using the Chi-square (χ^2) test on Cochran's Q statistics (with p-value) followed by I^2 statistics to determine the study's degree of heterogeneity. Standard Error (SE) was calculated using a standard formula for proportion calculation. A random-effect meta-analysis model was applied using the "mean" command specifying random due to the study's high degree of heterogeneity ($I^2 > 75\%$). The output has been illustrated using a forest plot [85].

3. Results

3.1. Descriptive Analysis

The literature search returned 65 articles (Figure 1, Supplementary Table S2) published from 1969 to 2019. These articles were from 11 out of 17 Middle East countries, such as Cyprus, Egypt, Iran, Iraq, Israel, Kuwait, Palestine, Qatar, Saudi Arabia, Syria, and Turkey (Figure 2). No report on rodent helminths was available from Bahrain, Jordan, Lebanon, Oman, United Arab Emirates, and Yemen. Cestodes were the most frequently reported (50 articles) helminths in the Middle Eastern rodents, followed by nematodes (49), and trematodes (14).

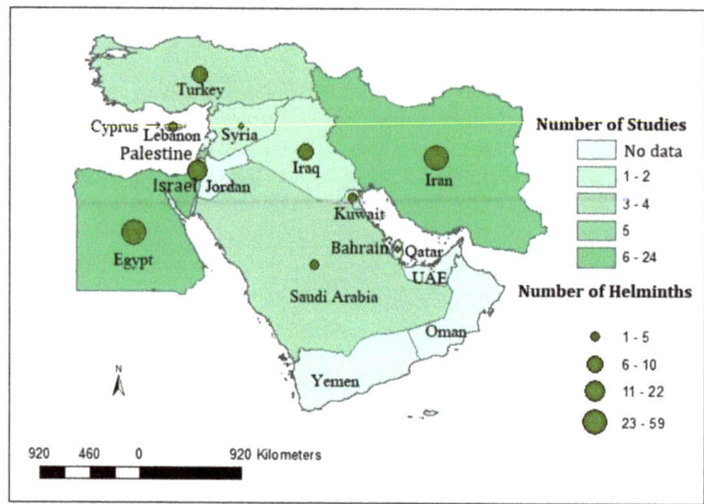

Figure 2. The map depicted the Middle East countries with the total number of studies and the number of helminths detected in rodents.

All 65 studies reported at least 9628 rodents (47% females and 53% males). A total of 44 rodent species from 6 families were listed (Supplementary Table S3). The analysis identified *Acomys dimidiatus*, *Jaculus jaculus*, *Meriones crassus*, *Mus musculus*, and *Rattus norvegicus*, *Rattus ratus* were widely distributed as these rodents were reported from where *Mus musculus* ($n = 1251$, 12.6%), *Rattus norvegicus* ($n = 3325$, 33.6%), and *Rattus rattus* ($n = 1694$, 17.1%) as the most common. Besides, three sub-species of *Rattus rattus*, such as *Rattus rattus alexandrines*, *Rattus rattus frugivorous*, and *Rattus rattus rattus*, are prevalent in the Middle East. Moreover, the review found some other rodents, which are important as zoonotic helminth carrier. These include *Acomys cahirinus*, *Acomys dimidiatus*, *Apodemus sylvaticus*, *Apodemus witherby*, *Arvicanthus niloticus*, *Calomyscus elburzensis*, *Cricetulus migratorius*, *Gerbillus cheesmani*, *Gerbillus gerbillus*, *Meriones libycus*, *Meriones persicus*, *Mesocricetus auratus*, *Microtus socialis*, *Microtus transcaspicus*, *Mus domesticus*, *Rhombomys opimus*, and *Tatera indica*. A total of 100 species of rodent helminths were identified. Based on the available data, the estimated pooled prevalence of the different types of parasites in rodents has been presented in Table 1. The random effect meta-analysis showed that the pooled prevalence of cestode ranged from 12.87% (95% CI: 5.17–20.57, $I^2\% = 80.6$, $p < 0.001$) in Turkey to 57.66% (95% CI: 34.63–80.70, $I^2\% = 85.6$, $p < 0.001$) in Saudi Arabia. The nematode prevalence was varying from 0.16% (95% CI: −0.15–0.47, $I^2\% = 0.0$) in Cyprus to 56.24% (95% CI: 11.40–101.1, $I^2\% = 96.7$, $p < 0.001$) in Turkey. Moreover, the prevalence of trematode ranged from 0.24% (95% CI: −0.11–0.59 $I^2\% = 0.0$, $p < 0.001$) in Iran to 15.83% (95% CI: 6.25–25.1, $I^2\% = 98.5$, $p < 0.001$) in Egypt.

Table 1. Estimated pooled prevalence of the rodent helminths in the Middle East countries.

Country	Parasite	Pooled Estimates (%)	95% CI	Heterogeneity Chi-Squared (χ^2)	$I^2\%$	*p*-Value
Cyprus	Nematode	0.160	−0.15–0.47	0.00	0	-
	Cestode	14.72	11.94–17.50	0.00	0	-
Egypt	Nematode	31.81	19.83–43.78	259.62	97.3	<0.001
	Cestode	27.49	23.72–31.26	17.18	70.9	<0.001
	Trematode	15.827	6.56–25.1	344.74	98.5	<0.001
Iran	Cestode	18.21	11.59–24.83	56.08	87.5	<0.001
	Nematode	33.93	16.52–51.35	154.53	96.8	<0.001
	Trematode	0.24	−0.11–0.59	0.19	0	<0.001
Palestine	Nematode	24.39	11.25–37.54	0.00	0	-
	Cestode	36.59	21.84–51.32	0.00	0	-
Qatar	Cestode	26.64	8.89–44.39	13.94	92.8	<0.001
Saudi Arabia	Cestode	57.66	34.63–80.70	6.74	85.6	<0.001
	Trematode	14.685	8.88–20.49	0.00	0	-
Turkey	Nematode	56.24	11.40–101.1	30.10	96.7	<0.001
	Cestode	12.87	5.17–20.57	10.34	80.6	<0.001

CI: confidence interval; I^2: inverse variance index; χ^2: Cochran's Q chi-square.

3.2. Rodent Cestodes in the Middle East Countries

Rodent cestodes information was available from all 11 Middle Eastern countries (Supplementary Table S3). A total of 21 rodent cestode species that belongs to 8 families have been reported in this review. Most of the cestodes were from Egypt and Iran (12 cestode species from each county). Out of 44 rodent species, *Mus musculus*, *Rattus norvegicus*, and *Rattus rattus* were frequently identified with the cestode infestation. Three species of cestodes have been frequently reported, viz: *Hymenolepis diminuta* (20 reports from 5 countries), *Hymenolepis nana* (30, 9), and *Cysticercus fasciolaris* (23, 4). Figure 3 shows the prevalence estimates from individual studies on cestodes in rodents of the Middle East countries, which ranged from 7.69 (95% CI: 2.22–13.17) to 68.57 (95% CI: 59.69–77.45) with an overall estimated prevalence 24.88 (95% CI: 19.99–29.77, $I^2\% = 94.9$, $p < 0.001$).

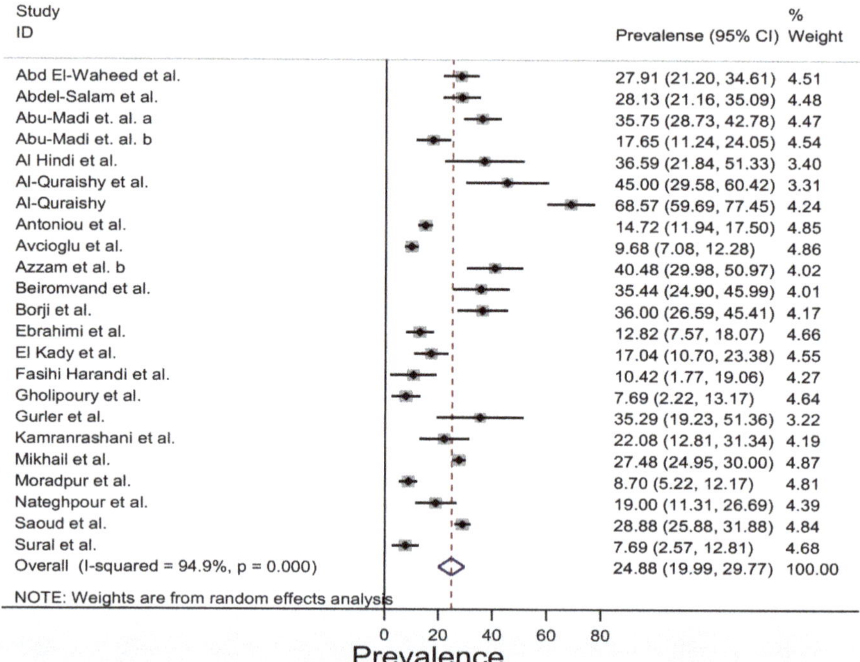

Figure 3. Forest plot of the prevalence estimates of cestode in rodents among the Middle East countries (the center dot representing point estimates whereas Gray Square representing the weight of each study to the meta-analysis).

3.3. Rodent Nematodes in the Middle East Countries

Rodent nematodes were studied in 8 countries in the Middle East, namely Cyprus, Egypt, Iran, Iraq, Israel, Kuwait, Palestine, and Turkey (Supplementary Table S3). Nematodes from 23 families represented the 56 nematode species in this region. Most of the rodent nematodes were reported from Egypt ($n = 24$) and Iran ($n = 31$) and the rodent species such as *Mus musculus*, *Rattus norvegicus*, *Rattus rattus*, *Meriones persicus*, *Acomys dimidiatus*, and *Tatera indica*. However, the nematodes, *Aspiculuris tetraptera*, *Capillaria hepatica*, *Syphacia obvelata*, *Streptopharagus kuntzi*, and *Trichuris muris* were most frequently reported and widely distributed. These nematodes were reported from three or more countries in the Middle East. Figure 4 shows the prevalence estimates from individual studies on nematodes in rodents of the Middle East countries, which ranged from 0.16 (95% CI: −0.15–0.47) to 79.41 (95% CI: 65.82–93.0) with an overall estimated prevalence of 32.71 (95% CI: 24.89–40.54, $I^2\% = 98.6$, $p < 0.001$).

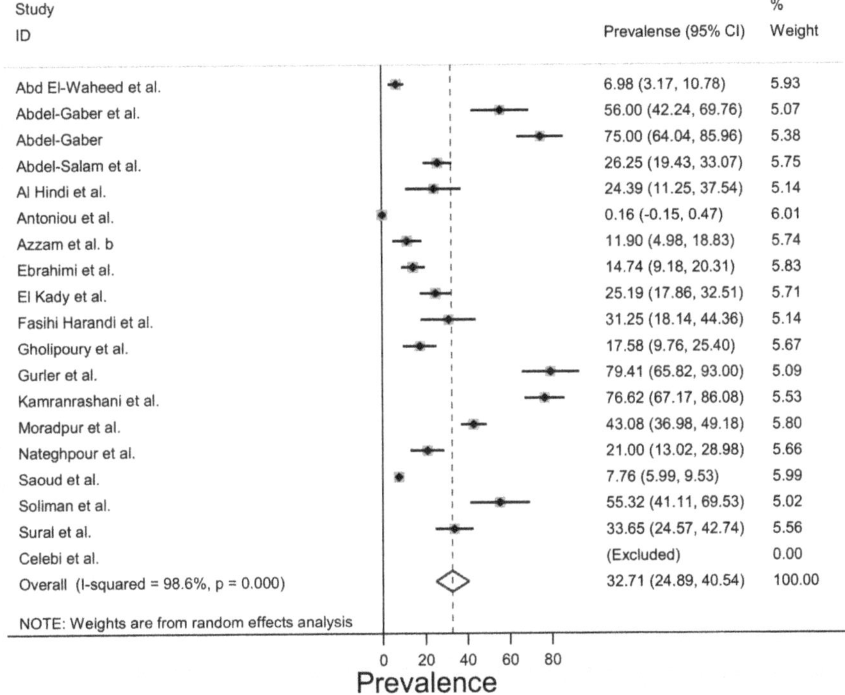

Figure 4. Forest plot of the prevalence estimates of nematode in rodents among the Middle East countries (the center dot representing point estimates whereas Gray Square representing the weight of each study to the meta-analysis).

3.4. Rodent Trematodes in the Middle East Countries

The reviewed studies reported rodent trematodes in Egypt, Iran, Israel, and Saudi Arabia (Supplementary Table S3). At least 23 trematode species from 11 families of trematodes were reported in the Middle Eastern rodents. Reports from Egypt ($n = 21$) were more descriptive of these trematodes. Moreover, *Fasciola* sp. was detected in Saudi Arabia, *Scaphiostomum* sp. in Israel, and *Notocotylus neyrai* and *Plagiorchis muris* were identified in Iran. The review found *Arvicanthus niloticus*, *Rattus norvegicus*, and *Rattus rattus* are three rodent species important for trematode infestation. Figure 5 shows the prevalence estimates from individual studies on trematode in rodents of the Middle East countries, which ranged from 0.20 (95% CI: −0.19–0.59) to 36.90 (95% CI: 26.59–47.22) with an overall estimated prevalence of 10.17 (95% CI: 6.7–13.65, $I^2\% = 98.3$, $p < 0.001$).

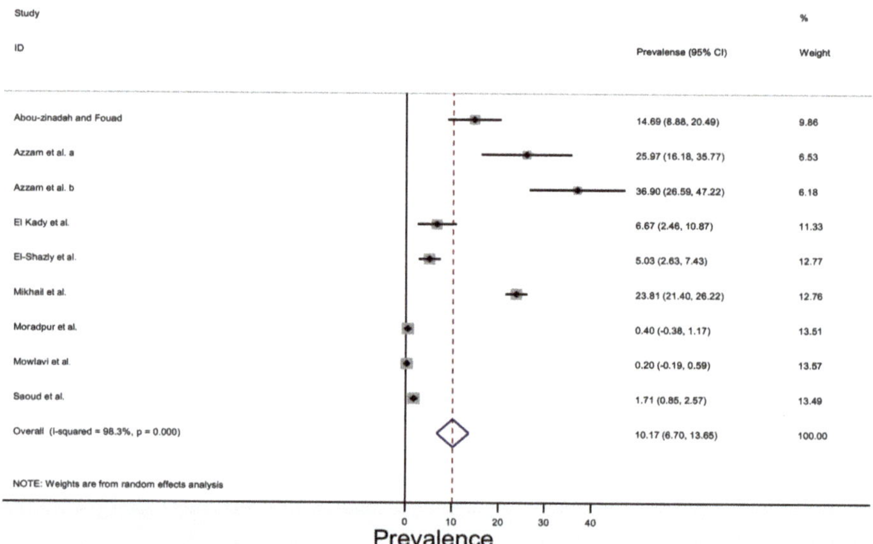

Figure 5. Forest plot of the prevalence estimates of trematode in rodents among the Middle East countries (the center dot representing point estimates, whereas Gray Square representing the weight of each study to the meta-analysis).

3.5. Zoonotic Importance of the Rodent Helminths in the Middle East Countries

Out of the 100 species of rodent helminths detected in this review, 22 species have zoonotic importance; 7 cestodes, 6 nematodes, and 9 trematodes. The zoonotic helminths, their hosts, and possible human infection sources have been illustrated in Table 2.

Table 2. Rodent-borne zoonotic helminths in the Middle East countries.

Parasites	Host	Source of Human Infection	Reference
Rodent-borne zoonotic cestodes:			
Raillieitina celebensis and *R. demerariensis.*	DH: rodent; IH: ant and beetle	Ingestion of food contaminated with infected insects	[5,6]
Hymenolepis diminuta and *H. nana*	DH: rodent; IH: *H. diminuta*: flea and beetle. *H. nana* does not require IH.	Consumption contaminated food with rodent feces containing parasitic egg	[6,86,87]
Mesocestoides sp.	DH: dog and cat; 1st IH: ant and mite, 2nd IH: rodent, bird, amphibian, and reptile	Consumption of undercooked meat of amphibians and reptiles containing infective larva (tetrathyridium)	[5,6]
Taenia taeniaeformis	DH: cat; IH: rodent	There is a report that *Taenia taeniaformis* can infect humans	[88]
Echinococcus multilocularis	DH: dog, fox; IH: rat	Ingestion of embryonated eggs	[86]

Table 2. Cont.

Parasites	Host	Source of Human Infection	Reference
	Rodent-borne zoonotic nematodes:		
Angiostrongylus cantonensis	DH: rat and mollusk; IH: snail, prawn, crab, and frog	Ingestion of uncooked IH or vegetables contaminated with infected larvae	[6]
Gongylonema pulchrum	DH: ruminant, pig, wild boar, non-human primate, carnivore, and rodent; IH: beetles and cockroaches	Ingestion of IH or drinking of water contaminated with infective larvae	[5,6]
Trichinella spp.	Pig, wild boar, and rodent	Ingestion of uncooked muscle with encysted larvae	[6]
Trichostrongylus spp.	Herbivorous animal	Consumption of food and water contaminated with animal feces containing infective larvae	[6]
Capillaria hepatica	Rat, carnivore, and humans	Consumption of food contaminated with feces containing embryonated eggs	[5]
Trichuris trichiura	Humans	Consumption of food contaminated with feces containing *Trichuris* egg.	[5]
	Rodent-borne zoonotic trematodes:		
Echinochasmus sp., *Echinoparyphium recurvatum*, and *Echoinostoma* sp.	DH: humans, rat, duck 1st IH: snail, 2nd IH: snail, amphibian, bivalve, fish	Ingestion of uncooked fish containing metacercariae	[2,89]
Fasciola hepatica	DH: herbivore; IH: snail	Ingestion of metacercariae contaminated vegetable	[6,34]
Haplorchis *pumilio*, *Pygidiopsis genata*, *Stictodora tridactyla*, *Prosthodendrium* spp., and *Plagiorchis muris*	DH: dog, cat, rat, duck, humans; 1st IH: snail, 2nd IH: fish	Eating uncooked fish harboring viable metacercariae	[88,90]
Schistosoma mansoni	DH: Vertebrate animal; IH: snail	Penetrate the DH skin	[6,86]

Note: DH: Definite host, IH: Intermediate host.

4. Discussion

This study reviewed the literature published in English on helminths-infested rodents in the Middle East region. The majority of the studies (47 of 65) were from Iran and Egypt, most likely due to their long history of rodent-borne zoonotic disease (like murine typhus, plague, tularemia) epidemics, which resulted in millions of death [4,91–94]. Thus, this topic became a central focus of public health research in these countries. The present review found three commensal rodent species: *Mus musculus*, *Rattus norvegicus*, and *Rattus rattus* to be more common and carrying most of the zoonotic helminths within this region. Previous literature described that these species occupy different habitats with higher population density than the other species and pose considerable risk to public health [4]. Although rodent cestodes were most frequently reported ($n = 50$) helminth in this review, the meta-analysis detected the overall rodent nematode prevalence was highest (32.7%) compared to cestodes (24.88%) and trematodes (10.17%) prevalence. Out of the 22 zoonotic helminths detected in this review, *Capillaria hepatica*, *H. diminuta*, *H. nana*, and *C. fasciolaris* have been found as widespread distribution. Furthermore, some non-zoonotic helminths such as *Aspiculuris tetraptera*, *Syphacia obvelata*, *Streptopharagus kuntzi*, and *Trichuris muris* were reported from three or more countries in this region.

Rodents have several beneficiary activities in ecology, such as soil aeration and water absorption ability, biotic recovery, and insect control [4,95]. In this regard, the presence of healthy rodents is essential for ecology [96]. Helminths infestation in rodents affects their own health and can subsequently alter the rodent-environment ecology to a considerable degree. Moreover, rodent helminths are important for humans, livestock, and pet animal health. Hymenolepiais is a major zoonotic rodent cestode [6]. Fascioliasis is hazardous for livestock health as well as for humans [6,97]. The definite host of *Taenia taeniaeformis* is the cat, where a stage of this cestode lifecycle (the cystic form, *Cysticercus fasciolaris*) is completed in rodents. An increase of *Cysticercus fasciolaris* in rodents can increase the health risk of cats [98]. Thus, rodent helminths have an impact on the ecology as well as humans and animal health.

Rodent-borne zoonotic helminths incur significant socioeconomic losses, although the zoonotic helminths' socioeconomic burden can differ from species to species [3]. *Hymenolepis diminuta* and *Hymenolepis nana* are major zoonotic cestodes [3,99]. *Trichostrongylus* sp. and *Trichuris trichiura* are generally considered as major nematode threats [100]. The socioeconomic burden caused by *Angiostrongylus cantonensis*, *Gongylonema pulchrum*, *Trichinella* sp., and *Capillaria hepatica* are likely to be very low [3].

There is an information gap on rodent-borne zoonotic helminths in the Middle East countries. Some zoonotic cases of helminths infestation were reported in rodents by some countries in the Middle East, but none involved humans who might have been infected with the same helminths. Humans hymenolepiasis were reported in Bahrain [101], Cyprus [102], Jordan [103], Oman [104], Palestine [105], Qatar [106], and Yemen [107]. The *Hymenolepis nana* is a common zoonotic helminth transmitted from rodents to humans and the prevalence ranged from 0.15% to 12.2% in some Middle East countries with prevalence of specific countries such as Jordan (1.8%) [103], Oman (5.9%) [104], Palestine (1.0%) [105], Qatar (0.15%) [106], and Yemen (12.2%) [107]. Egg of *Hymenolepis diminuta* was detected from soil samples of school playgrounds of Jordan [108]. There is no report of rodent hymenolepiasis within these countries. *Echinococcus* spp. is a major helminth for human health, which was detected in rodents of Egypt, Iran, and Turkey. Human cases of alveolar hydatid cysts were reported from Iran, Kuwait, Saudi Arabia, and Turkey [109,110].

The rodent lungworm, *Angiostrongylus cantonensis*, causes eosinophilic meningomyelitis in humans, reported in Israel [111]. Gongylonema infection is reported in humans [112] and dromedaries [113] from Iran. *Trichinella* was a widespread parasite infecting humans and other mammals, although the former makes for a poor host for said organism [6]. There are reports of humans trichinellosis from Iran [114], Israel [115], Lebanon [116], and Turkey [117]. Human cases of trichostrongyliasis infestation were reported in Egypt [118], Iran [119], Israel [120], and Turkey [121]. Eggs of *Trichostrongylus* sp. were detected from soil samples of public places of Jordan [108] and Iraq [122]. Humans reports of *Trichuris trichiura* are available from Bahrain [123], Egypt [118], Israel [120], Jordan [103], Oman [104], Palestine [105], Qatar [124], Saudi Arabia [125], Turkey [126], and Yemen [107]. *Trichuris muris*, the rodent whipworm, does not have any zoonotic importance. Eggs of *Trichuris* were found in the soil of the public place of Iraq [122]. *Trichuris trichiura* is not a rodent specific nematode. The report of *Trichuris trichiura* in Iranian rodents [67] may be a case of accidental infestation.

Schistosoma and *Fasciola* are two major humans trematodes globally [3]. The high prevalence of *Fasciola* was recorded in Egypt, Iran, and Yemen [127]. Human cases of schistosomiasis were noted in Egypt [3], Iran [128], Israel [129], Jordan [103], Saudi Arabia [130], Turkey [131], and Yemen [3,132], whereas *Heterophyes heterophyes* were reported from Egypt and Saudi Arabia [133,134]. There are non-humans (fish, dogs, and cats) reports of *Pygidiopsis genata*, *Haplorchis pumilio*, *Haplorchis yokogawai*, and *Heterophyes heterophyes* from Egypt, Iran, Iraq, Israel, Palestine, Kuwait, Saudi Arabia, Turkey, UAE, and Yemen [135–138]. However, these rodent trematodes in the current review were mostly reported from Egypt and Iran.

Based on the meta-analysis, the overall prevalence of rodent trematodes was less than that of nematodes and cestodes in the Middle East, which had received more emphasis in other similar reports [110,127]. Efficient management of water resources are important factor for prevalence of trematode prevalence [139,140]. The presence of deserts means shortage of surface water in some of the countries of Arabian Peninsula such as Bahrain, Kuwait, Oman, Qatar, Saudi Arabia, and United Arab Emirates [141], which may be the cause of shortage of aquatic intermediate hosts of trematodes in these countries. Therefore, rodent trematodes are less reported in these countries. More research should be conducted to find rodent-borne trematodes in the countries of this region.

The reviewed articles in the current study described some of the factors that can influence the population of rodent-borne helminths within the Middle East, necessitating a need to develop a plan of action to control rodent helminths. The abundance of rodent-borne helminths depends on the host organism's prevalence and its distribution [29,44]. An increase in the rodent population may increase the risk of humans getting infected by rodent parasites [142]. Rodents who inhabit animal farms have easy access to animal feed, and thus, they can be considered a potential vector and reservoir of animal and zoonotic diseases where animals serve as hosts [26]. *Hymenolepis diminuta* in the rodent are linked with some insects as intermediate hosts, such as *Xenopsylla astia*, which has a clear seasonal pattern. In Qatar, a research found that rodents are more infested with *Hymenolepis diminuta* in summer due to the *X. astia* abundance [24]. Several studies reported that the prevalence of rodent helminths is increased with rodent age [23,24,35,51]. Rodent helminths infestation can change with rodent host species [48]. The nematode, *Syphacia obvelata*, was reported to be most abundant in *Mus musculus* [64].

Rodent population control is a primary way to control rodent zoonotic diseases [143,144]. The other contributing factors, such as rodent species, seasons of the year, intermediate host, rodent control management in the residential areas, and animal farms, should also be considered on rodent related zoonoses control. As most of the rodent-related zoonotic helminths are linked to herbivores and carnivores [5,6,86,88], it is vital to manage dogs, cats, and livestock animals to avoid the spread of helminth infestation. Thus, One Health practice comes as a practical approach to control rodent-borne helminth prevalence [145]. "One Health is a collaborative, multisectoral, and transdisciplinary approach - working at the local, regional, national, and global levels—with the goal of achieving optimal health outcomes recognizing the interconnection between people, animals, plants, and their shared environment" [146]. One Health practice by linking veterinary, medical, ecology, entomology, parasitology, zoology fields, and local people are essential for rodent helminths prevention and control.

5. Conclusions

Rodent helminths in the Middle Eastern countries have been documented, which also highlighted rodent-borne zoonotic helminths. *Rattus norvegicus*, *Rattus rattus*, and *Mus musculus* were the most frequently reported rodents and infected with helminth parasites. Out of the 22 rodent-related zoonotic helminths, *Capillaria hepatica*, *H. diminuta*, *H. nana*, and *C. fasciolaris* were most frequent in this region. The current study illustrates that there is an information gap on the availability, diversity, and dynamics of rodent helminths and their interaction between humans and animals in the Middle East. Thus, the public health importance of rodent-borne helminth parasites is not fully recognized. However, rodent control should be the primary concentration by a One Health approach to control the spread of these helminths at the humans-animal-environmental interface in the countries of this region. We also suggest countrywide and detailed studies be conducted on rodent-borne helminths along with their impact on public health in this region.

Supplementary Materials: The following are available online at http://www.mdpi.com/2076-2615/10/12/2342/s1, Table S1: Prisma checklist, Table S2: Extracted data from the selected 65 studies, Table S3: Prevailing rodents and common cestodes, nematodes, and trematodes in the Middle East.

Author Contributions: Conceptualization, M.M.I., E.F. and Z.M.-K.; methodology, M.M.I., E.F. and M.M.H.; formal analysis, M.M.I. and M.M.H.; writing—original draft preparation, M.M.I., M.M.H. and Z.M.-K.; writing—review and editing, M.M.I., M.M.H., D.B., S.A.A., A.A., H.A.-R. and Z.M.-K.; visualization, D.B., E.F. and Z.M.-K.; supervision, Z.M.-K. All authors have read and agreed to the published version of the manuscript.

Funding: This research received no external funding.

Acknowledgments: The authors acknowledge the Qatar National Library for providing some full-text articles and partially paying the publication fee.

Conflicts of Interest: The authors declare no conflict of interest.

References

1. Macpherson, C.N.L.; Craig, P.S. *Parasitic Helminths and Zoonoses in Africa*; Springer: Eindhoven, The Netherlands, 1991.
2. Bruschi, F. *Helminth Infections and Their Impact on Global Public Health*; Springer: Wien, Austria, 2014.
3. Torgerson, P.R.; Macpherson, C.N. The socioeconomic burden of parasitic zoonoses: Global trends. *Vet. Parasitol.* **2011**, *182*, 79–95. [CrossRef] [PubMed]
4. Rabiee, M.H.; Mahmoudi, A.; Siahsarvie, R.; Krystufek, B.; Mostafavi, E. Rodent-borne diseases and their public health importance in Iran. *PloS Negl. Trop. Dis.* **2018**, *12*. [CrossRef] [PubMed]
5. Centers for Disease Control and Prevention. DPDx A-Z Index. Available online: https://www.cdc.gov/dpdx/az.html (accessed on 5 September 2020).
6. Dhaliwal, B.B.S.; Juyal, P.D. *Parasitic Zoonoses*; Springer: New Delhi, India, 2013; pp. 1–135.
7. Meerburg, B.G.; Singleton, G.R.; Kijlstra, A. Rodent-borne diseases and their risks for public health. *Crit. Rev. Microbiol.* **2009**, *35*, 221–270. [CrossRef] [PubMed]
8. World Population Review. The Middle East Population. Available online: http://worldpopulationreview.com/continents/the-middle-east-population/ (accessed on 21 November 2019).
9. Dabrowski, M.; Wulf, L. Economic Development, Trade and Investment in the Eastern and Southern Mediterranean Region. *SSRN Electron. J.* **2013**, 10. [CrossRef]
10. The World Bank. GDP Growth; All Countries and Economies. Available online: https://data.worldbank.org/indicator/NY.GDP.MKTP.KD.ZG (accessed on 21 November 2019).
11. The World Bank. Palestine's Economic Update—October 2019. Available online: https://www.worldbank.org/en/country/westbankandgaza/publication/economic-update-october-2019 (accessed on 21 November 2019).
12. Buliva, E.; Elhakim, M.; Tran Minh, N.N.; Elkholy, A.; Mala, P.; Abubakar, A.; Malik, S.M.M.R. Emerging and Reemerging Diseases in the World Health Organization (WHO) Eastern Mediterranean Region-Progress, Challenges, and WHO Initiatives. *Front. Public Health* **2017**, *5*, 276. [CrossRef] [PubMed]
13. Stewart, F. Root causes of violent conflict in developing countries. *BMJ* **2002**, *324*, 342–345. [CrossRef]
14. Bannazadeh Baghi, H.; Alinezhad, F.; Kuzmin, I.; Rupprecht, C.E. A Perspective on Rabies in the Middle East-Beyond Neglect. *Vet. Sci.* **2018**, *5*, 67. [CrossRef]
15. Hashemi Shahraki, A.; Carniel, E.; Mostafavi, E. Plague in Iran: Its history and current status. *Epidemiol. Health* **2016**, *38*. [CrossRef]
16. World Health Organization. Neglected Tropical Diseases. Available online: https://www.who.int/neglected_diseases/diseases/en/ (accessed on 12 September 2020).
17. Moher, D.; Liberati, A.; Tetzlaff, J.; Altman, D.G. Preferred reporting items for systematic reviews and meta-analyses: The PRISMA statement. *PLoS Med.* **2009**, *6*. [CrossRef]
18. Abd el-Wahed, M.M.; Salem, G.H.; el-Assaly, T.M. The role of wild rats as a reservoir of some internal parasites in Qalyobia governorate. *J. Egypt. Soc. Parasitol.* **1999**, *29*, 495–503.
19. Abdel-Gaber, R.; Abdel-Ghaffar, F.; Al Quraishy, S.; Morsy, K.; Saleh, R.; Mehlhorn, H. Morphological Re-Description and 18 S rDNA Sequence Confirmation of the Pinworm *Aspiculuris tetraptera* (Nematoda, Heteroxynematidae) Infecting the Laboratory Mice Mus musculus. *J. Nematol.* **2018**, *50*, 117–132. [CrossRef] [PubMed]
20. Abdel-Gaber, R. Syphacia obvelata (*Nematode, Oxyuridae*) infecting laboratory mice *Mus musculus* (*Rodentia, Muridae*): Phylogeny and host-parasite relationship. *Parasitol. Res.* **2016**, *115*, 975–985. [CrossRef]
21. Abdel-Salam, F.A.; Galal, A.A.; Ali, M.K. The Role of Rodents as a Reservoir of Zoonotic Intestinal Parasites at Sohag Governorate, Egypt. *Assiut Vet. Med. J.* **1994**, *30*, 14.
22. Abou-Zinadah, N.Y.; Fouad, M.A. Anti-Fasciola antibodies among rodents and sheep in Jeddah, Saudi Arabia. *J. Egypt. Soc. Parasitol.* **2005**, *35*, 711–716. [PubMed]
23. Abu-Madi, M.A.; Behnke, J.M.; Mikhail, M.; Lewis, J.W.; Al-Kaabi, M.L. Parasite populations in the brown rat Rattus norvegicus from Doha, Qatar between years: The effect of host age, sex and density. *J. Helminthol.* **2005**, *79*, 105–111. [CrossRef]
24. Abu-Madi, M.A.; Lewis, J.W.; Mikhail, M.; El-Nagger, M.E.; Behnke, J.M. Monospecific helminth and arthropod infections in an urban population of brown rats from Doha, Qatar. *J. Helminthol.* **2001**, *75*, 313–320. [CrossRef]

25. Al Hindi, A.I.; Abu-Haddaf, E. Gastrointestinal parasites and ectoparasites biodiversity of *Rattus rattus* trapped from Khan Younis and Jabalia in Gaza strip, Palestine. *J. Egypt. Soc. Parasitol.* **2013**, *43*, 259–268. [CrossRef]

26. Allymehr, M.; Tavassoli, M.; Manoochehri, M.H.; Ardavan, D. Ectoparasites and Gastrointestinal Helminths of House Mice (*Mus musculus*) from Poultry Houses in Northwest Iran. *Comp. Parasitol.* **2012**, *79*, 283–287. [CrossRef]

27. Al-Quraishy, S.; Gewik, M.M.; Abdel-Baki, A.A. The intestinal cestode Hymenolepis diminuta as a lead sink for its rat host in the industrial areas of Riyadh, Saudi Arabia. *Saudi J. Biol. Sci.* **2014**, *21*, 387–390. [CrossRef] [PubMed]

28. Al-Qureishy, S. Lead concentrations in some organs of the rat *Meriones libycus* and its parasite Hymenolepis diminuta from Riyadh City, KSA. *J. Egypt. Soc. Parasitol.* **2008**, *38*, 351–358.

29. Antoniou, M.; Psaroulaki, A.; Toumazos, P.; Mazeris, A.; Ioannou, I.; Papaprodromou, M.; Georgiou, K.; Hristofi, N.; Patsias, A.; Loucaides, F.; et al. Rats as Indicators of the Presence and Dispersal of Pathogens in Cyprus: Ectoparasites, Parasitic Helminths, Enteric Bacteria, and Encephalomyocarditis Virus. *Vector Borne Zoonotic Dis.* **2010**, *10*, 867–873. [CrossRef]

30. Arzamani, K.; Salehi, M.; Mobedi, I.; Adinezade, A.; Hasanpour, H.; Alavinia, M.; Darvish, J.; Shirzadi, M.R.; Mohammadi, Z. Intestinal Helminths in Different Species of Rodents in North Khorasan Province, Northeast of Iran. *Iran. J. Parasitol.* **2017**, *12*, 267–273.

31. Ashour, A.A.; Lewis, J.W. Gongylonema aegypti n. sp. (Nematoda: Thelaziidae) from Egyptian rodents. *Syst. Parasitol.* **1986**, *8*, 199–206. [CrossRef]

32. Avcioglu, H.; Guven, E.; Balkaya, I.; Kirman, R.; Bia, M.M.; Gulbeyen, H.; Kurt, A.; Yaya, S.; Demirtas, S. First detection of *Echinococcus multilocularis* in rodent intermediate hosts in Turkey. *Parasitology* **2017**, *144*, 1821–1827. [CrossRef]

33. Azzam, K.M.; Abd El-Hady, E.A.; El-Abd, N. Survey of Natural Infection with Echinostoma liei in Aquatic Snails and Wild Rodents in Egypt. *Egypt. J. Biol. Pest. Control.* **2015**, *25*, 427–432.

34. Azzam, K.M.; El-Abd, N.M.; Abd El-Hady, E.A. Survey of endoparasites of different rodent species in Egypt. *Egypt. J. Biol. Pest. Control.* **2016**, *26*, 815–820.

35. Behnke, J.M.; Barnard, C.J.; Mason, N.; Harris, P.D.; Sherif, N.E.; Zalat, S.; Gilbert, F.S. Intestinal helminths of spiny mice (*Acomys cahirinus dimidiatus*) from St Katherine's Protectorate in the Sinai, Egypt. *J. Helminthol.* **2000**, *74*, 31–43. [CrossRef]

36. Behnke, J.M.; Harris, P.D.; Bajer, A.; Barnard, C.J.; Sherif, N.; Cliffe, L.; Hurst, J.; Lamb, M.; Rhodes, A.; James, M.; et al. Variation in the helminth community structure in spiny mice (*Acomys dimidiatus*) from four montane wadis in the St Katherine region of the Sinai Peninsula in Egypt. *Parasitology* **2004**, *129*, 379–398. [CrossRef]

37. Beiromvand, M.; Akhlaghi, L.; Fattahi Massom, S.H.; Meamar, A.R.; Darvish, J.; Razmjou, E. Molecular Identification of *Echinococcus multilocularis* Infection in Small Mammals from Northeast, Iran. *PLoS Negl. Trop. Dis.* **2013**, *7*. [CrossRef] [PubMed]

38. Borji, H.; Khoshnegah, J.; Razmi, G.; Amini, H.; Shariatzadeh, M. A survey on intestinal parasites of golden hamster (*Mesocricetus auratus*) in the northeast of Iran. *J. Parasit. Dis.* **2014**, *38*, 265–268. [CrossRef]

39. Celebi, B.; Taylan Ozkan, A.; Babur, C. *Capillaria Hhepatica* in mouse (*Apodemus flavicollis*) from Giresun Province of Turkey. *Turk. Parazitol. Derg.* **2014**, *38*, 208–210. [CrossRef]

40. Ebrahimi, M.; Sharifi, Y.; Nematollahi, A. Assessment of gastrointestinal helminths among house mice (*Mus musculus*) caught in the north-west of Iran, with a special view on zoonotic aspects. *Comp. Clin. Pathol.* **2016**, *25*, 1047–1051. [CrossRef]

41. El Gindy, M.S.; Morsy, T.A.; Bebars, M.A.; Sarwat, M.A.; Arafa, M.A.; Salama, M.M. Rodents as reservoir of zoonotic intestinal helminths in Suez Canal Zone with the possible immunological changes. *J. Egypt. Soc. Parasitol.* **1987**, *17*, 259–273.

42. El Kady, G.A.; Gheneam, Y.M.; Bahgat, I.M. Zoonotic helminthes of commensal rodents in Talkha Center, Dakahlia Governorate. *J. Egypt. Soc. Parasitol.* **2008**, *38*, 863–872.

43. El Shazly, A.M.; Morsy, T.A.; el Kady, G.A.; Ragheb, D.A.; Handousa, A.E.; Ahmed, M.M.; Younis, T.A.; Habib, K.S. The helminthic parasites of rodents in Dakahlia Governorate, with reference to their Egyptian helminth fauna. *J. Egypt. Soc. Parasitol.* **1994**, *24*, 413–428.

44. Elshazly, A.M.; Awad, S.I.; Azab, M.S.; Elsheikha, H.M.; Abdel-Gawad, A.G.; Khalil, H.H.; Morsy, T.A. Helminthes of synanthropic rodents (*Rodentia: Muridae*) from Dakahlia and Menoufia, Egypt. *J. Egypt. Soc. Parasitol.* **2008**, *38*, 727–740. [PubMed]

45. El-Shazly, A.M.; Romia, S.A.; el Ganayni, G.A.; Abou-Zakham, A.A.; Sabry, A.H.; Morsy, T.A. Antibodies against some zoonotic parasites in commensal rodents trapped from Dakahlia Governorate, Egypt. *J. Egypt. Soc. Parasitol.* **1991**, *21*, 169–177. [PubMed]

46. Fair, J.M.; Schmidt, G.D.; Wertheim, G. New species of Andrya and Paranoplocephala (*Cestoidea: Anoplocephalidae*) from voles and mole-rats in Israel and Syria. *J. Parasitol.* **1990**, *76*, 641–644. [CrossRef]

47. Fasihi Harandi, M.; Madjdzadeh, S.M.; Ahmadinejad, M. Helminth parasites of small mammals in Kerman province, southeastern Iran. *J. Parasit. Dis.* **2016**, *40*, 106–109. [CrossRef]

48. Garedaghi, Y.; Afshin Khaki, A. Prevalence of Gastrointestinal and Blood Parasites of Rodents in Tabriz, Iran, with Emphasis on Parasitic Zoonoses. *Crescent J. Med. Biol. Sci.* **2014**, *1*, 9–12.

49. Gholipoury, M.; Rezai, H.R.; Namroodi, S.; Arab Khazaeli, F. Zoonotic and Non-zoonotic Parasites of Wild Rodents in Turkman Sahra, Northeastern Iran. *Iran. J. Parasitol.* **2016**, *11*, 350–357.

50. Greenberg, Z. Helminths of mammals and birds of israel i. helminths of acomys spp. (*Rodentia, Murinae*). *Isr. J. Ecol. Evol.* **1969**, *18*, 25–38. [CrossRef]

51. Gurler, A.T.; Beyhan, Y.E.; Bolukbas, C.S.; Acici, M.; Umur, S. Gastro-intestinal helminths of wild rats (brown rat-*Rattus norvegicus*, Berkenhout 1769) in Samsun, Turkey. *Ank. Univ. Vet. Fak. Derg.* **2011**, *58*, 289–290. [CrossRef]

52. Haridy, F.M.; Morsy, T.A.; Ibrahim, B.B.; Abdel Gawad, A.G.; Mazyad, S.A. Rattus rattus: A new host for fascioliasis. *J. Egypt. Soc. Parasitol.* **2003**, *33*, 647–648. [PubMed]

53. Hasson, R.H. Zoonotic &Nonzoonotic Endoparasites of Rodents from Some Districts in Baghdad. *Diyala J. Pure Sci.* **2010**, *6*, 11.

54. Jones, A.; El-Azazy, O.M.E. *Coelomotrema aegyptiaca* sp, nov., an unusual prosthogonimid trematode from *Rattus norvegicus* (berkenhout) in Egypt. *J. Nat. Hist.* **1986**, *20*, 707–712. [CrossRef]

55. Kamranrashani, B.; Kia, E.; Mobedi, I.; Mohebali, M.; Zarei, Z.; Mowlavi, G.; Hajjaran, H.; Abai, M.; Sharifdini, M.; Kakooei, Z.; et al. Helminth Parasites of Rhombomys opimus from Golestan Province, Northeast Iran. *Iran. J. Parasitol.* **2013**, *8*, 78–84.

56. Khajeh, A.; Mohammadi, Z.; Darvish, J.; Razmi, G.R.; Ghorbani, F.; Mohammadi, A.; Mobedi, I.; Shahrokhi, A.R. A survey on endoparasites in wild rodents of the Jaz Murian depression and adjacent areas, southeast of Iran. *J. Parasit. Dis.* **2018**, *42*, 589–597. [CrossRef]

57. Khalil, L.F.; Hassounah, O.; Behbehani, K. Helminth parasites of rodents in Kuwait with the description of a new species *Abbreviata kuwaitensis* (*Nematoda: Physalopteridae*). *Syst. Parasitol.* **1979**, *1*, 67–73. [CrossRef]

58. Kia, E.B.; Homayouni, M.M.; Farahnak, A.; Mohebai, M.; Shojai, S. Study of Endoparasites of Rodents and their Zoonotic Importance In Ahvaz, South West Iran. *Iran. J. Public Health* **2001**, *30*, 4.

59. Kia, E.; Shahryary-Rad, E.; Mohebali, M.; Mahmoudi, M.; Mobedi, I.; Zahabiun, F.; Zarei, Z.; Miahipoor, A.; Mowlavi, G.; Akhavan, A.; et al. Endoparasites of rodents and their zoonotic importance in germi, dashte mogan, ardabil province, Iran. *Iran. J. Parasitol.* **2010**, *5*, 15–20.

60. Meshkekar, M.; Sadraei, J.; Mahmoodzadeh, A.; Mobedi, I. Helminth Infections in *Rattus ratus* and *Rattus norvigicus* in Tehran, Iran. *Iran. J. Parasitol.* **2014**, *9*, 548–552.

61. Metwally, D.M.; Al-Enezy, H.A.; Al-Turaiki, I.M.; El-Khadragy, M.F.; Yehia, H.M.; Al-Otaibi, T.T. Gene-based molecular characterization of cox1 and pnad5 in Hymenolepis nana isolated from naturally infected mice and rats in Saudi Arabia. *Biosci. Rep.* **2019**, *39*. [CrossRef]

62. Mikhail, M.W.; Metwally, A.M.; Allam, K.A.; Mohamed, A.S. Rodents as reservoir host of intestinal helminthes in different Egyptian agroecosystems. *J. Egypt. Soc. Parasitol.* **2009**, *39*, 633–640.

63. Mirjalali, H.; Kia, E.B.; Kamranrashani, B.; Hajjaran, H.; Sharifdini, M. Molecular analysis of isolates of the cestode Rodentolepis nana from the great gerbil, Rhombomys opimus. *J. Helminthol.* **2016**, *90*, 252–255. [CrossRef]

64. Moradpour, N.; Borji, H.; Darvish, J.; Moshaverinia, A.; Mahmoudi, A. Rodents Helminth Parasites in Different Region of Iran. *Iran. J. Parasitol.* **2018**, *13*, 275–284.

65. Morsy, T.A.; Michael, S.A.; Bassili, W.R.; Saleh, M.S. Studies on rodents and their zoonotic parasites, particularly leishmania, in Ismailiya Governorate, A.R. Egypt. *J. Egypt. Soc. Parasitol.* **1982**, *12*, 565–585.

66. Mowlavi, G.; Mobedi, I.; Abedkhojasteh, H.; Sadjjadi, S.M.; Shahbazi, F.; Massoud, J. *Plagiorchis muris* (Tanabe, 1922) in *Rattus norvegicus* in Iran. *Iran. J. Parasitol.* **2013**, *8*, 486–490.

67. Nateghpour, M.; Motevalli-Haghi, A.; Akbarzadeh, K.; Akhavan, A.A.; Mohebali, M.; Mobedi, I.; Farivar, L. Endoparasites of wild rodents in Southeastern Iran. *J. Arthropod Borne Dis.* **2014**, *9*, 1–6.

68. Pakdel, N.; Naem, S.; Rezaei, F.; Chalehchaleh, A.A. A survey on helminthic infection in mice (*Mus musculus*) and rats (*Rattus norvegicus* and *Rattus rattus*) in Kermanshah, Iran. *Vet. Res. Forum Int. Q. J.* **2013**, *4*, 105–109.

69. Ranjbar, M.J.; Sarkari, B.; Mowlavi, G.R.; Seifollahi, Z.; Moshfe, A.; Abdolahi Khabisi, S.; Mobedi, I. Helminth Infections of Rodents and Their Zoonotic Importance in Boyer-Ahmad District, Southwestern Iran. *Iran. J. Parasitol.* **2017**, *12*, 572–579.

70. Sadighian, A.; Ghadirian, E.; Sadjadpour, E. Two new species of nematodes of lagomorphs and rodents from Iran. *J. Helminthol.* **1974**, *48*, 241–245. [CrossRef] [PubMed]

71. Saoud, M.F.; Ramadan, M.M.; Ashour, A.A.; Shahawy, A.A. On the helminth parasites of rodents in the Eastern Delta. I. General survey. *J. Egypt. Soc. Parasitol.* **1986**, *16*, 197–209.

72. Soliman, M.F.; Ibrahim, M.M.; Zalat, S.M. Gastrointestinal nematode community of spiny mice (Acomys dimidiatus) from St. Katherine, South Sinai, Egypt. *J. Parasit. Dis.* **2015**, *39*, 705–711. [CrossRef]

73. Sures, B.; Scheible, T.; Bashtar, A.R.; Taraschewski, H. Lead concentrations in *Hymenolepis diminuta* adults and *Taenia taeniaeformis* larvae compared to their rat hosts (*Rattus norvegicus*) sampled from the city of Cairo, Egypt. *Parasitology* **2003**, *127*, 483–487. [CrossRef] [PubMed]

74. Sursal, N.; Gokpinar, S.; Yildiz, K. Prevalence of intestinal parasites in hamsters and rabbits in some pet shops of Turkey. *Turk. Parazitol. Derg.* **2014**, *38*, 102–105. [CrossRef]

75. Wanas, M.Q.; Shehata, K.K.; Rashed, A.A. Larval occurrence of *Hydatigera taeniaeformis* Batsch (1786) (*Cestoda: Taeniidae*) in the liver of wild rodents in Egypt. *J. Egypt. Soc. Parasitol.* **1993**, *23*, 381–388.

76. Mohamed; Wanas, Q.A.; Shehata, K.K.; Rashed, A.A. Studies on the nematode parasites from Egyptian rodents. I. Spirurid nematodes. *J. Egypt. Soc. Parasitol.* **1994**, *23*, 851–863.

77. Wertheim, G.; Chabaud, A.G. Helminths of birds and mammals of Israel. VIII.—Skrjabinocapillaria rodentium n. sp. (*Nematoda Capillariidae*) from gerbillid and murid rodents. *Ann. De Parasitol. Hum. Comp.* **1979**, *54*, 65–68. [CrossRef]

78. Wertheim, G.; Giladi, M. Helminths of birds and mammals of Israel. VII.—*Pneumospirura rodentium* n. sp. (*Pneumospiruridae-Thelazioidea*). *Ann. Parasitol. Hum. Comp.* **1977**, *52*, 643–646. [CrossRef] [PubMed]

79. Wertheim, G. Helminths of Birds and Mammals from Israel: III. Helminths from Chromosomal Forms of the Mole-Rat, Spalax ehrenbergi. *J. Helminthol.* **1971**, *45*, 161–169. [CrossRef]

80. Yousefi, A.; Eslami, A.; Mobedi, I.; Rahbari, S.; Ronaghi, H. Helminth Infections of House Mouse (*Mus musulus*) and Wood Mouse (*Apodemus sylvaticus*) from the Suburban Areas of Hamadan City, Western Iran. *Iran. J. Parasitol.* **2014**, *9*, 511–518.

81. Yousif, F.; Ibrahim, A. The first record of *Angiostrongylus cantonensis* from Egypt. *Z. Parasitenkdunde* **1978**, *56*, 73–80. [CrossRef] [PubMed]

82. Zarei, Z.; Mohebali, M.; Heidari, Z.; Davoodi, J.; Shabestari, A.; Haghi, A.M.; Khanaliha, K.; Kia, E.B. Helminth infections of meriones persicus (Persian jird), *Mus musculus* (house mice) and cricetulus migratorius (grey ham-ster): A cross-sectional study in Meshkin-Shahr district, north-west Iran. *Iran. J. Parasitol.* **2016**, *11*, 213–220.

83. Nijsten, T.; Stern, R.S. How Epidemiology Has Contributed to a Better Understanding of Skin Disease. *J. Investig. Dermatol.* **2012**, *132*, 994–1002. [CrossRef] [PubMed]

84. Neta, G.; Brownson, R.C.; Chambers, D.A. Opportunities for Epidemiologists in Implementation Science: A Primer. *Am. J. Epidemiol.* **2018**, *187*, 899–910. [CrossRef]

85. Higgins, J.P.; Thompson, S.G. Quantifying heterogeneity in a meta-analysis. *Stat. Med.* **2002**, *21*, 1539–1558. [CrossRef]

86. Chatterjee, K.D. *Parasitology: Protozoology and Helminthology in Relation to Clinical Medicine: With Two Hundred Fourteen Illustrations*; CBS Publishers & Distributors: New Delhi, India, 2012.

87. Gárate, I.; Jimenez, P.; Flores, K.; Espinoza, B. Xenopsylla cheopis record as natural intermediate host of Hymenolepis diminuta in Lima, Peru. *Rev. Peru. Biol.* **2011**, *18*, 249–252. [CrossRef]

88. Stěrba, J.; Barus, V. First record of Strobilocercus fasciolaris (*Taenidae-larvae*) in man. *Folia Parasitol.* **1976**, *23*, 221–226.

89. Toledo, R.; Esteban, J.G. An update on human echinostomiasis. *Trans. R. Soc. Trop. Med. Hyg.* **2016**, *110*, 37–45. [CrossRef]

90. Boyce, K. Transmission Ecology of Gastrointestinal Trematodes of Small Mammals, Malham Tarn, North Yorkshire, UK. Ph.D. Thesis, University of Salford, Salford, UK, July 2013.

91. Loftis, A.D.; Reeves, W.K.; Szumlas, D.E.; Abbassy, M.M.; Helmy, I.M.; Moriarity, J.R.; Dasch, G.A. Surveillance of Egyptian fleas for agents of public health significance: Anaplasma, bartonella, coxiella, ehrlichia, rickettsia, and Yersinia pestis. *Am. J. Trop. Med. Hyg.* **2006**, *75*, 41–48. [CrossRef]

92. Trevisanato, S.I. The 'Hittite plague', an epidemic of tularemia and the first record of biological warfare. *Med. Hypotheses* **2007**, *69*, 1371–1374. [CrossRef] [PubMed]

93. Bishara, J.; Hershkovitz, D.; Yagupsky, P.; Lazarovitch, T.; Boldur, I.; Kra-Oz, T.; Pitlik, S. Murine typhus among Arabs and Jews in Israel 1991–2001. *Eur. J. Epidemiol.* **2004**, *19*, 1123–1126. [CrossRef] [PubMed]

94. Rosenthal, T.; Michaeli, D. Murine typhus and spotted fever in Israel in the seventies. *Infection* **1977**, *5*, 82–84. [CrossRef] [PubMed]

95. Khaghani, R. The economic and health impact of rodent in urban zone and harbours and their control methods. *Ann. Mil. Health Sci. Res.* **2007**, *4*, 1071–1078.

96. Morand, S.; Bordes, F. Parasite diversity of disease-bearing rodents of Southeast Asia: Habitat determinants and effects on sexual size dimorphism and life-traits. *Front. Ecol. Evol.* **2015**, *3*. [CrossRef]

97. Radostits, O.M.; Done, S.H. *Veterinary Medicine: A Textbook of the Diseases of Cattle, Sheep, Pigs, Goats, and Horses*; Elsevier Saunders: Philadelphia, PA, USA, 2007.

98. Rodan, I.; Sparkes, A.H. Chapter 8-Preventive Health Care for Cats. In *The Cat, Little, S.E., Ed.*; W.B. Saunders: Saint Louis, MO, USA, 2012; pp. 151–180. [CrossRef]

99. Dorny, P.; Praet, N.; Deckers, N.; Gabriel, S. Emerging food-borne parasites. *Vet. Parasitol.* **2009**, *163*, 196–206. [CrossRef]

100. Youn, H. Review of zoonotic parasites in medical and veterinary fields in the Republic of Korea. *Korean J. Parasitol.* **2009**, *47*, S133–S141. [CrossRef]

101. Amin, O.M. Pathogenic micro-organisms and helminths in sewage products, Arabian Gulf, country of Bahrain. *Am. J. Public Health* **1988**, *78*, 314–315. [CrossRef]

102. Berger, S. *Infectious Diseases of Cyprus*; Incorporated Gideon e-book series; Global Infectious Diseases and Epidemiology Network Informatics: Winconsin, WI, USA, 2019.

103. Al-Lahham, A.B.; Abu-Saud, M.; Shehabi, A.A. Prevalence of Salmonella, Shigella and intestinal parasites in food handlers in Irbid, Jordan. *J. Diarrhoeal Dis Res.* **1990**, *8*, 160–162.

104. Patel, P.K.; Khandekar, R. Intestinal parasitic infections among school children of the Dhahira Region of Oman. *Saudi Med. J.* **2006**, *27*, 627–632.

105. Astal, Z. Epidemiological survey of the prevalence of parasites among children in Khan Younis governorate, Palestine. *Parasitol. Res.* **2004**, *94*, 449–451. [CrossRef]

106. Abu-Madi, M.A.; Behnke, J.M.; Boughattas, S.; Al-Thani, A.; Doiphode, S.H.; Deshmukh, A. Helminth infections among long-term-residents and settled immigrants in Qatar in the decade from 2005 to 2014: Temporal trends and varying prevalence among subjects from different regional origins. *Parasites Vectors* **2016**, *9*, 153. [CrossRef]

107. Alharbi, R.A.; Alwajeeh, T.S.; Assabri, A.M.; Almalki, S.S.R.; Alruwetei, A.; Azazy, A.A. Intestinal parasitoses and schistosome infections among students with special reference to praziquantel efficacy in patients with schistosomosis in Hajjah governorate, Yemen. *Ann. Parasitol.* **2019**, *65*, 217–223. [CrossRef]

108. Abo-Shehada, M.N. Prevalence of Toxocara ova in some schools and public grounds in northern and central Jordan. *Ann. Trop Med. Parasitol.* **1989**, *83*, 73–75. [CrossRef]

109. Geramizadeh, B.; Baghernezhad, M. Hepatic Alveolar Hydatid Cyst: A Brief Review of Published Cases from Iran in the Last 20 Years. *Hepat Mon.* **2016**, *16*. [CrossRef]

110. Al-Aboody, M.S.; Omar, M.A.; Alsayeqh, A.F. Epizootiology of zoonotic parasites in Middle East: A comprehensive review. *Ann. Parasitol.* **2020**, *66*, 125–133. [CrossRef] [PubMed]

111. Fellner, A.; Hellmann, M.A.; Kolianov, V.; Bishara, J. A non-travel related case of Angiostrongylus cantonensis eosinophilic meningomyelitis acquired in Israel. *J. Neurol. Sci.* **2016**, *370*, 241–243. [CrossRef]

112. Molavi, G.H.; Massoud, J.; Gutierrez, Y. Human Gongylonema infection in Iran. *J. Helminthol.* **2006**, *80*, 425–428. [CrossRef]

113. Mirzayans, A.; Halim, R. Parasitic infection of Camelus dromedarius from Iran. *Bull. Soc. Pathol Exot. Fil.* **1980**, *73*, 442–445.

114. Moazeni, M.; Khamesipour, F.; Anyona, D.N.; Dida, G.O. Epidemiology of taeniosis, cysticercosis and trichinellosis in Iran: A systematic review. *Zoonoses Public Health* **2019**, *66*, 140–154. [CrossRef]

115. Eisenman, A.; Einat, R. A family outbreak of trichinosis acquired in Israel. *Harefuah* **1992**, *122*, 702–704, 751. [PubMed]

116. Haim, M.; Efrat, M.; Wilson, M.; Schantz, P.M.; Cohen, D.; Shemer, J. An outbreak of Trichinella spiralis infection in southern Lebanon. *Epidemiol. Infect.* **1997**, *119*, 357–362. [CrossRef]

117. Turk, M.; Kaptan, F.; Turker, N.; Korkmaz, M.; El, S.; Ozkaya, D.; Ural, S.; Vardar, I.; Alkan, M.Z.; Coskun, N.A.; et al. Clinical and laboratory aspects of a trichinellosis outbreak in Izmir, Turkey. *Parasite* **2006**, *13*, 65–70. [CrossRef]

118. El Shazly, A.M.; Awad, S.E.; Sultan, D.M.; Sadek, G.S.; Khalil, H.H.; Morsy, T.A. Intestinal parasites in Dakahlia governorate, with different techniques in diagnosing protozoa. *J. Egypt. Soc. Parasitol.* **2006**, *36*, 1023–1034.

119. Ghanbarzadeh, L.; Saraei, M.; Kia, E.B.; Amini, F.; Sharifdini, M. Clinical and haematological characteristics of human trichostrongyliasis. *J. Helminthol.* **2019**, *93*, 149–153. [CrossRef]

120. Greenberg, Z.; Giladi, L.; Bashary, A.; Zahavi, H. Prevalence of intestinal parasites among Thais in Israel. *Harefuah* **1994**, *126*, 507–509, 563.

121. Akdemir, C.; Helvaci, R. Evaluation of parasitology laboratory results of a group of people older than 15 years of age in Kutahya. *Turk. Parazit. Derg.* **2007**, *31*, 129–132.

122. Nooraldeen, K. Contamination of public squares and parks with parasites in Erbil city, Iraq. *Ann. Agric. Environ. Med.* **2015**, *22*, 418–420. [CrossRef]

123. Mukhtar, A. Intestinal parasites in the state of Bahrain. *Indian J. Pathol. Microbiol.* **1995**, *38*, 341–344. [PubMed]

124. Abu-Madi, M.A.; Behnke, J.M.; Ismail, A. Patterns of infection with intestinal parasites in Qatar among food handlers and housemaids from different geographical regions of origin. *Acta Trop.* **2008**, *106*, 213–220. [CrossRef]

125. Amer, O.S.O.; Al-Malki, E.S.; Waly, M.I.; AlAgeel, A.; Lubbad, M.Y. Prevalence of Intestinal Parasitic Infections among Patients of King Fahd Medical City in Riyadh Region, Saudi Arabia: A 5-Year Retrospective Study. *J. Parasitol. Res.* **2018**, *2018*. [CrossRef] [PubMed]

126. Yentur Doni, N.; Yildiz Zeyrek, F.; Simsek, Z.; Gurses, G.; Sahin, I. Risk Factors and Relationship between Intestinal Parasites and the Growth Retardation and Psychomotor Development Delays of Children in Sanliurfa, Turkey. *Turk. Parazitol. Derg.* **2015**, *39*, 270–276. [CrossRef]

127. Hotez, P.J.; Savioli, L.; Fenwick, A. Neglected tropical diseases of the Middle East and North Africa: Review of their prevalence, distribution, and opportunities for control. *PLoS Negl. Trop. Dis.* **2012**, *6*. [CrossRef]

128. Rollinson, D.; Knopp, S.; Levitz, S.; Stothard, J.R.; Tchuem Tchuente, L.A.; Garba, A.; Mohammed, K.A.; Schur, N.; Person, B.; Colley, D.G.; et al. Time to set the agenda for schistosomiasis elimination. *Acta Trop.* **2013**, *128*, 423–440. [CrossRef]

129. Hornstein, L.; Lederer, G.; Schechter, J.; Greenberg, Z.; Boem, R.; Bilguray, B.; Giladi, L.; Hamburger, J. Persistent Schistosoma mansoni infection in Yemeni immigrants to Israel. *Isr. J. Med. Sci.* **1990**, *26*, 386–389. [PubMed]

130. Taha, H.A.; Soliman, M.I.; Banjar, S.A. Intestinal parasitic infections among expatriate workers in Al-Madina Al-Munawarah, Kingdom of Saudi Arabia. *Trop Biomed.* **2013**, *30*, 78–88.

131. Kiremit, M.C.; Cakir, A.; Arslan, F.; Ormeci, T.; Erkurt, B.; Albayrak, S. The bladder carcinoma secondary to schistosoma mansoni infection: A case report with review of the literature. *Int. J. Surg. Case. Rep.* **2015**, *13*, 76–78. [CrossRef]

132. Al-Mekhlafi, A.M.; Abdul-Ghani, R.; Al-Eryani, S.M.; Saif-Ali, R.; Mahdy, M.A. School-based prevalence of intestinal parasitic infections and associated risk factors in rural communities of Sana'a, Yemen. *Acta Trop.* **2016**, *163*, 135–141. [CrossRef]

133. El-Shazly, A.M.; el-Nahas, H.A.; Soliman, M.; Sultan, D.M.; Abedl Tawab, A.H.; Morsy, T.A. The reflection of control programs of parasitic diseases upon gastrointestinal helminthiasis in Dakahlia Governorate, Egypt. *J. Egypt. Soc. Parasitol.* **2006**, *36*, 467–480.

134. Chai, J.Y.; Seo, B.S.; Lee, S.H.; Hong, S.J.; Sohn, W.M. Human infections by *Heterophyes heterophyes* and *H. dispar* imported from Saudi Arabia. *Kisaengchunghak Chapchi* **1986**, *24*, 82–86. [CrossRef]

135. Chai, J.Y.; Jung, B.K. Foodborne intestinal flukes: A brief review of epidemiology and geographical distribution. *Acta Trop.* **2019**, *201*. [CrossRef]

136. El-Gayar, A.K. Studies on some trematode parasites of stray dogs in Egypt with a key to the identification of intestinal trematodes of dogs. *Vet. Parasitol.* **2007**, *144*, 360–365. [CrossRef]

137. El-Azazy, O.M.; Abdou, N.E.; Khalil, A.I.; Al-Batel, M.K.; Majeed, Q.A.; Henedi, A.A.; Tahrani, L.M. Potential Zoonotic Trematodes Recovered in Stray Cats from Kuwait Municipality, Kuwait. *Korean J. Parasitol.* **2015**, *53*, 279–287. [CrossRef]

138. Khalil, M.I.; El-Shahawy, I.S.; Abdelkader, H.S. Studies on some fish parasites of public health importance in the southern area of Saudi Arabia. *Rev. Bras. Parasitol. Vet.* **2014**, *23*, 435–442. [CrossRef] [PubMed]

139. Steinmann, P.; Keiser, J.; Bos, R.; Tanner, M.; Utzinger, J. Schistosomiasis and water resources development: Systematic review, meta-analysis, and estimates of people at risk. *Lancet Infect. Dis.* **2006**, *6*, 411–425. [CrossRef]

140. El-Sayed, H.F.; Rizkalla, N.H.; Mehanna, S.; Abaza, S.M.; Winch, P.J. Prevalence and epidemiology of *Schistosoma mansoni* and *S. haematobium* infection in two areas of Egypt recently reclaimed from the desert. *Am. J. Trop. Med. Hyg.* **1995**, *52*, 194–198. [CrossRef]

141. Odhiambo, G.O. Water scarcity in the Arabian Peninsula and socio-economic implications. *Appl. Water Sci.* **2017**, *7*, 2479–2492. [CrossRef]

142. Bordes, F.; Blasdell, K.; Morand, S. Transmission ecology of rodent-borne diseases: New frontiers. *Integr. Zool.* **2015**, *10*, 424–435. [CrossRef]

143. Eisen, R.J.; Enscore, R.E.; Atiku, L.A.; Zielinski-Gutierrez, E.; Mpanga, J.T.; Kajik, E.; Andama, V.; Mungujakisa, C.; Tibo, E.; MacMillan, K.; et al. Evidence that rodent control strategies ought to be improved to enhance food security and reduce the risk of rodent-borne illnesses within subsistence farming villages in the plague-endemic West Nile region, Uganda. *Int. J. Pest. Manag.* **2013**, *59*, 259–270. [CrossRef]

144. Brown, L.M.; Laco, J. Rodent Control and Public Health: A Description of Local Rodent Control Programs. *J. Environ. Health* **2015**, *78*, 28–29.

145. Kaplan, B.; Kahn, L.H.; Monath, T.P.; Woodall, J. 'ONE HEALTH' and parasitology. *Parasites Vectors* **2009**, *2*, 36. [CrossRef] [PubMed]

146. Centers for Disease Control and Prevention. One Health Basics. Available online: https://www.cdc.gov/onehealth/basics/ (accessed on 2 October 2020).

Publisher's Note: MDPI stays neutral with regard to jurisdictional claims in published maps and institutional affiliations.

 ***animals***

Article

Identification of *Sarcocystis* spp. in One-humped Camels (*Camelus dromedarius*) from Riyadh and Dammam, Saudi Arabia, via Histological and Phylogenetic Approaches

Dina M. Metwally [1,2,*], Tahani T. Al-Otaibi [3], Isra M. Al-Turaiki [4], Manal F. El-Khadragy [5,6] and Reem A. Alajmi [1]

1   Department of Zoology, College of Science, King Saud University, Riyadh 11451, KSA; ralajmi@ksu.edu.sa
2   Department of Parasitology, Faculty of Veterinary Medicine, Zagazig University, Zagazig 44519, Egypt
3   Department of Biology, Al-Nairiyah University College, University of Hafr Al-Batin, Hafr Al-Batin 31991, Saudi Arabia; ttaalotaibi@gmail.com
4   Department of Information Technology, College of Computer and Information Sciences, King Saud University, Riyadh 11451, Saudi Arabia; ialturaiki@ksu.edu.sa
5   Department of Biology, Faculty of Science, Princess Nourah Bint Abdelrahman University, Riyadh 84428, Saudi Arabia; manalelkhadragy@yahoo.com
6   Department of Zoology and Entomology, Faculty of Science, Helwan University, Cairo 11795, Egypt
*   Correspondence: dhasanin@ksu.edu.sa

Received: 5 May 2020; Accepted: 22 June 2020; Published: 28 June 2020

Simple Summary: In the present work, we explored the existence of *Sarcocystis* spp. in samples of camels obtained from abattoirs in Riyadh, Saudi Arabia. We examined tissues of the tongue, heart, esophagus, diaphragm, and skeletal muscle by macroscopic assessments, optical microscopy of tissues, optical microscopy of digested sediment, Transmission Electron Microscopy (TEM), and Polymerase chain reaction (PCR) followed by gene sequencing. The results identified *Sarcocystis cameli* (*S. cameli*) and *S. camelicanis*. *Sarcocystis* spp. were detected in Saudi Arabian camels by molecular analysis. *S. levinei* and *S. miescheriana* were most closely related.

Abstract: *Sarcocystis* (*S.*) spp. are intracellular protozoan parasites that infect birds and animals, resulting in substantial commercial losses. *Sarcocystis* spp. have an indirect life cycle; canines and felines are known to act as final hosts, and numerous domestic and wild animals act as intermediate hosts. The presence of sarcocysts in camel meat may diminish its commercial quality. There is limited knowledge regarding the taxonomy and diagnosis of *Sarcocystis* spp. that infect camels in Saudi Arabia. In this study, transmission electron microscopy (TEM) revealed *S. cameli* and *S. camelicanis* (*camelicanis*) in *Camelus* (*C.*) *dromedarius*. This is the first report of *S. camelicanis* in Saudi Arabia and is considered a significant finding. Based on cytochrome c oxidase subunit I gene (COX1) sequences, two samples of *Sarcocystis* spp. isolated from *C. dromedarius* in Riyadh and Dammam were grouped with *S. levinei* hosted by *Bubalus bubalis* in India, *S. rangi* hosted by *Rangifer tarandus* in Norway, *S. miescheriana* hosted by *Sus scrofa* in Italy and *S. fayeri* hosted by *Equus caballus* in Canada. The sequences obtained in this study have been deposited in GenBank.

Keywords: *Sarcocystis* spp.; COX1; Camelus dromedarius

---

1. Introduction

Sarcocystis (*S.*) is a genus of intracellular coccidian parasites that was first identified in 1843 [1]. More than 200 species have been identified and were demonstrated to infect a wide range of domestic and wild animals, resulting in significant losses in farm animals worldwide [2,3]. *Sarcocystis* spp. accomplish their life cycles in two hosts, a final and an intermediate host, and are known to have a high level of host specificity with regard to their intermediate hosts as opposed to their final hosts [4]. The asexual stages of *Sarcocystis* development occur in intermediate hosts (herbivorous animals, such as sheep, cattle and camels, and primates, such as humans, poikilothermic animals and birds), where sarcocysts generally become visible in skeletal muscles and in the striated muscles of the heart, diaphragm and esophagus [5,6] but are rarely found in the smooth muscle of the intestine and the central nervous system [4–7]. The final hosts (e.g., canids, felids, marsupials and primates) become infected by eating intermediate host tissue infected with mature sarcocysts. Next, a stage of sexual reproduction occurs in the final host, after which oocysts/sporocysts are expelled into the environment through feces to be eaten by appropriate intermediate hosts.

The one-humped camel *Camelus dromedarius* is broadly distributed in the hot, arid regions of the Middle East, South Asia, Africa, the Canary Islands, and Central Australia. Compared to beef, lamb, and ostrich meat, camel meat is favored in several countries due to its reduced fat and cholesterol contents [8,9]. However, the presence of sarcocysts in camel meat may decrease its value for human utilization because camels act as intermediate hosts for at least two *Sarcocystis* spp.; furthermore, such infection is a general phenomenon in the one-humped camel due to its worldwide distribution [10–12]. The presence of sarcocysts in the musculature decreases the economic value of the musculature, especially if macroscopic cysts are detected, as their presence leads to condemning the product for consumption [13,14]. Microscopic cysts, on the other hand, may cause serious pathological conditions in infected animals, especially in the case of acute forms, and result in heavy production losses [15–17].

Infected camels generally exhibit subclinical infections, although *Sarcocystis* spp. can produce extensive pathology or death in these animals [12,18].

The diagnosis of acute sarcocystosis in camels is an onerous task because of the absence of an industrial standard indicator test, asymptomatic features of the infection, and the presence of microscopic sarcocysts. Muscle squash, pepsin digestion, trypsin digestion and histopathological examination have been used for the analysis of microscopic sarcocysts in camels [11,12,18,19]. Ultrastructural analysis of the cyst wall is quite useful in the identification of sarcocysts in camels. Transmission electron microscopy (TEM) has demonstrated that *S. cameli* has a primary cyst wall that appears as a thin wall with finger-like villar protrusions and rows of knob-like projections. *S. ippeni* was defined in Egypt [20], wherein the authors of that study observed cyst walls harboring cone-like villar protrusions. *S. camelicanis* was also defined in Egypt [19], wherein the authors found that the primary cyst wall appeared as a thick layer and had finger-like protrusions with a blunt apex. Furthermore, another study defined *S. miescheri* in Egypt [10], in which the primary cyst wall appeared as a thick electron-dense layer with spine-like protrusions.

Although molecular analysis might be an alternative technique for the identification of *Sarcocystis* spp., there is a limitation of such data for camels. The first molecular identification of *S. cameli* sarcocysts was performed in camels in Iran, wherein the 18S rRNA gene fragment was amplified from bradyzoite DNA using conventional polymerase chain reaction (PCR), followed by sequencing (GenBank: GU074011.1) and restriction fragment length polymorphism (RFLP) analysis [21,22]. However, the RFLP technique was found to be more expensive than DNA sequencing or electron microscopy techniques. In another study, the 18S rRNA gene was amplified from microscopic sarcocysts in camels, although no phylogenetic analysis was reported [12]. Therefore, in the present study, a molecular marker, cytochrome oxidase subunit I (COX1), was used to identify different *Sarcocystis* spp. infecting camels from Riyadh and Dammam, Saudi Arabia.

2. Materials and Methods

2.1. Sample Collection

Between February and October 2018, veterinarians collected samples of tissues (diaphragm, skeletal muscle, cardiac muscle, tongue, and esophagus) during postmortem investigations of slaughtered animals in the West Abattoir and Dammam Slaughterhouses in Riyadh and Dammam, Saudi Arabia, respectively. Tissue samples were collected from 37 (27 in Riyadh and 10 in Dammam) male camels aged >5 years and transported to the laboratory in boxes containing ice packs. This study (IRB number: KSU-SE-18-33) secured the approval of the Institutional Committee of Postgraduate Studies and Research at King Saud University (Saudi Arabia).

2.2. Macroscopic Examination

On the same day as the tissues were collected, a macroscopic analysis was performed. In this process, a scalpel was used to make as many as five transverse incisions on organs, such as the heart and the tongue, for the purpose of revealing the cysts (macroscopic). A macroscopic analysis of the entire esophagus was also performed to examine the external and internal walls; the lumen was exposed after longitudinal sectioning [23].

2.3. Microscopic Examination

Microscopic evaluation of cysts was performed using the squashing method [24]. Distinct fragments of each tissue with a thickness of approximately 5 mm were squashed robustly from both slides, after which an optical microscope was used to analyze the specimen. For all tissues, the procedure was performed in triplicate. *Sarcocystis* spp. were also examined by light microscopy and DNA analyses.

2.4. Digestion Method

For all tissues, approximately 20 g were minced before digestion for a period of 30 min in 100 mL at a temperature of 40 °C; the digestion solution comprised pepsin (1.3 g), HCl (3.5 mL) and NaCl (2.5 g) in distilled water (500 mL) [25]. After digestion, the mixture was centrifuged at 3500× g for 3 min, followed by Giemsa staining and optical microscopy evaluation [26].

2.5. Histopathological Examination

Small specimens of muscles and organs were collected, fixed in 10% neutral buffered formalin, serially dehydrated in increasing concentrations of ethanol (30%, 70% and 95% absolute) and embedded in paraffin. Three 5-μm-thick sections of each of the abovementioned organs and muscles were prepared, stained with hematoxylin and eosin and examined under an ECLIPSE NI-4 (Nikon, Tokyo, Japan) [27].

2.6. Transmission Electron Microscopy (TEM)

Six *Sarcocystis* spp. cysts embedded in the tissues were collected from organs, fixed in 0.1 M sodium cacodylate buffer (pH 7.4) supplemented with 3% glutaraldehyde solution for 4 h at 4 °C and stored at 4 °C until processing. After fixation, the samples were washed in 0.1 M sodium cacodylate buffer, fixed with 2% osmium tetroxide for 24 h and rewashed four to five times in the buffer (10–15 min each) [19]. The samples were serially dehydrated in increasing concentrations of acetone (30%, 40%, 50%, 70%, 90% and 100%) and blocked with buffer supplemented with 1% phosphotungstic acid and 1% uranyl acetate. Next, the 100% acetone solution was replaced with Polybed resin, followed by paraffin embedding and polymerization in an oven at 60 °C [28]. Moderately thin sections were prepared to observe *Sarcocystis* spp. cysts under a microscope (CX31, Olympus Corporation, Tokyo, Japan). Ultrathin sections were stained with uranyl acetate and lead citrate and examined using a JEM-1400 transmission electron microscope (JEOL, Tokyo, Japan) at 80 kV.

2.7. Molecular Analysis

2.7.1. DNA Extraction and PCR Amplification

From all the microscopic *Sarcocystis* isolates, six isolates were selected and washed five times in distilled water (sterile). gDNA from tissue was extracted using a DNA Mini Kit (QIAGEN GmbH, Hilden, Germany) (Cat. No. 51304) according to the manufacturer's instructions. Amplification of the

COX1 gene was performed using the primer pairs mentioned in Table 1 [28,29] and a thermocycler (Veriti® 96-well Thermal Cycler, Model 9902, Biosystem). This procedure was conducted in a mixture (20 µL) containing 4 µL of master mix (5×), 12 µL of RNase-free water and DNA template (2 µL). The PCR program consisted of denaturation at 95 °C for 5 min, followed by 40 cycles of denaturation for 45 s at 95 °C, annealing for 45 s at 54 °C and extension for 10 min at 72 °C. The PCR products were analyzed by 1.5% agarose gel electrophoresis.

Table 1. List of primers used for the amplification of cytochrome c oxidase subunit I (COX1) gene in *Sarcocystis* spp.

Gene	Primers	Sequences	References
COX1	SF1	5′-ATG GCG TAC AAC AAT CAT AAA GAA-3′	[28–30]
	SR9	5′-ATA TCC ATA CCR CCA TTG CCC AT-3′	

2.7.2. DNA Sequencing

The aforementioned PCR products were first purified and then sequenced in the reverse and forward directions (only) using a Genetic Analyzer at the Central Laboratory of King Saud University.

The sequences were analyzed using Geneious Prime Build [31]. All sequences were truncated slightly using the error probability method with a limit of 0.05 at both ends. A Basic Local Alignment Search Tool (BLAST) search was performed to identify related sequences. Multiple sequence alignments were generated using CLUSTAL Omega [32]. The maximum likelihood trees were constructed using PhyML 3.3 with 100 bootstraps [33].

2.8. Statistical Analysis

Statistical analysis was performed using the Statistical Package for Social Sciences (SPSS) software (version 17, SPSS, Inc., Chicago, IL, USA). The lengths and widths of at least five cysts from each organ (diaphragm, skeletal muscle, cardiac muscle, tongue, and esophagus) were determined by light microscopy and expressed as the mean sizes and amplitudes of variation.

3. Results

3.1. Macroscopic Examination

No macroscopic cysts were detected by the naked eye during the inspection of carcasses or collected samples.

3.2. Microscopic Examination

A compression technique with a light microscope was used for microscopic examination, in which 15 of 37 slaughtered camels (40.54%) were found to be positive for microscopic cysts. In particular, the numbers of positive cases for diaphragm, skeletal muscle, cardiac muscle, tongue, and esophagus samples were 14/37 (37.83%), 10/37 (27.01%), 9/37 (24.32%), 8/37 (21.62%), and 3/37 (8.10%), respectively. Thin-walled sarcocysts and thick-walled sarcocysts were observed. In contrast to the tissue squash method, the pepsin–hydrochloric acid digestion technique revealed a greater number of animals that were positive.

3.2.1. Thin-Walled Microscopic Sarcocysts

Unstained sarcocysts were found to have thin walls, to be elongated and spindle-like in shape, and to be present within the muscle fibers. The sarcocysts measured 197.9–405.6 µm in length (mean 301.75 µm) and 57.3–125.6 µm in breadth (mean 91.45 µm) (Figure 1A). Stained sarcocysts are shown in Figure 1B. Histopathological sections revealed that the cysts measured 83.50–135.50 µm in length (mean 109.5 µm). The cyst wall consisted of two layers, an outer striated layer and an inner homogenous layer. The cystic cavity was divided into several compartments by fine trabeculae originating from the cyst wall (Figure 1C). Ultrastructural analysis of the cyst wall showed that *S. cameli* sarcocysts (shown in Figure 1D) have an outer cyst wall (Ocw) that is in close contact with the cyst wall, ground substance, metrocytes, and bradyzoites or merozoites. The primary cyst wall (Pcw) displayed irregular folded nonbranched finger-like villar protrusions with fibrillar elements originating from the ground substance at a distance of 0.54 µm below the primary cyst wall and protruding into it. The villar protrusions measured 0.93 µm in length and 0.77 µm in width at the base and 0.35 µm at the apex. The distance between each villar protrusion measured 1.7 µm. The finger-like process (Flp) was generally surrounded by numerous scattered host cell mitochondria. The ground substance was located directly under the primary cyst wall with a diameter of 1.27–1.7 µm (mean: 1.48 µm). The space between the primary cyst wall (Pcw) and the contents of the cyst was primarily fine, showing dense homogenous granules and fibrillar elements folding into the primary cyst wall. The ground substance extended into the interior of the cyst, forming a thin septum separating the entire cyst into compartments enclosing the metrocytes and bradyzoites or merozoites.

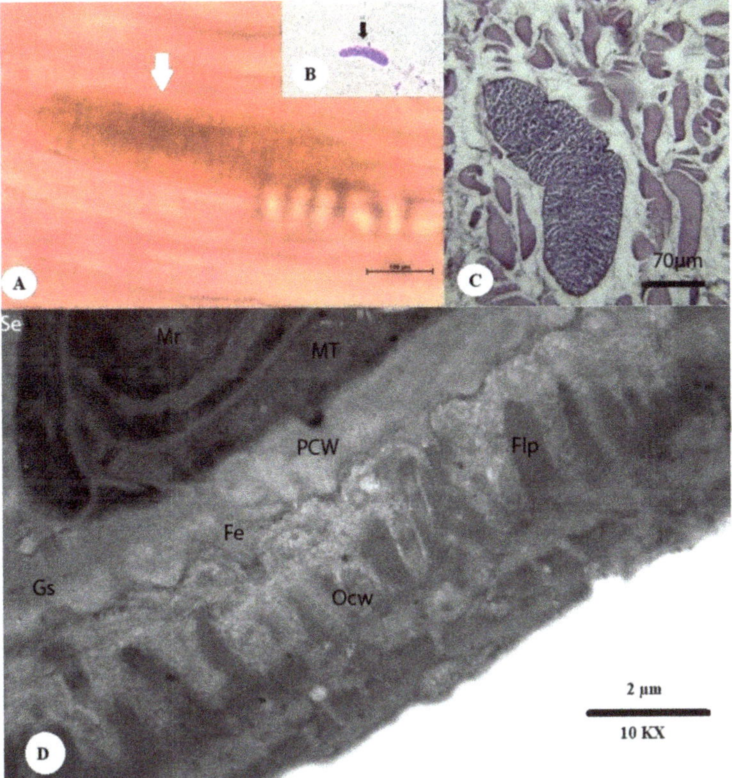

Figure 1. Morphology of a sarcocyst of *S. cameli* from the esophageal muscles of a camel (*Camelus dromedarius*). (**A**) Microscopic thin-walled *S. cameli* (white arrow) (bar = 100 µm). (**B**) Cysts stained with Giemsa after pepsin–hydrochloric acid digestion (black arrow). (**C**) Histopathological section of camel esophageal muscles showing the thin cyst wall (CW) of *S. cameli* (H&E) (bar = 70 µm). The thin and smooth cyst wall and the clearly visible septum are shown. (**D**) The cyst wall in detail, containing an outer cyst wall (Ocw), a primary cyst wall (Pcw), ground substance (Gs), septae (Se), merozoites (Mr), fibrillar elements (Fe), metrocytes (Mt), and finger-like process (Flp) (10,000×).

3.2.2. Thick-walled Microscopic Sarcocysts

The sarcocysts appeared fusiform or spindle-shaped, measuring 151–449 µm in length (mean 300 µm) and 65–140 µm in breadth (mean 102.5 µm) (Figure 2A). Cysts after pepsin–hydrochloric acid digestion are depicted in Figure 2B. Histopathological sections demonstrated that the cysts had a thick wall composed of two layers: an outer striated layer and an inner smooth layer. The cyst cavity was divided into compartments by a narrow septum originating from the wall in the periphery of the cyst (Figure 2C). Ultrastructural analysis (Figure 1D) of the cyst wall revealed an outer cyst wall connected to the primary cyst wall (Pcw), ground substance (Gs), metrocytes, and bradyzoites for *S. camelicanis*.

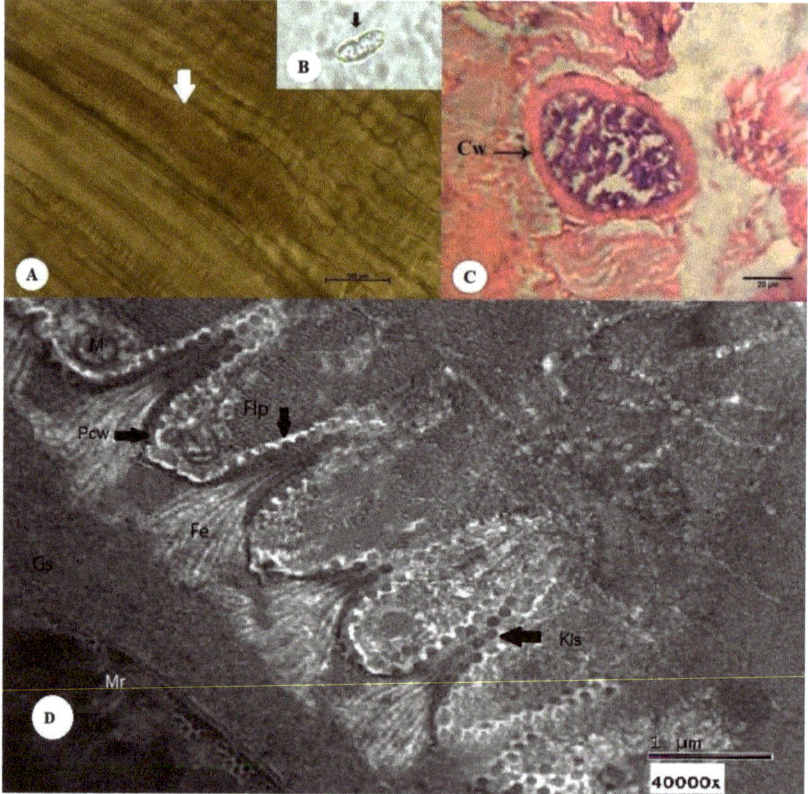

Figure 2. Morphology of a sarcocyst of *S. camelicanis* from the skeletal muscles of a camel (*Camelus dromedarius*). (**A**) Microscopic thick-walled *S. camelicanis* (white arrow) (bar = 100 µm). (**B**) Cysts after pepsin–hydrochloric acid digestion (small black arrow). (**C**) A histopathological section of the skeletal muscles of a camel shows the thick cyst wall (Cw) (black arrow) of *S. camelicanis* and a clearly visible septum (H&E; bar = 20 µm). (**D**) A *Sarcocystis camelicanis* thick-walled cyst showing the primary cyst wall (Pcw) (black arrow), ground substance (Gs), fibrillar elements (Fe), finger-like process (Flp), knoblike structures (Kls) (black arrow), Mitochondria (M), and merozoites (Mr) (40,000×).

The primary cyst wall (Pcw) is a thick, dense layer adjacent to the Gs. Fibrillar elements (Fe) originating from the Gs at 0.60 μm below the Pcw aggregated towards the Pcw and embedded into it, forming finger-like protrusions (Flp). The cyst measured 2.15–2.91 μm (mean 2.53 μm) in length and 0.52–0.60 μm (mean 0.56 μm) in breadth. Each Flp carried characteristic multiple numerous knob-like structures (Kls), which were spherical in shape with a diameter of 0.10–0.14 μm (mean 0.12 μm); their number varied from 19 to 25 on each protrusion. The distance between each Flp was 0.71–0.89 μm (mean 0.80 μm), and they were generally surrounded by numerous dispersed host cell mitochondria (M). The Gs was found to be 1.25–1.60 μm (mean 1.42 μm) below the Pcw and in between the Pcw and metrocytes. It appeared as a homogeneous substance and extended to the interior of the cyst by a septum separating the entire cyst into a number of compartments enclosing the metrocytes, merozoites and other structures.

3.3. Genetic Characteristics

Genomic DNA extracted from four *Sarcocystis* spp. isolates, D7S (thin-walled cyst), D10S (thick-walled cyst), R10C (thick-walled cyst) and R16C (thin-walled cyst), were used for the amplification of 1000-bp COX1 sequences.

The first two were isolated from the skeletal muscle of a Dammam camel, and the last two were isolated from the cardiac muscle of a Riyadh camel. These sequences were 692, 513, 194, and 205 nucleotides in length, respectively, and shared a pairwise identity of 63.1%. All sequences obtained in this study have been deposited in GenBank under accession numbers MK948444 (D7S), MK948443 (D10S), MK948442 (R10C), and MK948441 (R16C).

BLAST results showed that the D7S sequence had similarity to *S. levinei* hosted by *Bubalus bubalis* (heart tissue) in India, accession numbers MH255774–MH255777, with 86.5% pairwise identity and 89.6% coverage. The D7S sequence also displayed similarity to *S. rangi* hosted by *Rangifer tarandus* in Norway, accession numbers KC209662–KC209666, with 85.2% pairwise identity and 87.57% coverage.

Furthermore, the D10S sequence exhibited similarity to *S. miescheriana*, accession number MH404202, hosted by *Sus scrofa* in Italy, with 80.7% pairwise identity and 43% coverage. The D10S sequence was also similar to *S. fayeri*, accession number LC171854, hosted by *Equus caballus* in Canada, with 79.1% pairwise identity and 43% coverage. The phylogenetic tree with bootstrap proportions is illustrated in Figure 3. *Toxoplasma gondii* (JX473253) was used as an outgroup. For R10C and R16C, '194 and 205' are very short sequences inappropriate for analysis. The tree shows that MK948444 (D7S) and MK948442 (R10C) are placed in a clade with *S. levinei*. MK948443 (D10S) is grouped with *Sarcocystis miescheriana*. MK948441 (R16C) is grouped with *Sarcocystis fayeri*.

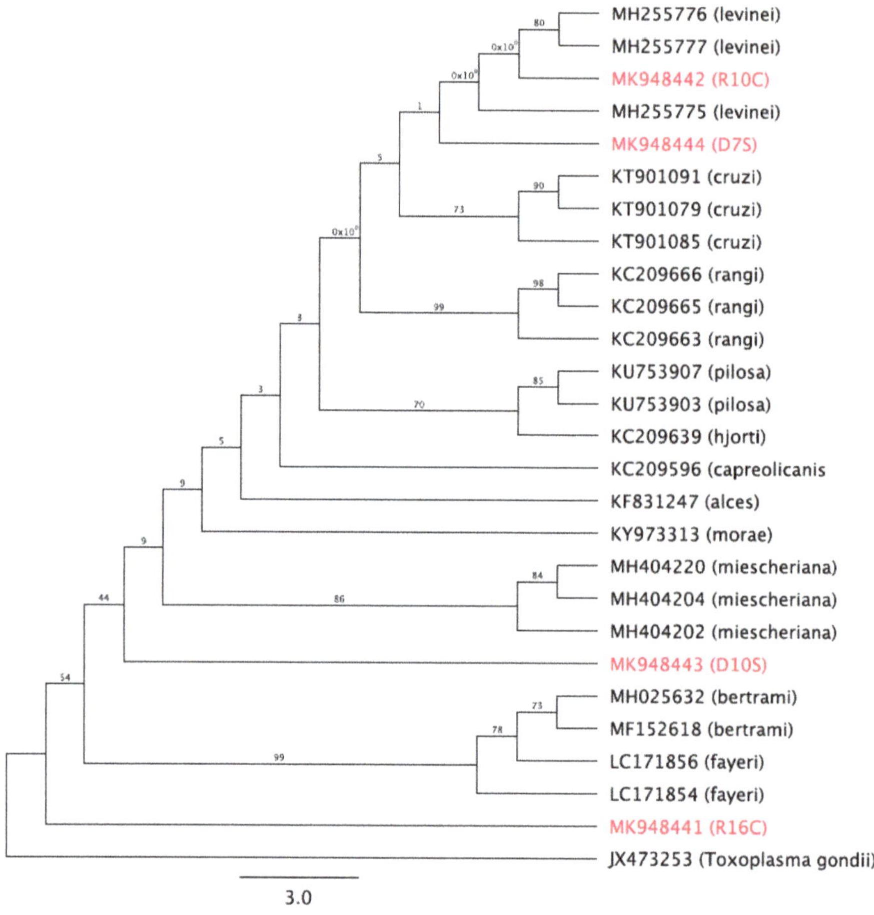

Figure 3. Genetic relationships of our samples isolated from camels in Saudi Arabia (highlighted in pink) with other *Sarcocystis* spp. retrieved from GenBank based on the COX1 region.

4. Discussion

To establish the occurrence of *Sarcocystis* spp., this study examined the esophagus, tongue, diaphragm, skeletal muscle, and heart tissue from 37 camel specimens (27 from Riyadh and 10 from Dammam). The results showed that the overall prevalence rates of *Sarcocystis* spp. were 40.74% (11/27) in Riyadh specimens and 40% (4/10) in Dammam specimens.

In the present study, macroscopic cysts were not detected. However, macroscopic cysts have been reported to be less common in some esophageal samples [34,35]. In this study, the squashing and pepsin-hydrochloric acid digestion methods accompanied by light microscopy were used, wherein the technique of pepsin-hydrochloric acid digestion revealed a greater number of positive animals. Cysts from esophageal and skeletal muscle samples appeared in a cylindrical and spindle-like shape.

Cysts were found to have thin and smooth or thick walls divided internally into several compartments by trabeculae. Thin-walled cysts have been identified in the esophagus and diaphragm of camels slaughtered at Al-Ahsa Abattoir, Saudi Arabia [36]; in the esophagus, diaphragm, heart, shoulder, and masseter muscles of one-humped camels slaughtered in southern Ethiopia [37]; in the esophagus, diaphragm, heart, skeletal muscles, and tongue of one-humped camels slaughtered in Iran [18], and in the esophagus of one-humped camels slaughtered in Egypt [38].

Thick-walled cysts were found in the esophagus of one-humped camels slaughtered in Egypt [10]; in the esophagus, diaphragm, heart, skeletal muscles, and tongue of one-humped camels slaughtered in Egypt [19]; in the esophagus of one-humped camels slaughtered in Egypt [39], and in the esophagus, skeletal muscles, and tongue of one-humped camels slaughtered in Egypt [40]. Similarly, both thin- and thick-walled *Sarcocystis* cysts have been suggested to be at different stages in the same parasite and have been termed *S. cameli* [41]. In addition, thick-walled cysts were found repeatedly and termed *S. cameli*, whereas thin-walled cysts were unnamed [18,25,36].

Sarcocystis spp. isolated from *C. dromedarius* from Dammam and Riyadh grouped alongside *S. levinei* hosted by *B. bubalis* (heart tissue) in India (accession numbers MH255774–MH255777) [42], *S. rangi* hosted by *R. tarandus* in Norway (accession numbers KC209662–KC209666) [43], *S. miescheriana* hosted by *S. scrofa* in Italy (accession number MH404202) [44], and *S. fayeri* hosted by *E. caballus* in Canada (accession number LC171854) [45].

In the present study, four partial sequences of the COX1 gene were analyzed. Two samples, R10C (thick-walled cyst) and D7S (thin-walled cyst), isolated from *C. dromedarius* in Riyadh and Dammam, Saudi Arabia, respectively, were identified by histological and TEM approaches as *Sarcocystis* spp. The phylogenetic tree shows that these two samples are related to *S. mansoni* hosted by South American camelids, *S. levinei* hosted by water buffaloes, and *S. miescheriana* hosted by pigs, suggesting that camels are receptors of *Sarcocystis* spp. from other intermediate hosts.

The main limitation of this study lies in the other two samples, D10S and R16C. These two samples were identified by histological and TEM approaches as *Sarcocystis* spp. (D10S, thick-walled cyst; and R16C, thin-walled cyst). However, in the phylogenetic tree, they were not grouped in a clade with R10C and D7S. Through a deeper analysis of the raw data, it was found that the samples were of lower quality than R10C and D7S. This low quality affected the process of phylogenetic tree construction.

5. Conclusions

In this study, our results highlight the necessity to reinforce a more in-depth phylogenetic analysis and identification of *Sarcocystis* spp. with more taxa and different molecular markers. For instance, amplification of the 18S rRNA gene should be performed to further distinguish among the closely related *Sarcocystis* spp.

Author Contributions: Conceptualization: D.M.M. Data curation: I.M.A.-T. Formal analysis: T.T.A.-O. Funding acquisition: D.M.M. Investigation: D.M.M. and T.T.A.-O. Methodology: T.T.A.-O. Project administration: D.M.M. and R.A.A. Resources: T.T.A.-O. and M.F.E.-K. Software: I.M.A.-T. Supervision: D.M.M. and R.A.A. Validation: D.M.M. Visualization: M.F.E.-K. Writing: original draft: D.M.M. Writing, reviewing and editing: D.M.M., R.A.A., M.F.E.-K., T.T.A.-O. and I.M.A.-T. All authors have read and agreed to the published version of the manuscript.

Funding: The authors extend their appreciation to the Deanship of Scientific Research at King Saud University for funding this work through research group No. RG-1441-381.

Conflicts of Interest: The authors declare no conflict of interest. The funders had no role in the design of the study; in the collection, analyses, or interpretation of data; in the writing of the manuscript, or in the decision to publish the results

References

1. Miescher, F. Über eigenthümliche Schläuche in den Muskeln einer Hausmaus. *Ber. Verhandl Nat. Gesell Basel* **1843**, *5*, 198–202.

2. Tenter, A.M. Current research on Sarcocystis species of domestic animals. *Int. J. Parasitol.* **1995**, *25*, 1311–1330. [CrossRef]

3. Dubey, J. Foodborne and waterborne zoonotic sarcocystosis. *Food Waterborne Parasitol.* **2015**, *1*, 2–11. [CrossRef]

4. Fayer, R. Sarcocystis spp. Human Infections. *Clin. Microbiol. Rev.* **2004**, *17*, 894–902. [CrossRef]

5. Hidron, A.; Vogenthaler, N.; Santos-Preciado, J.I.; Rodriguez-Morales, A.J.; Franco-Paredes, C.; Rassi, A. Cardiac Involvement with Parasitic Infections. *Clin. Microbiol. Rev.* **2010**, *23*, 324–349. [CrossRef]
6. Dubey, J.P.; Calero-Bernal, R.; Rosenthal, B.M.; Speer, C.A.; Fayer, R. *Sarcocystosis of Animals and Humans*; CRC Press: Boca Raton, FL, USA, 2015.
7. Miller, M.; Barr, B.; Nordhausen, R.; James, E.; Magargal, S.; Murray, M.; Conrad, P.; Toy-Choutka, S.; Jessup, D.; Grigg, M. Ultrastructural and molecular confirmation of the development of Sarcocystis neurona tissue cysts in the central nervous system of southern sea otters (Enhydra lutris nereis). *Int. J. Parasitol.* **2009**, *39*, 1363–1372. [CrossRef]
8. Kadim, I.; Mahgoub, O.; Purchas, R. A review of the growth, and of the carcass and meat quality characteristics of the one-humped camel (Camelus dromedaries). *Meat Sci.* **2008**, *80*, 555–569. [CrossRef] [PubMed]
9. Kadim, I.T.; Mahgoub, O.; Mbaga, M. Potential of camel meat as a non-traditional high quality source of protein for human consumption. *Anim. Front.* **2014**, *4*, 13–17. [CrossRef]
10. Mandour, A.M.; Rabie, S.A.; Mohammed, N.; Hussein, N.M. On the presence of Sarcocystis miescheri sp. nov. in camels of Qena Governorate. *Egypt. Acad. J. Boil. Sci. E. Med Èntomol. Parasitol.* **2011**, *3*, 1–7. [CrossRef]
11. Hamidinejat, H.; Hekmatimoghaddam, S.; Jafari, H.; Sazmand, A.; Molayan, P.H.; Derakhshan, L.; Mirabdollahi, S. Prevalence and distribution patterns of Sarcocystis in camels (Camelus dromedarius) in Yazd province, Iran. *J. Parasit. Dis.* **2012**, *37*, 163–165. [CrossRef]
12. Omer, S.A.; Alzuraiq, A.A.; Mohammed, O.B. Prevalence and molecular detection of Sarcocystis spp. infection in the dromedary camel (Camelus dromedarius) in Riyadh city, Saudi Arabia. *Biomed. Res.* **2017**, *28*, 4962–4965.
13. Oryan, A.; Moghaddar, N.; Gaur, S.N. The distribution pattern of Sarcocystis species, their transmission and pathogenesis in sheep in Fars Province of Iran. *Veter. Res. Commun.* **1996**, *20*, 243–253. [CrossRef]
14. Dubey, J.P.; Lindsay, D.S.; Speer, C.A.; Fayer, R.; Livingston, C.W. Sarcocystis arieticanis and Other Sarcocystis Species in Sheep in the United States. *J. Parasitol.* **1988**, *74*, 1033. [CrossRef]
15. Fayer, R. Economic losses to Sarcocystis. *Natl. Wool. Grow.* **1976**, *66*, 22–28.
16. Munday, B. The effect of Sarcocystis ovicanis on growth rate and haematocrit in lambs. *Veter. Parasitol.* **1979**, *5*, 129–135. [CrossRef]
17. Munday, B. Effects of different doses of dog-derived Sarcocystis sporocysts on growth rate and haematocrit in lambs. *Veter. Parasitol.* **1986**, *21*, 21–24. [CrossRef]
18. Valinezhad, A.; Oryan, A.; Ahmadi, N. Sarcocystis and Its Complications in Camels (Camelus dromedarius) of Eastern Provinces of Iran. *Korean J. Parasitol.* **2008**, *46*, 229–234. [CrossRef]
19. Abdel-Ghaffar, F.; Mehlhorn, H.; Bashtar, A.-R.; Al-Rasheid, K.; Sakran, T.; El-Fayoumi, H. Life cycle of Sarcocystis camelicanis infecting the camel (Camelus dromedarius) and the dog (Canis familiaris), light and electron microscopic study. *Parasitol. Res.* **2009**, *106*, 189–195. [CrossRef]
20. Entzeroth, R.; A Ghaffar, F.; Chobotar, B.; Scholtyseck, E. Fine structural study of Sarcocystis sp. from Egyptian camels (Camelus dromed arius). *Acta Veter. Acad. Sci. Hung.* **1981**, *29*, 335–339.
21. Motamedi, G.R.; Dalimi, A.; Nouri, A.; Aghaeipour, K. Ultrastructural and molecular characterization of Sarcocystis isolated from camel (Camelus dromedarius) in Iran. *Parasitol. Res.* **2010**, *108*, 949–954. [CrossRef] [PubMed]
22. Eslampanah, M.; Motamedi, G.R.; Dalimi, A.; Noori, A.; Habibi, G.R.; Aghaeepour, K.; Niroumand, M. Study of camel and goat Sarcocystis by electron microscopic and PCR-RFLP. *Vet. Res. Biol. Prod.* **2016**, *29*, 77–84.
23. Bittencourt, M.V.; Meneses, I.D.S.; Ribeiro-Andrade, M.; De Jesus, R.F.; De Araújo, F.R.; Gondim, L.F. Sarcocystis spp. in sheep and goats: Frequency of infection and species identification by morphological, ultrastructural, and molecular tests in Bahia, Brazil. *Parasitol. Res.* **2016**, *115*, 1683–1689. [CrossRef] [PubMed]
24. Odening, K.; Stolte, M.; Bockhardt, I. On the diagnostics of Sarcocystis in cattle: Sarcocysts of a species unusual for Bos taurus in a dwarf zebu. *Veter. Parasitol.* **1996**, *66*, 19–24. [CrossRef]
25. Dubey, J.; Speer, C.; Charleston, W. Ultrastructural differentiation between sarcocysts of Sarcocystis hirsuta and Sarcocystis hominis. *Veter. Parasitol.* **1989**, *34*, 153–157. [CrossRef]
26. Hamidinejat, H.; Moetamedi, H.; Alborzi, A.; Hatami, A. Molecular detection of Sarcocystis species in slaughtered sheep by PCR–RFLP from south-western of Iran. *J. Parasit. Dis.* **2013**, *38*, 233–237. [CrossRef]
27. Carleton, H.M. *Histological Technique for Normal and Pathological Tissues and the Identification of Parasites*, 3rd ed.; Oxford University Press: London, UK, 1957.

28. Gjerde, B. Phylogenetic relationships among Sarcocystis species in cervids, cattle and sheep inferred from the mitochondrial cytochrome c oxidase subunit I gene. *Int. J. Parasitol.* **2013**, *43*, 579–591. [CrossRef]

29. Gjerde, B. Sarcocystis species in red deer revisited: With a re-description of two known species as Sarcocystis elongata n. sp. and Sarcocystis truncata n. sp. based on mitochondrial cox1 sequences. *Parasitology* **2013**, *141*, 441–452. [CrossRef]

30. Metwally, D.M.; Al-Damigh, M.A.; Al-Turaiki, I.; El-Khadragy, M.F. Molecular Characterization of Sarcocystis Species Isolated from Sheep and Goats in Riyadh, Saudi Arabia. *Animals* **2019**, *9*, 256. [CrossRef] [PubMed]

31. Geneious Prime. Available online: http://www.geneious.com/ (accessed on 7 April 2020).

32. Sievers, F.; Wilm, A.; Dineen, D.; Gibson, T.J.; Karplus, K.; Li, W.; López, R.; McWilliam, H.; Remmert, M.; Soeding, J.; et al. Fast, scalable generation of high-quality protein multiple sequence alignments using Clustal Omega. *Mol. Syst. Boil.* **2011**, *7*, 539. [CrossRef] [PubMed]

33. Guindon, S.; Lefort, V.; Anisimova, M.; Hordijk, W.; Gascuel, O.; Dufayard, J.-F. New Algorithms and Methods to Estimate Maximum-Likelihood Phylogenies: Assessing the Performance of PhyML 3.0. *Syst. Boil.* **2010**, *59*, 307–321. [CrossRef] [PubMed]

34. Wahba, A.A. Morphobiological Studies on Cryptosporidia and Sarcocystis in Animals and Their Impact on Their Health. Ph.D. Thesis, Faculty of Veterinary Medicine, Zagazig University, Zagazig, Egypt, 1994.

35. Fatani, A.; Hilali, M.A.; Al-Atiya, S.; Al-Shami, S. Prevalence of Sarcocystis in camels (Camelus dromedarius) from Al-Ahsa, Saudi Arabia. *Veter. Parasitol.* **1996**, *62*, 241–245. [CrossRef]

36. Al-Quraishy, S.A.; Bashtar, A.R.; Al-Rasheid, K.; Abdel-Ghaffar, F. Prevalence and ultrastructure of Sarcocystis species infecting camels (Camelus dromedarius) slaughtered in Riyadh city Saudi Arabia. *Saudi J. Biol. Sci.* **2004**, *11*, 135–141.

37. Woldemeskel, M.; Gumi, B. Prevalence of sarcocysts in one-humped camel (Camelus dromedarius) from Southern Ethiopia. *J. Vet. Med. Ser. B* **2001**, *48*, 223–226. [CrossRef] [PubMed]

38. Hilali, M.; Mohamed, A. The dog (Canis familiaris) as the final host of Sarcocystis cameli (Mason, 1910). *Trop. Parasitol.* **1980**, *31*, 213–214.

39. Abd-Elmalek, B.S.; Abed, G.H.; Mandour, A.M. Study on Sarcosystis sp. By Light and Electron Microscopy in Camel Muscles at Assiut Governorate. *J. Veter. Sci. Technol.* **2015**, *6*. [CrossRef]

40. Mason, F.E. Sarcocysts in the camel in Egypt. *J. Comp. Pathol. Ther.* **1910**, *23*, 168–176. [CrossRef]

41. Moré, G.; Regensburger, C.; Gos, M.L.; Pardini, L.; Verma, S.K.; Ctibor, J.; Serrano-Martínez, E.; Dubey, J.P.; Venturini, M.C. Sarcocystis masoni, n sp. (Apicomplexa: Sarcocystidae), and redescription of Sarcocystis aucheniae from llama (Lama glama), guanaco (Lama guanicoe) and alpaca (Vicugna pacos). *Parasitology* **2016**, *143*, 617–626.

42. Singh, D.; Kumar, B. UP Pandit Deen Dayal Upadhyaya Pashu-Chikitsa Vigyan Vishwavidyalaya Evam Go Anusandhan Sansthan. Ph.D. Thesis, Mathura, Mathura, India, 2018.

43. Gjerde, B.; Luzón, M.; Alunda, J.M.; De La Fuente, C. Morphological and molecular characteristics of six Sarcocystis spp. from red deer (Cervus elaphus) in Spain, including Sarcocystis cervicanis and three new species. *Parasitol. Res.* **2017**, *116*, 2795–2811. [CrossRef]

44. Gazzonis, A.L.; Gjerde, B.; Villa, L.; Minazzi, S.; Zanzani, S.A.; Riccaboni, P.; Sironi, G.; Manfredi, M.T. Prevalence and molecular characterisation of Sarcocystis miescheriana and Sarcocystis suihominis in wild boars (Sus scrofa) in Italy. *Parasitol. Res.* **2019**, *118*, 1271–1287. [CrossRef]

45. Murata, R.; Suzuki, J.; Hyuga, A.; Shinkai, T.; Sadamasu, K. Molecular identification and characterization of Sarcocystis spp. in horsemeat and beef marketed in Japan. *Parasite* **2018**, *25*, 27. [CrossRef]

Article

Ancylostoma ceylanicum: The Neglected Zoonotic Parasite of Community Dogs in Thailand and Its Genetic Diversity among Asian Countries

Doolyawat Kladkempetch [1], Sahatchai Tangtrongsup [2,3] and Saruda Tiwananthagorn [4,5,*

1 Master's Degree Program in Veterinary Science, Faculty of Veterinary Medicine, Chiang Mai University, Chiang Mai 50100, Thailand; doolyawat_kl@cmu.ac.th
2 Department of Companion Animal and Wildlife Clinic, Faculty of Veterinary Medicine, Chiang Mai University, Chiang Mai 50100, Thailand; sahatchai.t@cmu.ac.th
3 Research Center of Producing and Development of Products and Innovations for Animal Health and Production, Faculty of Veterinary Medicine, Chiang Mai University, Chiang Mai 50100, Thailand
4 Department of Veterinary Biosciences and Veterinary Public Health, Faculty of Veterinary Medicine, Chiang Mai University, Chiang Mai 50100, Thailand
5 Excellent Center in Veterinary Bioscience, Faculty of Veterinary Medicine, Chiang Mai University, Chiang Mai 50100, Thailand
* Correspondence: saruda.t@cmu.ac.th; Tel.: +66-53-948-046 or +66-95-446-5955

Received: 3 November 2020; Accepted: 17 November 2020; Published: 19 November 2020

Simple Summary: Ancylostomiasis is a zoonotic disease caused by the *Ancylostoma* hookworm infection of dogs, cats, and other wildlife species and is frequently found in Asia and tropical regions. The present study had confirmed the species of hookworm in dogs and soil environments collected from temple communities. In addition, we investigated the association of hookworm-contaminated soil in temple areas to increase awareness and establish a regimen to prevent zoonotic hookworm transmission. Lastly, we analyzed the genetic diversity and evolution of hookworm in dogs and soil environments among Thai *A. ceylanicum* and other Asian populations and disclosed insight into their fundamental genetic relationship.

Abstract: *Ancylostoma ceylanicum* is a zoonotic helminth that is commonly found in domestic dogs and cats throughout Asia but is largely neglected in many countries. This study aimed to confirm the species of hookworm in dogs and soil environments and investigate the evolutionary analyses of *A. ceylanicum* among Thai and Asian populations. In a total of 299 dog fecal samples and 212 soil samples from 53 temples, the prevalence rates of hookworm infection by microscopic examination were 26.4% (79/299) and 10.4% (22/212) in dog and soil samples, respectively. A PCR-RFLP targeting the ITS region was then utilized to identify the hookworm species. In dogs, *A. ceylanicum* was the main hookworm species, and the rates of *A. ceylanicum* and *A. caninum* infections were 96.6% and 3.5%, respectively. The genetic characterization and diversity indices of the *A. ceylanicum cox1* gene among Thai and Asian populations were evaluated. Nine haplotypes were identified from Thai *A. ceylanicum*, in which the haplotype diversity and the nucleotide diversity were 0.4436 and 0.0036, respectively. The highest nucleotide diversity of Chinese *A. ceylanicum* populations suggested that it could be the ancestor of the populations. Pairwise fixation indices indicated that Thai *A. ceylanicum* was closely related to the Malaysian population, suggesting a gene flow between these populations. The temples with hookworm-positive dogs were associated with the presence of hookworm-contaminated soil, as these levels showed an approximately four-fold increase compared with those in temples with hookworm-negative dogs (OR = 4.38, 95% CI: 1.55–12.37). Interestingly, the genotypes of *A. ceylanicum* in the contaminating soil and infecting dogs were identical. Therefore, increased awareness and concern from the wider public communities with regard to the responsibility of temples and municipal offices to provide proper deworming programs to community dogs should be strongly endorsed to reduce the risk of the transmission of this zoonotic disease. In addition, parasitic examination and treatment should be strongly implemented before dogs are imported and exported worldwide.

Keywords: *Ancylostoma ceylanicum*; community dogs; ITS region; *cox1*; Thailand; population diversity

1. Introduction

Hookworms are soil-transmitted gastrointestinal nematodes of the Ancylostomatidae family that infect dogs, cats, and humans. These nematodes are endemic in tropical and subtropical regions of the world, including parts of China, Southeast Asia, Australia, and Africa [1–4]. Hookworms can infect the host by the fecal-oral route and through the skin penetration of the third-stage larvae [2]. Hookworm infection causes intestinal blood loss, anemia, malnutrition, and dermatitis [5]. Common species of hookworm infection in dogs include *Ancylostoma caninum*, *A. ceylanicum*, *A. braziliense*, and *Uncinaria stenocephala*, while *A. braziliense*, *A. tubaeforme*, *A. ceylanicum*, and *U. stenocephala* are commonly found in cats [6,7]. Recent studies have shown that ancylostomiasis in humans results from *Necator americanus*, *A. duodenale*, and *A. ceylanicum* [5,8,9]. Among *Ancylostoma* species, *A. ceylanicum* is the only species of animal hookworms known to produce patent infections in humans and is an important cause of zoonotic ancylostomiasis in Asia and Southeast Asia [9–14].

In general, the epidemiological study revealed that animal-derived *Ancylostoma* species are emerging due to human and animal interaction [15,16]. In Thailand, free-roaming community dogs have been widespread and come in close contact with people in some areas, such as temples and rural communities. The presence of hookworm larvae in soil reported by George et al. [5] confirmed that humans and dogs are at risk of infection from hookworm-contaminated soil. Additionally, a previous report in Malaysia from Ngui et al. [15] indicated that people who are not wearing shoes are at risk of hookworm infection from hookworm-contaminated soil. The opportunity for exposure to hookworm is high in children, farmers, and especially Buddhist monks, who do not wear shoes based on their practices and Buddhist tradition in Thai temples. Information regarding whether the soil in the area is contaminated with hookworm is, therefore, important for public health to promote awareness and establish a regimen to prevent zoonotic hookworm transmission.

The conventional method for hookworm diagnosis is coprological examination. However, this method cannot differentiate hookworm species due to the similar egg size and morphology. Morphological identification of adult hookworms is achievable, but the process is time-consuming and labor-intensive and requires personal skill [16]. Alternatively, molecular techniques have been developed for the identification of hookworms at the species level, such as polymerase chain reaction (PCR) assays and PCR with restriction fragment length polymorphism (PCR-RFLP) assays targeting the internal transcribed spacer (ITS) region [16]. The ITS region is a conserved region with a low mutation rate and has no intraspecific variation [17]. Thus, the ITS region is suitable as a genetic marker for hookworm species identification [18]. Recently, the cytochrome *c* oxidase subunit 1 (*cox1*) gene has been utilized to evaluate the evolution and relationship of hookworm species among the population [19,20], given its high intraspecific sequence variation from maternal inheritance and high evolutionary rates [21].

In Northern Thailand, a few reports have assessed the prevalence of hookworm infection in humans, domestic dogs, and cats. By microscopic examination, hookworm infection rates of approximately 21.3% in dogs and 13.9% in cats [22] and 12% in dogs and 4.5% in cats [23] have been reported in the lower northern region and in the Chiang Mai province of the upper northern region, respectively. The prevalence of hookworm infection in humans ranged from 0.6% to 13.4% [24,25]. However, the species of hookworm in Upper Northern Thailand have never been clarified. The present study, therefore, aimed to identify the species of hookworm in dogs and investigate the presence of hookworm contamination in the soil environment of community temples where there are high risks of communicable pathogen transmission. Furthermore, the genetic characteristics of *Ancylostoma* hookworm and the evolutionary genetic relationships among the Thai hookworm and the populations of neighboring Asian countries have been elucidated.

2. Materials and Methods

2.1. Study Area and Sample Size

The study area is located in the upper northern region of Thailand. Four provinces, including Chiang Mai (18°47′26″ N, 98°59′14″ E), Chiang Rai (20°01′05″ N, 99°40′22″ E), Lampang (18°17′21″ N, 99°29′26″ E), and Phayao (19°11′30″ N, 99°52′47″ E), were chosen and considered as high dog population density provinces (Figure 1). Sampling sites were 53 community temples that were chosen based on convenience and distribution from at least 3 districts in each province. The inclusion criteria for temple selection were (i) a human population density of more than 30 persons per sq. km. [26,27], (ii) no regular deworming program in dogs within 3 months, and (iii) ease of access and sample transportation to the laboratory. The sample size of 299 dog fecal samples for the detection of hookworm infection was calculated using the Win episcope 2.0 program [28] based on an estimated hookworm prevalence of 12.8% in dogs [29] and an error rate of 5% with a 95% level of confidence. The stratified sampling method was used to generate the number of samples per province based on dog population data [30].

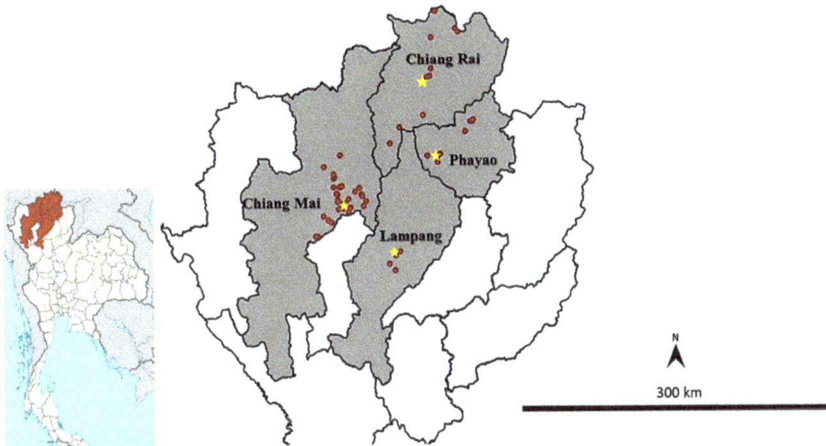

Figure 1. Map of the study areas in Thailand. The gray color indicates four study provinces, red dots indicate the location of sampling sites (temples), and stars indicate the capital city of each province.

All the owners signed an informed consent form, and the animal use was approved by the Ethics Committee of the Laboratory Animal Center, Faculty of Veterinary Medicine, Chiang Mai University (S41/2561), on 18 January 2019.

2.2. Sample Collection and Examination

2.2.1. Fecal and Soil Sample Collection

Samples were collected between June and September 2019. Two hundred and ninety-nine fecal samples were collected from the rectum of dogs or fresh fecal samples found on the ground. All the collected fresh feces were normal in appearance (medium to dark brown in color, soft to firm in texture) based on the WALTHAM fecal scoring system [31]. The dog ID, date of collection, temple name, district, province, cleaning housing program, and type of food were recorded. Soil environment samples were collected at the temple in the morning (08:00–10:00). For soil sample collection, the sites of sample collection were defined by 4 areas in each temple: (i) temple courtyard, (ii) dog dwelling area, (iii) human activity area (such as sand pagoda), and (iv) under the big tree. In the collection site that was not dirt ground, the soil sample for that site was obtained from the adjacent ground within a 2 m radius of the area. Soil samples were collected from moist areas within the selected sampling sites, and approximately 50 g (wet weight) was obtained by scraping the surface layer to gather a 0.09 square meter area using a metal spade. The soil samples were kept separately in a sealed plastic bag.

All the samples were stored on ice and transported to the Parasitology Laboratory, Faculty of Veterinary Medicine, Chiang Mai University. Samples were refrigerated at 4 °C upon arrival.

2.2.2. Microscopic Examinations of Fecal and Soil Samples

Fresh fecal samples were examined under a light microscope after simple floatation using zinc sulfate solution (specific gravity: 1.18) for the presence of eggs. The remainder of the fecal sample was stored at −20 °C until further molecular analysis. The Baermann technique was used to determine the presence of hookworm-like larvae in the soil samples [32]. Twenty grams of soil samples were primarily filtered through a tea strainer to remove any debris and then subjected to another filtration step using 2 sieve filters with pore sizes of 0.4 and 0.2 mm. The filtrated soil sample was cultured individually using tap water. Hookworm-like larvae containing sediments were stored at −20 °C until further molecular analysis could be carried out.

2.3. DNA Extraction

Genomic DNA (gDNA) samples from hookworm-positive samples were extracted using the NucleoSpin® DNA stool kit (Macherey-Nagel GmbH, Düren, Germany) according to the manufacturer's instructions. In the elution step, gDNA samples were eluted with 100 µL of elution buffer. The extracted DNA concentration and quality were determined using a spectrophotometer (Beckman Coulter DU® 730 Life Sciences, Pasadena, CA, USA) at wavelengths of 260 and 280 nm. All the gDNA samples were kept at −20 °C until the next step.

2.4. Molecular Analyses

2.4.1. Identification of Hookworm Species Using a Polymerase Chain Reaction-Restriction Fragment Length Polymorphism (PCR-RFLP) Assay Based on ITS Region

The PCR-RFLP assay, which was established by Traub et al. [16], was conducted to identify the species of hookworm in the positive dog fecal and soil samples. A forward primer RTGHF1 (5′-CGTGCTAGTCTTCAGGACTTTG-3′) and a reverse primer RTABCR1 (5′-CGGGAATTGCTATAAGCAAGTGC-3′) were used to amplify the 545-bp region of the internal transcribed spacer (ITS1, 5.8S, and ITS2). PCR amplification reactions were performed in a 25 µL reaction volume containing 10 ng of template gDNA (2–4 µL), 0.2 µM of each primer (0.5 µL of 10 µM), and 12.5 µL of 2× PCR Master Mix (KOD One™ PCR Mastermix; Toyobo, Japan). The thermocycling conditions were as follows: initial denaturation at 94 °C for 2 min, 45 cycles of amplification (94 °C for 30 s, 62 °C for 30 s, and 72 °C for 30 s), and a final extension at 72 °C for 5 min. The reaction was performed on a T100™ Thermal Cycler (Bio-Rad, Hercules, CA, USA). The amplified PCR products were digested with Hinf1 to differentiate *A. caninum* from *A. ceylanicum* and *U. stenocephala*. The digestions were performed at 37 °C for 16 h, using 1 unit of restriction enzyme and 1 µg of DNA (3–5 µL of PCR product) in a total reaction volume of 20 µL. The PCR products were analyzed using 2% agarose gels electrophoresis. Distilled water (DW) and the gDNA of *A. caninum* and *A. ceylanicum* were used as a negative and positive control, respectively.

2.4.2. PCR Assay and Nucleotide Analysis Based on *A. ceylanicum* and *A. caninum cox1* Gene

The mitochondrial *cox1* gene served as the target to assess the genetic characteristics and evolutionary analysis. The previously identified *A. ceylanicum*- and *A. caninum*-positive samples were successively subjected to PCR for the *cox1* gene using a set of primers described by Inpankaew et al. [33], including the forward primer AceyCOX1F (5′-GCTTTTGGTATTGTAAGACAG-3′) and the reverse primer AceyCOX1R (5′-CTAACAACATAATAAGTATCATG-3′), amplifying a 377-bp amplicon. The reactions were performed in a 50 μL reaction mixture volume, containing 10 ng of template gDNA (2–4 μL), 0.2 μM of each primer (1 μL of 10 μM), and 25 μL of 2× PCR Master Mix (KOD One™ PCR Mastermix; Toyobo, Japan). The PCR conditions were set as follows: initial denaturation at 94 °C for 2 min, 45 cycles of amplification (94 °C for 30 s, 55 °C for 30 s, and 68 °C for 45 s), and a final extension at 72 °C for 5 min. The reaction was performed using a T100™ Thermal Cycler (Bio-Rad, Hercules, CA, USA). The DW and gDNA of *A. caninum* and *A. ceylanicum* were used as negative and positive controls, respectively. The PCR products were purified using a NucleoSpin® PCR Clean-up Kit (Macherey-Nagel GmbH, Düren, Germany). Purified PCR products were submitted for fluorescent dye-terminator sequencing by Bio Basic, Inc. (Singapore, Singapore) in the sense and antisense directions using the PCR primer set described above. For nucleotide analysis, the *cox1* amplicons of 32 samples (29 *A. ceylanicum* samples from dogs, 2 *A. ceylanicum* samples from soil, and 1 *A. caninum* sample from dog) were selected based on the geographic distribution, covering all positive provinces.

2.5. Data Analyses

2.5.1. Prevalence of Hookworm Infection/Contamination

Positive fecal samples from the zinc sulfate flotation technique and positive soil samples from the Baermann technique were considered as positive for hookworm detection. The prevalence of hookworm infection in dogs and contamination in soil samples were estimated using the formula prevalence = (number of positives samples divided by the number of samples in dogs/soil examined) × 100. The association between hookworm-positive results and sampling areas was assessed using Fisher's exact test. The significance was defined as $p < 0.05$.

2.5.2. Phylogenetic Analysis

The *cox1* sequences were edited and manually aligned. The consensus sequences were generated using the BioEdit Sequence Alignment Editor [34] and then compared with the nucleotide sequences from GenBank using a Basic Local Alignment Search Tool (BLAST) analysis. Phylogenetic analysis was performed using MEGA X [35]. Multiple sequences were aligned using ClustalW (https://www.genome.jp/tools-bin/clustalw), and phylogenetic analysis was performed using a maximum likelihood (ML) method based on the Kimura 2-Parameter model. The consensus tree was obtained after a 1000-replication bootstrap analysis. The *cox1* sequences of *A. ceylanicum* from other areas of Thailand and other countries, including Malaysia, Cambodia, China, Japan, Papua New Guinea, Australia, and Tanzania, and *A. caninum* from Australia (NC012309) were used as reference sequences for tree construction (Table S1). In addition, the sequence of *A. duodenale* from China (accession number: NC003415) was used as an outgroup species.

2.5.3. Population Genetic Analysis

The DnaSP 6 program [36] was used for the haplotype analysis and calculations to determine the haplotype diversity (Hd), nucleotide diversity (π), and the number of variable sites (S). The gene flow among populations was analyzed by fixation index (F_{ST}) statistics using the Arlequin program Version 3.5 [37]. The gene flow of *A. ceylanicum* was analyzed using the reference sequences from Asian countries, including Thailand, China, Malaysia, and Cambodia (Table S2). The Network program [38] was used to generate a median-joining (MJ) network of *cox1* haplotypes and reference haplotypes and was subsequently drawn manually. The references haplotype and their frequency obtained from the NCBI database (https://www.ncbi.nlm.nih.gov/genbank/) are described above.

3. Results

3.1. Prevalence and Distribution of Hookworm Infection

A total of 299 fecal samples and 212 soil samples from 53 temples were examined microscopically, and the hookworm species were identified by PCR-RFLP targeting the ITS region. The prevalence rate of hookworm in dogs was 26.4% (79/299), as assessed by microscopic examination. PCR-RFLP successfully identified the species of hookworm in 58 samples, and the rates of *A. ceylanicum* and *A. caninum* infection were 96.55% (56/58) and 3.45% (2/58), respectively (Figure 1). On the other hand, the prevalence of hookworm-like larvae in soil was 10.38% (22/212), as assessed by microscopic examination (Table 1). However, PCR-RFLP only identified the species in 8 of 22 samples, and all the samples were identified as *A. ceylanicum* (Table 1).

Table 1. Summary of hookworm infection in dog fecal samples and soil samples in Thailand, as demonstrated by the microscopic results and subsequent PCR-RFLP targeting the ITS region.

Province	Fecal Samples										Soil Samples								
	Positive Microscopic			N_2	*A. ceylanicum*		*A. caninum*		n/a	Positive Microscopic			N_4	*A. ceylanicum*		*A. caninum*		n/a
	N_1	n	%		n	%	n	%	n	N_3	n	%		n	%	n	%	n
Chiang Mai	159	57	35.85 [a]	42	40	95.24	2	4.76	15	120	16	13.33	3	3	100	0	0	13
Chiang Rai	70	6	8.57 [b]	5	5	100	0	0	1	48	1	2.08	1	1	100	0	0	0
Lampang	20	9	45.00 [a]	5	5	100	0	0	4	16	3	18.75	2	2	100	0	0	1
Phayao	50	7	14.00 [b]	6	6	100	0	0	1	28	2	7.14	2	2	100	0	0	0
Total	299	79	26.42	58	56	96.55	2	3.45	21	212	22	10.38	8	8	100	0	0	14

N_1: total number of examined dog fecal samples assessed by microscopic examination; N_2: number of positive dog samples successfully assessed with PCR-RFLP; N_3: total number of examined soil samples by microscopic examination; N_4: number of positive soil samples successfully assessed with PCR-RFLP; n/a: species unidentified samples; superscripts (a,b) indicate test results of Chi-square on prevalence among four provinces (any measurements with shared superscript letters are not significantly different from each other at $p < 0.05$).

Considering the population, the temple-level prevalence of positive dogs and positive soils was 47.17% (25/53) and 26.42% (14/53), respectively (Table S3). A total of 10 temples were positive for hookworm both in dog and soil samples (Table S3). Although not statistically significant, the hookworm-positive results in dogs were speculated to be associated with the hookworm-positive results in the soil ($p = 0.06$). Regarding the different areas in each temple, we found that the soil from dog-dwelling areas had the highest rates of hookworm contamination (15.09%, 8/53), followed by the soil from human activity areas, such as sand pagoda (11.32%; 6/53); the areas under big trees (9.43%; 5/53); and the temple courtyard (5.66%, 3/53) (Table S4). However, the frequency of hookworm-contaminated soil in each sampling site in the temple exhibited no significant differences ($p > 0.05$; Fisher's exact).

3.2. Hookworm Species Confirmation and Genetic Characteristics of the A. ceylanicum Mitochondrial cox1 Gene

The nucleotide BLAST of partial *cox1* sequences (315 bp) of 31 *A. ceylanicum* sequences exhibited a high identity (98.07–100%) with Thai *A. ceylanicum* (GenBank accession no. KF896595). An *A. caninum* sequence from a dog exhibited a 96.71% identity with *A. caninum* from Australia (GenBank accession no. AJ407962). The obtained *A. ceylanicum cox1* gene sequences from dog and soil isolates were classified into nine haplotypes, which were referred to as Acy-COX1-TH01 to Acy-COX1-TH09. The sequences of *A. ceylanicum* of dogs and soil were deposited in GenBank (DDBJ/EMBL/GenBank database accession no. LC533318-LC533327). The sequence of *A. caninum* was deposited as LC533328 (Table S5).

Genetic variation among nine haplotypes of Thai *A. ceylanicum*, as assessed by multiple alignment analysis, revealed 12 different variable sites, including 2 transversions and 10 transitions (Table 2).

Table 2. Multiple alignment of the partial *cox1* from 9 haplotypes of *A. ceylanicum* in this study.

Haplotype Name	Nucleotide Position on the *A. ceylanicum cox1* Gene											
	48	72	87	88	90	105	147	150	198	199	219	264
Acy-COX1-TH01	A	C	A	G	A	A	G	A	G	G	A	G
Acy-COX1-TH02	G	.	.	.	.	.	A	.	.	.	.	.
Acy-COX1-TH03	.	T	.	.	.	.	.	.	.	.	.	.
Acy-COX1-TH04	.	.	.	.	.	G	.	.	A	A	.	.
Acy-COX1-TH05	.	.	.	C	.	.	.	.	.	.	.	.
Acy-COX1-TH06	.	T	G	.	G	.	A	.	.	.	G	A
Acy-COX1-TH07	.	.	.	.	.	.	.	T	.	.	.	.
Acy-COX1-TH08	.	.	G	.	.	.	.	.	.	.	.	.
Acy-COX1-TH09	G	.	.	.	.	.	.	.	.	.	.	.

Dots indicated the matching of nucleotide position the *A. ceylanicum* sequence.

The distribution and frequencies of *A. ceylanicum* haplotypes in Upper Northern Thailand were analyzed and revealed that the Acy-COX1-TH01 haplotype was detected at the highest frequency at 74.2% (23/31 sequences). This haplotype was commonly observed in all four provinces and found in both fecal and soil samples. As displayed in Figure 2, the sequence of two soil isolates (Acy-COX1-TH01-soil) was close to Acy-COX1-TH01-dog. Six haplotypes are present in Chiang Mai (Acy-COX1-TH01 to TH06), 3 haplotypes are present in Lampang (Acy-COX1-TH01, TH08, and TH09), whereas only one or two haplotypes were observed in Chiang Rai (Acy-COX1-TH01 and TH07) and Phayao (Acy-COX1-TH01) provinces (Table S5). Regarding the relationship among the human and animal isolates, Thai Acy-COX1-TH01 was close to the sequences of *A. ceylanicum* from Malaysian people (MK792828), dog (MK792824), and cat (MK792825) that belong to the same haplotype MA02 (Figure 2 and Table S1). Similarly, Acy-COX1-TH03 was close to the sequences of *A. ceylanicum* from human (KC247737) and dog (KC247734) that belong to the haplotype Malaysian MA01. Regarding the relationship between *A. ceylanicum* and geographic distribution, a phylogenetic tree demonstrated that *A. ceylanicum* from Thailand and all reference countries, including Malaysia, Cambodia, China, Japan, Papua New Guinea, and one country from the African continent (Tanzania), were located in the same clade, with the exception of one human sequence in Australia (AJ407937) (Figure 2).

3.3. Population Analyses of the A. ceylanicum Mitochondrial cox1 Gene among Asian Countries

The haplotype network displayed a star-like pattern and exhibited a wide dispersal around the predominant haplotype Acy-COX1-TH01 (in this study). This haplotype was identical to haplotypes MA02 from Malaysia, CA01 from Cambodia, and CH01 haplotype from China (Figure 3). The second most abundant haplotype was Acy-COX1-TH03, which was placed in the same group as haplotype MA01 from Malaysia and haplotype CA03 from Cambodia. Acy-COX1-TH06 was the most distinct haplotype, which includes four nucleotide-substitutions and was delineated from the haplotype CA02 from Cambodia.

The genetic diversity of the Thai *A. ceylanicum* population has been analyzed and compared pairwise to the populations of Cambodia, Malaysia, and China. The Thai *A. ceylanicum cox1* population (32 sequences; 31 from this study and 1 sequence from Inpankaew et al. [33]) displayed the lowest genetic diversity—i.e., a haplotype diversity (Hd) of 0.4435 and a nucleotide diversity (π) of 0.0036 (Table 3).

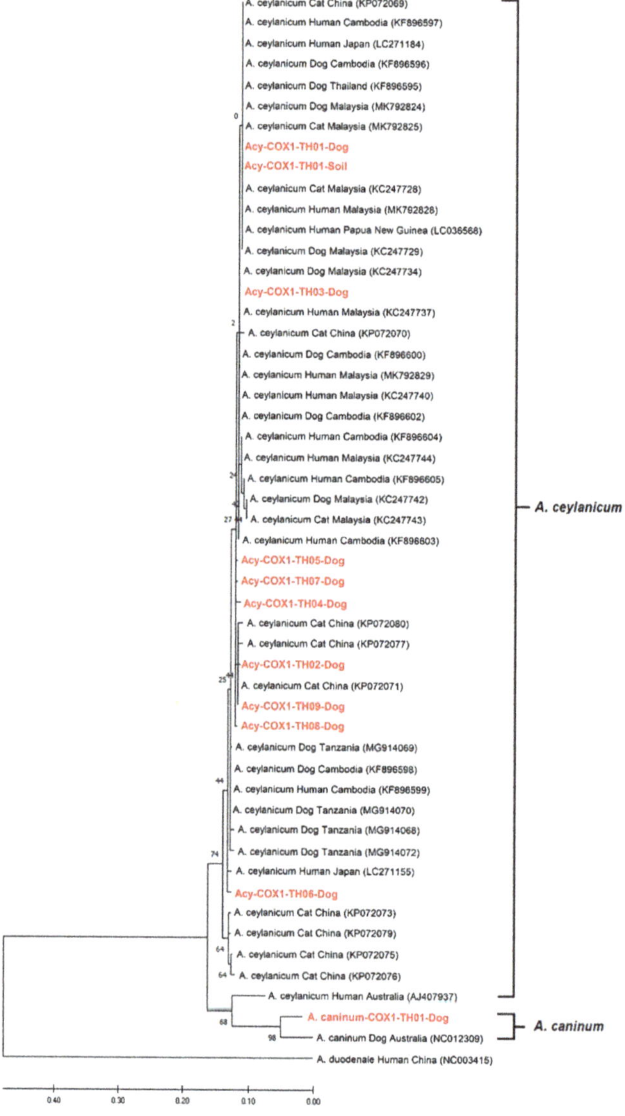

Figure 2. The evolutionary relationship of *A. ceylanicum* and *A. caninum* isolates from dogs and soil based on the 259-bp fragments of the mitochondrial cytochrome *c* oxidase subunit 1 (*cox1*) gene. The tree was constructed using a maximum likelihood method based on the Kimura 2-Parameter using the MEGA X software version 10.0.5. The number in each branch indicates the percentage of 1000 bootstrap replications. Sequences obtained from GenBank are indicated by their accession numbers.

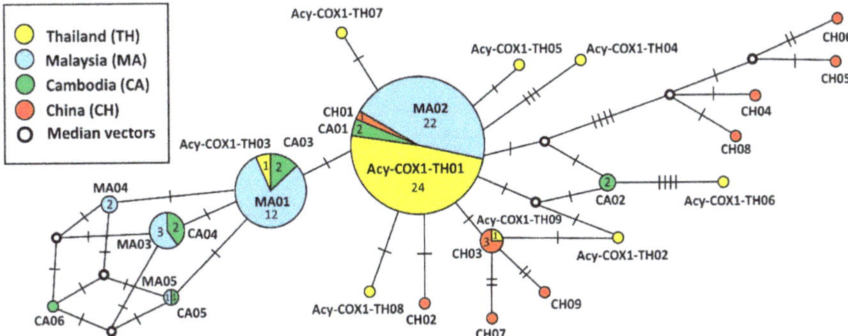

Figure 3. The haplotype network map of *A. ceylanicum* based on a fragment of the *cox1* gene from Thailand, Cambodia, China, and Malaysia. Circle size is scaled to frequencies of each haplotype. The hash marks indicate nucleotide substitutions among adjacent haplotypes.

Table 3. Summary of the genetic diversity of the 4 populations of *A. ceylanicum* based on the nucleotide sequences of the partial mitochondrial *cox1* gene.

Populations	N	Diversity			
		S	h	Hd ± SD	π ± SD
Thailand	32	11	9	0.4435 ± 0.1105	0.0036 ± 0.0028
Cambodia	10	6	6	0.9111 ± 0.0620	0.0088 ± 0.0059
Malaysia	40	4	5	0.6141 ± 0.0593	0.0043 ± 0.0006
China	12	17	9	0.9394 ± 0.0577	0.0201 ± 0.0118

N: numbers of *A. ceylanicum* sequences used; S: numbers of variable sites; h: numbers of haplotypes; Hd: haplotype diversity; SD: standard deviation; π: nucleotide diversity.

The highest haplotype diversity and nucleotide diversity were observed in the Chinese *A. ceylanicum* population (Hd = 0.9394; π = 0.0201), followed by the Cambodian and Malaysian populations. Pairwise (F_{ST}) indices between the Thai *A. ceylanicum* population and other reference populations were 0.2644, 0.2507, and 0.1526 for the Chinese, Cambodian, and Malaysian populations, respectively. In addition, the highest F_{ST} value of 0.3577 was observed between the Chinese and Malaysian populations. All the F_{ST} values were statistically significant (Table 4).

Table 4. The pairwise fixation index (F_{ST}) among four *A. ceylanicum* populations based on the partial mitochondrial *cox1* gene.

Populations	Thailand	Cambodia	Malaysia	China
Thailand	-			
Cambodia	0.2507	-		
Malaysia	0.1526	0.0999	-	
China	0.2644	0.1939	0.3577	-

All values were statistically significant ($p < 0.001$).

4. Discussion

The major hookworm species in the community dogs in Upper Northern Thailand was *A. ceylanicum*. We demonstrated an approximately 50% temple-level prevalence of hookworm infection in dogs, and these positive dogs were commonly living in the community temples and freely roaming in the surrounding communities. The prevalence of hookworm infection at the individual dog level in this study (26.4%) showed an approximately two-fold increase compared with previous studies in Chiang

Mai province (approximately 12%), as reported by Tiwananthagorn et al. (unpublished results) and Tangtrongsup et al. [23], and greater than that reported by Pumidonming et al. [22] in Lower Northern Thailand (21.3%). These differences are probably due to the different study areas and the inclusion criteria of no regular deworming program. The proportion of *A. ceylanicum* infection in dogs from our study in Upper Northern Thailand (96.6%) was higher than the previous report in the Lower Northern region, which reported a rate of 82.1% [22]. Recently, various studies have reported the rate of *A. ceylanicum* in hookworm infection in the bordering countries of Thailand: 94.4% of dogs in Cambodia, 52% of dogs in Malaysia [39], 77.8% of dogs in Laos [40], and 42.7% of dogs in China [7]. The failure of PCR amplification in some fecal and soil samples was noticed in the present study and was also noted in other studies. This failure could be due to the low intensity of infection, resulting in a reduced number of hookworm eggs, and some PCR inhibitors in the fecal and soil samples interfere with the amplification reaction [15,41]. Therefore, developing a more robust PCR assay to improve the diagnosis of hookworm infection is needed for the efficient control of this zoonosis.

The temples with hookworm-infected dogs are prone to have hookworm-contaminated soil. Here, we confirmed the presence of *A. ceylanicum* larvae in the soil. Moreover, the *A. ceylanicum* genotypes in dogs and soil were identical, suggesting the shedding of hookworm from dogs to the soil. The frequently detected areas in the positive temples included the dog's dwelling areas; human activity areas, such as the sand pagoda; and the areas under the big tree. These areas are the places that community people commonly access for the Buddhist rituals, ceremonies, commerce, and recreation. In addition, allowing dogs to roam freely with indiscriminate defecating areas in the Thai temples can lead to the widespread contamination of infective *A. ceylanicum* hookworm larvae in the environment and possible transmission to humans who occasionally go barefoot during these activities. As suggested in rural Malaysia, close contact with community dogs and cats was associated with *A. ceylanicum* hookworm infection in humans [42]. Previous reports worldwide have reported soil contaminated with hookworm in the public parks of Brazil and Malaysia [43,44], a tribal area of India [5], dog parks in Portugal [45], and organic farms in the Philippines [46]. Therefore, good hygienic practices of shoe-wearing and temple ground cleaning programs should be encouraged to reduce the risk of transmission [47]. In addition, reducing the number of free-roaming dog population size through a sterilization campaign or by sheltering free-roaming dogs should be implemented [48].

The molecular epidemiologic data, phylogenetic relationship, and population analysis obtained from the characterization of the *cox1* gene of *A. ceylanicum* hookworms strongly support the distribution of this parasite in the country, within the Asian continent, and possibly worldwide. A common haplotype of Thailand (Acy-COX1-TH01), which was observed in all four provinces of Upper Northern Thailand, suggested the circulation of *A. ceylanicum* in the region. Genetic differentiation analysis showed that Thai *A. ceylanicum* had the lowest haplotype diversity and nucleotide diversity, indicating minimal differences between haplotypes in Thailand. However, further studies in other regions of Thailand are necessary. The pairwise fixation index showed that Thai *A. ceylanicum* was mostly close to Malaysian *A. ceylanicum*, suggesting gene flow between these populations. In addition, the sharing haplotype of the *cox1* gene of *A. ceylanicum* among the countries (Acy-COX1-TH01 and MA02 and CA01 and CH01; or Acy-COX1-TH03 and MA01 and CA03) advocated the transmission and widespread geographic distribution among these countries [49]. The haplotype relationship between Malaysia and Cambodia populations revealed a very complex evolutionary pattern with some median vectors that indicated many lost or potential ancestors [50]. Interestingly, the Chinese *A. ceylanicum* exhibited the highest haplotype and nucleotide diversities, suggesting that the Chinese *A. ceylanicum* might be an ancestor of the populations in Asian countries. Our hypothesis is consistent with a previous study on the history of domestic dogs, in which China was the origin of the dog species [51]. Animals potentially migrated to accompany human colonization to other Asian countries and subsequently carried the parasites and allowed the parasites to colonize new habitats [51–53]. Hence, the transportation of companion animals should strictly follow the guidelines and regulations of each county to prevent the spreading of zoonotic hookworm.

Based on the phylogenetic tree, the sequences of *A. ceylanicum* from humans (MK792828-29), dogs (MK792824), and cats (MK792825) in Malaysia belong to the same haplotype MA02, confirming that transmission from dogs to humans has occurred. This species was also found in wildlife, such as Asian golden cat, leopard cat, civet, and wild canids [9,54,55]. The biological, epidemiological, and pathophysiological characteristics of *A. ceylanicum* in humans and other wildlife species in Thailand and their genetic diversity with isolates from companion animals deserve further investigation.

Integrated control programs intended for combining chemotherapeutic interventions by deworming programs with improvements in community hygiene and animal health programs will aid in curbing this potentially opportunistic zoonosis. For the deworming program of community dogs in Thailand, an oral dose of 10 mg/kg BW of pyrantel pamoate and the subcutaneous injection of 0.2–0.4 mg/kg BW ivermectin have been regularly administered for deworming; however, the efficacy of these regimens against *A. ceylanicum* in dogs remains unclear. Currently, many commercial drugs against *A. ceylanicum* infection in dogs have been tested at the laboratory level and exhibited greater than 99% efficacy within 3–7 days, including a combination of 15 mg/kg BW of febantel and 14.4 mg/kg BW of pyrantel embonate (Drontal® Plus, Bayer Animal health) [56], a spot-on of 2.5 mg/kg BW moxidectin (Advocate®, Bayer Animal health) [57], and 0.5 mg/kg BW milbemycin oxime (NexGard Spectra®, Boehringer Ingelheim) [58]. The clinical field assessment of the efficacy of anthelminthic agents against *A. ceylanicum* infection warrants further investigation.

5. Conclusions

In conclusion, this study provided the data, genetic characteristics, and evolutionary relationship of *A. ceylanicum* in Thailand and other Asian countries. We confirmed that *A. ceylanicum* was the predominant hookworm species in Northern Thailand, and the data fully supported the transmission dynamics from dogs to soil environments. Therefore, the awareness of *A. ceylanicum* transmission from dogs and soil in the temple should be raised. A proper schedule for the deworming of community dogs and the regular cleaning of the temple ground are suggested to reduce the risk of transmission of this zoonotic disease.

Supplementary Materials: The following are available online at http://www.mdpi.com/2076-2615/10/11/2154/s1: Table S1: Reference *Ancylostoma cox1* gene sequences used for phylogenetic analysis, Table S2: List of *A. ceylanicum* haplotypes based on a 259-bp fragment of the *cox1* gene by countries for population analysis, Table S3: Summary of hookworm infection in dog fecal samples and soil samples obtained from 53 temples as assessed by microscopic examination, Table S4: Hookworm detection in the soil based on the site of collection in temples in Thailand, Table S5: Haplotypes of *Ancylostoma* hookworms based on the *cox1* gene in Thailand based on sampling location.

Author Contributions: Conceptualization, S.T. (Saruda Tiwananthagorn); methodology, D.K., S.T. (Sahatchai Tangtrongsup), and S.T. (Saruda Tiwananthagorn); software, D.K., S.T. (Sahatchai Tangtrongsup), and S.T. (Saruda Tiwananthagorn); validation, S.T. (Sahatchai Tangtrongsup), and S.T. (Saruda Tiwananthagorn); formal analysis, D.K., S.T. (Sahatchai Tangtrongsup), and S.T. (Saruda Tiwananthagorn); investigation, D.K. and S.T. (Saruda Tiwananthagorn); resources, D.K. and S.T. (Saruda Tiwananthagorn); data curation, D.K., S.T. (Sahatchai Tangtrongsup), and S.T. (Saruda Tiwananthagorn); writing—original draft preparation, D.K., S.T. (Sahatchai Tangtrongsup), and S.T. (Saruda Tiwananthagorn); writing—review and editing, S.T. (Sahatchai Tangtrongsup) and S.T. (Saruda Tiwananthagorn); visualization, S.T. (Saruda Tiwananthagorn); supervision, S.T. (Sahatchai Tangtrongsup) and S.T. (Saruda Tiwananthagorn); project administration, S.T. (Saruda Tiwananthagorn); funding acquisition, S.T. (Saruda Tiwananthagorn). All authors have read and agreed to the published version of the manuscript.

Funding: This study was supported by the Excellent Center of Veterinary Bioscience, Faculty of Veterinary Medicine, Chiang Mai University and the research project "A multicenter study of dog and cat parasites in East and Southeast Asia; R000017642".

Acknowledgments: We would like to thank all the community dog owners for their valuable participation and the excellent assistance of Mrs. Duanghatai Sripakdee, Parasitological laboratory of Faculty of Veterinary Medicine, Chiang Mai University.

Conflicts of Interest: The authors declare no conflict of interest.

References

1. Liu, Y.; Shen, Y.; Zhang, T. Survey of the relationship between hookworm disease and socioeconomic factors. *Zhongguo Ji Sheng Chong Xue Yu Ji Sheng Chong Bing Za Zhi* **1999**, *17*, 399. [PubMed]

2. Bethony, J.; Brooker, S.; Albonico, M.; Geiger, S.M.; Loukas, A.; Diemert, D.; Hotez, P.J. Soil-transmitted helminth infections: Ascariasis, trichuriasis, and hookworm. *Lancet* **2006**, *367*, 1521–1532. [CrossRef]

3. Merino-Tejedor, A.; Nejsum, P.; Mkupasi, E.M.; Johansen, M.V.; Olsen, A. Molecular identification of zoonotic hookworm species in dog faeces from Tanzania. *J. Helminthol.* **2019**, *93*, 313–318. [CrossRef] [PubMed]

4. Gordon, C.A.; Kurscheid, J.; Jones, M.K.; Gray, D.J.; McManus, D.P. Soil-transmitted helminths in tropical Australia and Asia. *Trop. Med. Infect. Dis.* **2017**, *4*, 56. [CrossRef] [PubMed]

5. George, S.; Levecke, B.; Kattula, D.; Velusamy, V.; Roy, S.; Geldhof, P.; Sarkar, R.; Kang, G. Molecular identification of hookworm isolates in humans, dogs and soil in a tribal area in Tamil Nadu, India. *PLoS Negl. Trop. Dis.* **2016**, *10*, e0004891. [CrossRef]

6. Bowman, D.D.; Montgomery, S.P.; Zajac, A.M.; Eberhard, M.L.; Kazacos, K.R. Hookworms of dogs and cats as agents of cutaneous larva migrans. *Trends Parasitol.* **2010**, *26*, 162–167. [CrossRef] [PubMed]

7. Liu, Y.J.; Zheng, G.C.; Zhang, P.; Alsarakibi, M.; Zhang, X.H.; Li, Y.W.; Liu, T.; Ren, S.N.; Chen, Z.X.; Liu, Y.L.; et al. Molecular identification of hookworms in stray and shelter dogs from Guangzhou city, China using ITS sequences. *J. Helminthol.* **2015**, *89*, 196–202. [CrossRef]

8. Smout, F.A.; Skerratt, L.F.; Butler, J.R.A.; Johnson, C.N.; Congdon, B.C.; Thompson, R.C.A. The hookworm *Ancylostoma ceylanicum*: An emerging public health risk in Australian tropical rainforests and Indigenous communities. *One Health* **2017**, *3*, 66–69. [CrossRef]

9. Zibaei, M.; Nosrati, M.R.C.; Shadnoosh, F.; Houshmand, E.; Karami, M.F.; Rafsanjani, M.K.; Majidiani, H.; Ghaffarifar, F.; Cortes, H.C.E.; Dalvand, S.; et al. Insights into hookworm prevalence in Asia: A systematic review and meta-analysis. *Trans. R. Soc. Trop. Med. Hyg.* **2020**, *114*, 141–154. [CrossRef]

10. Palmer, C.S.; Traub, R.J.; Robertson, I.D.; Hobbs, R.P.; Elliot, A.; While, L.; Rees, R.; Thompson, R.C. The veterinary and public health significance of hookworm in dogs and cats in Australia and the status of *A. ceylanicum*. *Vet. Parasitol.* **2007**, *145*, 304–313. [CrossRef]

11. Traub, R.J. *Ancylostoma ceylanicum*, a re-emerging but neglected parasitic zoonosis. *Int. J. Parasitol.* **2013**, *43*, 1009–1015. [CrossRef] [PubMed]

12. Schar, F.; Inpankaew, T.; Traub, R.J.; Khieu, V.; Dalsgaard, A.; Chimnoi, W.; Chhoun, C.; Sok, D.; Marti, H.; Muth, S.; et al. The prevalence and diversity of intestinal parasitic infections in humans and domestic animals in a rural Cambodian village. *Parasitol. Int.* **2014**, *63*, 597–603. [CrossRef] [PubMed]

13. Kaya, D.; Yoshikawa, M.; Nakatani, T.; Tomo-Oka, F.; Fujimoto, Y.; Ishida, K.; Fujinaga, Y.; Aihara, Y.; Nagamatsu, S.; Matsuo, E.; et al. *Ancylostoma ceylanicum* hookworm infection in Japanese traveler who presented chronic diarrhea after return from Lao People's Democratic Republic. *Parasitol. Int.* **2016**, *65*, 737–740. [CrossRef] [PubMed]

14. Yoshikawa, M.; Ouji, Y.; Hirai, N.; Nakamura-Uchiyama, F.; Yamada, M.; Arizono, N.; Akamatsu, N.; Yoh, T.; Kaya, D.; Nakatani, T.; et al. *Ancylostoma ceylanicum*, novel etiological agent for traveler's diarrhea-report of four Japanese patients who returned from southeast Asia and Papua New Guinea. *Trop. Med. Health* **2018**, *46*, 6. [CrossRef] [PubMed]

15. Ngui, R.; Ching, L.S.; Kai, T.T.; Roslan, M.A.; Lim, Y.A. Molecular identification of human hookworm infections in economically disadvantaged communities in Peninsular Malaysia. *Am. J. Trop. Med. Hyg.* **2012**, *86*, 837–842. [CrossRef]

16. Traub, R.J.; Robertson, I.D.; Irwin, P.; Mencke, N.; Thompson, R.C. Application of a species-specific PCR-RFLP to identify *Ancylostoma* eggs directly from canine faeces. *Vet. Parasitol.* **2004**, *123*, 245–255. [CrossRef]

17. Romstad, A.; Gasser, R.B.; Nansen, P.; Polderman, A.M.; Chilton, N.B. *Necator americanus* (Nematoda: Ancylostomatidae) from Africa and Malaysia have different ITS-2 rDNA sequences. *Int. J. Parasitol.* **1998**, *28*, 611–615. [CrossRef]

18. Gasser, R.B.; Stewart, L.E.; Speare, R. Genetic markers in ribosomal DNA for hookworm identification. *Acta Trop.* **1996**, *62*, 15–21. [CrossRef]

19. Blouin, M.S. Molecular prospecting for cryptic species of nematodes: Mitochondrial DNA versus internal transcribed spacer. *Int. J. Parasitol.* **2002**, *32*, 527–531. [CrossRef]
20. Hu, M.; Chilton, N.B.; Gasser, R.B. The mitochondrial genomics of parasitic nematodes of socio-economic importance: Recent progress, and implications for population genetics and systematics. *Adv. Parasitol.* **2004**, *56*, 133–212. [CrossRef]
21. Avise, J.C.; Arnold, J.; Ball, R.M.; Bermingham, E.; Lamb, T.; Neigel, J.E.; Reeb, C.A.; Saunders, N.C. Intraspecific phylogeography: The mitochondrial DNA bridge between population genetics and systematics. *Annu. Rev. Ecol. Syst.* **1987**, *18*, 489–522. [CrossRef]
22. Pumidonming, W.; Salman, D.; Gronsang, D.; Abdelbaset, A.E.; Sangkaeo, K.; Kawazu, S.I.; Igarashi, M. Prevalence of gastrointestinal helminth parasites of zoonotic significance in dogs and cats in lower Northern Thailand. *J. Vet. Med. Sci.* **2017**, *78*, 1779–1784. [CrossRef]
23. Tangtrongsup, S.; Scorza, A.V.; Reif, J.S.; Ballweber, L.R.; Lappin, M.R.; Salman, M.D. Seasonal distributions and other risk factors for *Giardia duodenalis* and *Cryptosporidium* spp. infections in dogs and cats in Chiang Mai, Thailand. *Prev. Vet. Med.* **2020**, *174*, 104820. [CrossRef] [PubMed]
24. Piangjai, S.; Sukontason, K.; Sukontason, K.L. Intestinal parasitic infections in hill-tribe schoolchildren in Chiang Mai, northern Thailand. *Southeast Asian J. Trop. Med. Public Health* **2003**, *34* (Suppl. 2), 90–93.
25. Yanola, J.; Nachaiwieng, W.; Duangmano, S.; Prasannarong, M.; Somboon, P.; Pornprasert, S. Current prevalence of intestinal parasitic infections and their impact on hematological and nutritional status among Karen hill tribe children in Omkoi District, Chiang Mai Province, Thailand. *Acta Trop.* **2018**, *180*, 1–6. [CrossRef] [PubMed]
26. Energy Policy and Plannning Office, Ministry of Energy. The Provinces and Administrative Areas of Thailand. Available online: http://www.e-report.energy.go.th/area.html (accessed on 31 May 2019).
27. Royal Thai Government Gazette, The Secretariat of the Cabinet. Thailand Population. 2018. Available online: http://www.ratchakitcha.soc.go.th/DATA/PDF/2562/E/036/T_0032.PDF (accessed on 31 May 2019).
28. Thrusfield, M.; Ortega, C.; de Blas, I.; Noordhuizen, J.P.; Frankena, K. WIN EPISCOPE 2.0: Improved epidemiological software for veterinary medicine. *Vet. Rec.* **2001**, *148*, 567–572. [CrossRef] [PubMed]
29. Tiwananthagorn, S.; Chiang Mai University, Chiang Mai, Thailand. Unpublished Work. 2018.
30. Bureau of Disease Control and Veterinary Services, Department of Livestock Development. *Dog Population in Thailand*. 2016. Available online: http://dcontrol.dld.go.th/webnew/index.php/th/news-menu/2018-07-04-04-12-47/rabies/360-dogpop2016 (accessed on 31 May 2019).
31. Moxham, G. The WALTHAM faeces scoring system—A tool for veterinarians and pet owners: How does your pet rate? *Walth. Focus* **2001**, *11*, 24–25.
32. Gibbons, L.M.; Jones, A.; Khalil, L.F. *Manual for the 8th International Training Course on Identification of Helminth Parasites of Economic Importance*; CABI Institue of Parasitology, Wallingford: Oxon, UK, 1996.
33. Inpankaew, T.; Schar, F.; Dalsgaard, A.; Khieu, V.; Chimnoi, W.; Chhoun, C.; Sok, D.; Marti, H.; Muth, S.; Odermatt, P.; et al. High prevalence of *Ancylostoma ceylanicum* hookworm infections in humans, Cambodia, 2012. *Emerg. Infect. Dis.* **2014**, *20*, 976–982. [CrossRef]
34. Hall, T. BioEdit: A user-friendly biological sequence alignment editor and analysis program for Window 95/98/NT. *Nucleic Acids Symp. Ser.* **1996**, *41*, 95–98.
35. Kumar, S.; Stecher, G.; Li, M.; Knyaz, C.; Tamura, K. MEGA X: Molecular evolutionary genetics analysis across computing platforms. *Mol. Biol. Evol.* **2018**, *35*, 1547–1549. [CrossRef]
36. Rozas, J.; Ferrer-Mata, A.; Sanchez-DelBarrio, J.C.; Guirao-Rico, S.; Librado, P.; Ramos-Onsins, S.E.; Sanchez-Gracia, A. DnaSP 6: DNA sequence polymorphism analysis of large data sets. *Mol. Biol. Evol.* **2017**, *34*, 3299–3302. [CrossRef] [PubMed]
37. Excoffier, L.; Lischer, H.E. Arlequin suite ver 3.5: A new series of programs to perform population genetics analyses under Linux and Windows. *Mol. Ecol. Resour.* **2010**, *10*, 564–567. [CrossRef] [PubMed]
38. Bandelt, H.J.; Forster, P.; Rohl, A. Median-joining networks for inferring intraspecific phylogenies. *Mol. Biol. Evol.* **1999**, *16*, 37–48. [CrossRef] [PubMed]

39. Mahdy, M.A.; Lim, Y.A.; Ngui, R.; Siti Fatimah, M.R.; Choy, S.H.; Yap, N.J.; Al-Mekhlafi, H.M.; Ibrahim, J.; Surin, J. Prevalence and zoonotic potential of canine hookworms in Malaysia. *Parasit Vectors* **2012**, *5*, 88. [CrossRef]

40. Conlan, J.V.; Khamlome, B.; Vongxay, K.; Elliot, A.; Pallant, L.; Sripa, B.; Blacksell, S.D.; Fenwick, S.; Thompson, R.C. Soil-transmitted helminthiasis in Laos: A community-wide cross-sectional study of humans and dogs in a mass drug administration environment. *Am. J. Trop. Med. Hyg.* **2012**, *86*, 624–634. [CrossRef]

41. Schrader, C.; Schielke, A.; Ellerbroek, L.; Johne, R. PCR inhibitors—occurrence, properties and removal. *J. Appl. Microbiol.* **2012**, *113*, 1014–1026. [CrossRef]

42. Ngui, R.; Mahdy, M.A.; Chua, K.H.; Traub, R.; Lim, Y.A. Genetic characterization of the partial mitochondrial cytochrome oxidase *c* subunit I *(cox1)* gene of the zoonotic parasitic nematode, *Ancylostoma ceylanicum* from humans, dogs and cats. *Acta Trop.* **2013**, *128*, 154–157. [CrossRef]

43. Marques, J.P.; Guimaraes Cde, R.; Boas, A.V.; Carnauba, P.U.; Moraes, J. Contamination of public parks and squares from Guarulhos (Sao Paulo State, Brazil) by *Toxocara* spp. and *Ancylostoma* spp. *Rev. Inst. Med. Trop. Sao Paulo* **2012**, *54*, 267–271. [CrossRef]

44. Tun, S.; Ithoi, I.; Mahmud, R.; Samsudin, N.I.; Kek Heng, C.; Ling, L.Y. Detection of helminth eggs and identification of hookworm species in stray cats, dogs and soil from Klang Valley, Malaysia. *PLoS ONE* **2015**, *10*, e0142231. [CrossRef]

45. Ferreira, A.; Alho, A.M.; Otero, D.; Gomes, L.; Nijsse, R.; Overgaauw, P.A.M.; Madeira de Carvalho, L. Urban dog parks as sources of canine parasites: Contamination rates and pet owner behaviours in Lisbon, Portugal. *J. Env. Public Health* **2017**, *2017*, 5984086. [CrossRef]

46. Paller, V.G.V.; Babia-Abion, S. Soil-transmitted helminth (STH) eggs contaminating soils in selected organic and conventional farms in the Philippines. *Parasite Epidemiol. Control.* **2019**, *7*, e00119. [CrossRef]

47. Traub, R.J.; Inpankaew, T.; Sutthikornchai, C.; Sukthana, Y.; Thompson, R.C. PCR-based coprodiagnostic tools reveal dogs as reservoirs of zoonotic ancylostomiasis caused by *Ancylostoma ceylanicum* in temple communities in Bangkok. *Vet. Parasitol.* **2008**, *155*, 67–73. [CrossRef]

48. Smith, L.M.; Hartmann, S.; Munteanu, A.M.; Dalla Villa, P.; Quinnell, R.J.; Collins, L.M. The effectiveness of dog population management: A systematic review. *Animals* **2019**, *9*, 1020. [CrossRef] [PubMed]

49. Posada, D.; Crandall, K.A. Intraspecific gene genealogies: Trees grafting into networks. *Trends Ecol. Evol.* **2001**, *16*, 37–45. [CrossRef]

50. Guo, L.; Sun, B.; Sang, F.; Wang, W.; Lu, Z. Haplotype distribution and evolutionary pattern of miR-17 and miR-124 families based on population analysis. *PLoS ONE* **2009**, *4*, e7944. [CrossRef] [PubMed]

51. Wang, G.D.; Zhai, W.; Yang, H.C.; Wang, L.; Zhong, L.; Liu, Y.H.; Fan, R.X.; Yin, T.T.; Zhu, C.L.; Poyarkov, A.D.; et al. Out of southern East Asia: The natural history of domestic dogs across the world. *Cell Res.* **2016**, *26*, 21–33. [CrossRef] [PubMed]

52. Witt, K.E.; Judd, K.; Kitchen, A.; Grier, C.; Kohler, T.A.; Ortman, S.G.; Kemp, B.M.; Malhi, R.S. DNA analysis of ancient dogs of the Americas: Identifying possible founding haplotypes and reconstructing population histories. *J. Hum. Evol.* **2015**, *79*, 105–118. [CrossRef] [PubMed]

53. Loeurng, V.; Ichikawa-Seki, M.; Wannasan, A.; Sothyra, T.; Chaisowwong, W.; Tiwananthagorn, S. Genetic characterization of cambodian *Fasciola gigantica* and dispersal direction of the species in Asia. *Vet. Parasitol.* **2019**, *273*, 45–51. [CrossRef]

54. Biocca, E.; Chabaud, A. Redescription of *Seuratum mucronatum* (Nematoda-Cucullanidae). *Ann. Parasitol. Hum. Comp.* **1951**, *26*, 85–92. [CrossRef]

55. Smout, F.A.; Thompson, R.C.; Skerratt, L.F. First report of *Ancylostoma ceylanicum* in wild canids. *Int. J. Parasitol. Parasites Wildl.* **2013**, *2*, 173–177. [CrossRef]

56. Taweethavonsawat, P.; Chungpivat, S.; Satranarakun, P.; Traub, R.J.; Schaper, R. Efficacy of a combination product containing pyrantel, febantel and praziquantel (Drontal® Plus Flavour, Bayer Animal Health) against experimental infection with the hookworm *Ancylostoma ceylanicum* in dogs. *Parasitol. Res.* **2010**, *106*, 533–537. [CrossRef] [PubMed]

57. Taweethavonsawat, P.; Chungpivat, S.; Satranarakun, P.; Traub, R.J.; Schaper, R. Experimental infection with *Ancylostoma ceylanicum* in dogs and efficacy of a spot on combination containing imidacloprid 10% and moxidectin 2.5% (Advocate®/Advantage Multi, Bayer Animal Health). *Parasitol. Res.* **2010**, *106*, 1499–1502. [CrossRef] [PubMed]

58. Tielemans, E.; Lebon, W.; Dumont, P.; Taweethavonsawat, P.; Larsen, D.; Rehbein, S. Efficacy of afoxolaner plus milbemycin oxime chewable tablets (NexGard Spectra®, Merial) against adult *Ancylostoma ceylanicum* hookworm, in dogs. *Vet. Parasitol.* **2017**, *238*, 87–89. [CrossRef] [PubMed]

Publisher's Note: MDPI stays neutral with regard to jurisdictional claims in published maps and institutional affiliations.

animals

Article

Geographic Spatial Distribution Patterns of *Dirofilaria immitis* and *Brugia pahangi* Infection in Community Dogs in Chiang Mai, Thailand

Manusvee Kaikuntod [1], Orapun Arjkumpa [2], Doolyawat Kladkempetch [3], Shinya Fukumoto [4], Kriangkrai Thongkorn [1], Chavalit Boonyapakorn [1], Veerasak Punyapornwithaya [5,6,*] and Saruda Tiwananthagorn [3,7,*]

1   Department of Companion Animals and Wildlife Clinic, Faculty of Veterinary Medicine, Chiang Mai University, Chiang Mai 50100, Thailand; manusvee.k@gmail.com (M.K.); kktk23@hotmail.com (K.T.); chavalitboonyapakorn@gmail.com (C.B.)
2   Phra Nakhon Si Ayutthaya Provincial Livestock Office, Thanu Subdistrict, Uthai District, Phra Nakhon Si Ayutthaya 13000, Thailand; arjkumpa@hotmail.com
3   Department of Veterinary Biosciences and Veterinary Public Health, Faculty of Veterinary Medicine, Chiang Mai University, Chiang Mai 50100, Thailand; doolyawat_kl@cmu.ac.th
4   National Research Center for Protozoan Diseases, Obihiro University of Agriculture and Veterinary Medicine, Obihiro 080-8555, Hokkaido, Japan; fukumoto@obihiro.ac.jp
5   Department of Food Animal Clinic, Faculty of Veterinary Medicine, Chiang Mai University, Chiang Mai 50100, Thailand
6   Research Group for Veterinary Public Health, Faculty of Veterinary Medicine, Chiang Mai University, Chiang Mai 50100, Thailand
7   Research Center of Producing and Development of Products and Innovations for Animal Health and Production, Faculty of Veterinary Medicine, Chiang Mai University, Chiang Mai 50100, Thailand
*   Correspondence: veerasak.p@cmu.ac.th (V.P.); saruda.t@cmu.ac.th (S.T.); Tel.: +66-53-948-023 (V.P.); +66-53-948-046 or +66-95-446-5955 (S.T.)

Citation: Kaikuntod, M.; Arjkumpa, O.; Kladkempetch, D.; Fukumoto, S.; Thongkorn, K.; Boonyapakorn, C.; Punyapornwithaya, V.; Tiwananthagorn, S. Geographic Spatial Distribution Patterns of *Dirofilaria immitis* and *Brugia pahangi* Infection in Community Dogs in Chiang Mai, Thailand. *Animals* **2021**, *11*, 33. https://dx.doi.org/10.3390/ani11010033

Received: 24 November 2020
Accepted: 23 December 2020
Published: 26 December 2020

Publisher's Note: MDPI stays neutral with regard to jurisdictional claims in published maps and institutional affiliations.

Simple Summary: Filariasis is emerging as a public health concern for humans, dogs, cats, and other wildlife species, and is frequently found in southeast Asian countries. The present study confirmed the species of filarial nematodes in free-roaming dogs from temple communities. Two species were found: *Dirofilaria immitis* infection and, for the first time, *Brugia pahangi*. The occurrence of the two species was comparable. Geographic spatial distribution revealed the abundance of *D. immitis* and *B. pahangi* in the central areas at altitudes less than 400 m. However, at higher altitudes between 400 and 800 m, we found a significantly higher number of *B. pahangi* infections than *D. immitis* infections. In conclusion, *D. immitis* and *B. pahangi* were the most common filarial infections found in community dogs in Northern Thailand. Dogs might be an important reservoir for *B. pahangi* in that region. The population dynamics of the mosquito vector of *B. pahangi* across altitudinal gradients merits further study.

Abstract: Filariasis is emerging as a public health concern in tropical and subtropical areas. Filariasis is an endemic problem commonly found in southeast Asian countries. Using the PCR-restriction fragment length polymorphism (PCR-RFLP) of the ITS1 region with *Vsp* I, the overall prevalence rates of *Dirofilaria immitis* (12.2% (41/337); 95% confidence interval: 9.1–16.1%) and *Brugia pahangi* (8.3% (28/337); 95% confidence interval: 5.8–11.8%) were determined based on 337 free-roaming community dogs from 20 districts in Northern Thailand. Microfilaremia was found in only 6.2% of dogs (21/337). Co-infection with *D. immitis* and *B. pahangi* was observed in two dogs. Of the 215 blood samples examined using a Canine Heartworm Ag Kit, only 3.72% (eight dogs) were *D. immitis* antigen positive. Among these eight, six dogs had occult *D. immitis* infections. In terms of geographic distribution, we found the abundance of *D. immitis* and *B. pahangi* in the central areas at altitudes less than 400 m to be 12.1% and 10.3%, respectively. In contrast, at higher altitudes between 400 and 800 m, a significantly higher number of *B. pahangi* compared with *D. immitis* infected individuals were observed at 14.29% and 4.1%, respectively. In conclusion, *D. immitis* and *B. pahangi* were the most common filarial infections found in community dogs in Northern Thailand. Dogs might be an important reservoir of *B. pahangi* in that region. Increasing awareness and concern and including proper deworming programs for community dogs should be endorsed to reduce the transmission risk. Additionally, the population dynamics of the mosquito vector of *B. pahangi* across altitudinal gradients deserved further investigation.

Keywords: *Brugia pahangi*; *Dirofilaria immitis*; PCR-RFLP; community dogs; spatial distribution; altitude

1. Introduction

Filarial nematode infection is an important vector-borne disease in tropical countries. According to the World Health Organization (WHO) roadmap (WHO 2020), the goal of eliminating filariasis is expected to be achieved by 2030. Filarial infections are currently common in companion animals worldwide [1]. Specifically, in Southeast Asia, lymphatic filariasis caused by *Brugia malayi* (Brugian or Malayan filariasis), *Wuchereria bancrofti* (bancroftian filariasis) and *Brugia timori* is considered a significant human health problem. Filariasis in dogs is caused by various species, e.g., *Dirofilaria* spp., *Acanthocheilonema* spp., and *Brugia* spp. [2–4]. Typically, the life cycle of the filarial worm requires a bloodsucking insect such as mosquitoes as a transmission vector, known as an intermediate host. The ingested microfilariae (L1) from the dog develop into the infective stage larvae (L3) in the vector. However, the worm has a species–specific target organ within the final host [5].

Canine heartworm disease, *D. immitis* infection, is one of the serious types of filariasis observed in veterinary practice. Adult worms commonly live in pulmonary arteries and congestive heart failure can occur in severe cases. *Dirofilaria repens* and *Acanthocheilonema reconditum* are other [6] filarial worms usually located in subcutaneous tissues or subconjunctival areas. *Brugia* spp., such as *Brugia pahangi* and *Brugia malayi*, are known to cause lymphatic filariasis (LF), whereas *B. malayi* causes a serious lymphatic obstruction in humans. Whether *B. pahangi* is a causative agent of human disease in the natural environment is not yet known [7]. However, microfilaria in the blood, as well as signs and symptoms of LF, were detected in experimentally infected human volunteers [8]. A clinical description of lymphatic filariasis caused by natural infection with *B. pahangi* in Malaysia was recently published [9]. Dogs and cats are reservoir hosts of *B. malayi*, but they are the essential hosts of *B. pahangi* [10].

Currently, several methods are employed to diagnose filarial infections. The basic conventional technique generally uses microscopic examination to detect microfilariae. The microhematocrit centrifugation technique, or Woo's test, is easy to perform, quick, and inexpensive [11,12], but it cannot identify the species of the parasites. An alternative approach for the identification of filarial worms is DNA technology. Polymerase chain reaction-restriction fragment length polymorphism (PCR-RFLP) analysis is a molecular technique that involves cutting the specific target of the PCR product with a restriction endonuclease. According to Nuchprayoon et al. [13], differentiation of a wide variety of species of filarial worms was accomplished using PCR-RFLP analysis of the internal transcribed spacer 1 (ITS1). Rishniw et al. [3] reported that using the pan-filarial primers DIDR-F1 and DIDR-R1 in a single PCR test could discriminate among six discordant microfilaria species, including *D. immitis*, *D. repens*, *A. reconditum*, *B. malayi*, and *B. pahangi*. The species–specific (for *D. immitis*) PCR-targeting cytochrome *c* oxidase subunit 1 (*cox1*) gene has been used to confirm *D. immitis* infection.

In tropical countries, including Thailand, the endemic areas for *D. immitis*, *B. malayi*, and *B. pahangi* have been reported, including different species of filarial infection [14,15]. *D. immitis* infection was found throughout Thailand and has a higher prevalence in stray dogs than in pet dogs [16]. *Brugia* spp. infection was reported to be found mostly in the southern part of Thailand [17,18]. In the Chiang Mai province in Northern Thailand, canine filariasis showed a prevalence of 18.2% in 2008; however, this study lacked reliable filarial species identification [19]. In addition, *Brugia* infection in dogs has never been confirmed in Northern Thailand. The WHO recommended using a mapping method for delimiting areas requiring mass drug treatment to save resources [20]; however, to date, no spatial clustering of vector-borne diseases of dogs has been detected in this region. The present study was aimed at updating the occurrence of filarial infection in free-roaming community dogs in the province of Chiang Mai, confirming filarial species using molecular techniques and documenting the geographic distribution of *D. immitis* and *B. pahangi* using spatial mapping in an effort to augment disease control efforts in animals and humans, thus reducing public health concerns.

2. Materials and Methods

2.1. Study Area, Sample Collection, and Ethical Concerns

The study area was in the province of Chiang Mai in the upper northern region of Thailand, which is considered the province with the highest dog population density in the nation (278,943 dogs in Chiang Mai of 645,368 total dogs in the region, Bureau of Disease Control and Veterinary Services 2016). The study area was divided into three zones (northern, central, and southern; Figure 1A), encompassing 146 community temples as well as households distributed over 20 of 25 districts and 72 subdistricts to include a wide range of geographical features from north to south of Chiang Mai. The 20 districts included six districts in the northern part of the province (Chiang Dao, Chai Prakan, Fang, Mae Rim, Mae Taeng, and Phrao), 10 districts in the central part (Doi Saket, Hang Dong, Mae On, Mae Wang, Mueang Chiang Mai, Samoeng, San Kamphaeng, San Pa Tong, San Sai, and Saraphi), and four districts in the southern part (Chom Thong, Doi Lo, Doi Tao, and Hot). GPS location was determined (Map Plus TM version 2.4, mobile application) for each sampling site, including coordinates (latitude and longitude) and altitude (meters). Samples were collected from June to December 2019. According to a government announcement (Thai Meteorological Department 2018), this sampling period could be defined as spanning the rainy (June to October) and cold seasons (November to February).

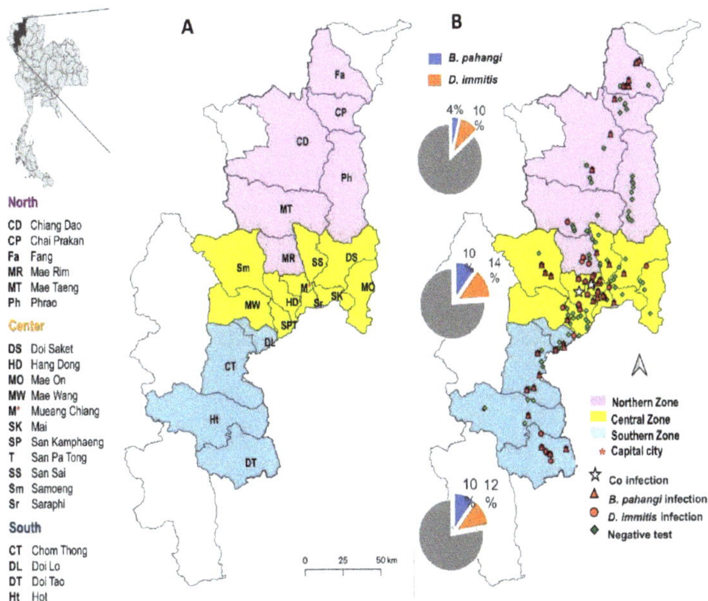

Figure 1. Study sites and geographic distribution of canine filariasis in three zones of Chiang Mai, Thailand. (**A**). The study districts; (**B**). Distribution of *Brugia pahangi* and *Dirofilaria immitis* positives and the proportion of prevalence of filarial nematode species in each zone.

A total of 337 blood samples were collected from the community dogs (the dogs owned by specific owners, able to be free ranging in the community, and taken care of by people in the communities), including 168 males and 169 females. Inclusion criteria included age >7 months and having not received heartworm prevention medication. From each dog, 2 mL of blood was collected from either the cephalic or saphenous vein which was kept in EDTA tubes. The first 0.5 mL of blood was used for Woo's examination for the presence of microfilaria. The remaining 1.5 mL was kept in a freezer at −20 °C for further molecular analysis.

All dog owners (temple master or dog keepers) signed an informed consent form, and the treatment of the animals was approved by the Animal Care and Use Committee, Faculty of Veterinary Medicine, Chiang Mai University (R15/2562) on 12 June 2019.

2.2. Laboratory Examination

2.2.1. Detection of Circulating Microfilaria and *D. immitis* Antigens

The whole blood samples were examined for total circulating microfilariae using the microhematocrit centrifugation technique (Woo's technique) [11]. In addition, the packed cell volume (PCV) was also measured as a percent. Measurement of the severity of anemia in this study followed the guidelines for classification of severity of anemia from the World Small Animal Veterinary Association (WSAVA) 2005 [21]. A total of 215 whole blood samples were randomly selected for additional study to detect the adult female *D. immitis* antigen using a Thinka Canine Heartworm Ag Test kit (Thinka CHW, Arkray, Kyoto, Japan) following the manufacturer's instructions. For quality control, the blood samples were examined within 48 h after blood collection.

2.2.2. Molecular Techniques and Sequencing

Genomic DNA (gDNA) was extracted from 200 μL anticoagulated blood samples. A commercial DNA extraction kit (Nucleospin® Blood, Macherey-Nagel, Duren, Germany) was used following the manufacturer's instructions.

Various diagnostic tools have been developed for the precise diagnosis of filaria infection. PCR-RFLP—which is based on a diagnostic method previously designed by Nuchprayoon et al. [13] and targets the ITS1 region (primer: ITS1-F and ITS1-R; Table 1)—was conducted. PCR amplification reactions were performed in a 20 μL reaction volume containing 100 ng gDNA (3–5 μL), 0.2 μM of each primer (0.4 μL of 10 μM), and 10 μL of 2 × Quick Taq® HS DyeMix (TOYOBO, Osaka, Japan). PCR products were analyzed on a 1.5% agarose gel electrophoresis. The positive ITS1 PCR products were then digested with five units of Vsp I (SibEnzyme, Novosibirsk, Russia) as a restriction endonuclease, according to the manufacturer's protocol. One unit of Vsp I was used to digest 1 μg of the ITS1 PCR product for 1 h at 37 °C in a total reaction volume of 50 μL. DNA fragment analysis was conducted by 1.5% agarose gel electrophoresis, stained with RedSafe™ nucleic acid staining solution (iNtRON Biotechnology, Gyeonggi-do, Korea), and visualized using a GelMax™ Imager (Ultra-Violet Products, Cambridge, UK).

Two other PCR techniques were used to confirm the presence of filarial DNA in the suspicious samples, including PCR of pan-filarial primers (DIDR-F1 and DIDR-R1) to discriminate between *A. reconditum* and *Amphiachyris dracunculoides*, and primers specific for *D. immitis* (DI COI-F1 and DI COI-R1) to confirm *D. immitis* DNA in suspected occult infections. The PCR procedure was performed following previously published methods [3] and using the same reagent previously described for the PCR of ITS1. Distilled water (DW) as a negative control and the gDNA of *D. immitis*, *B. pahangi*, and *B. malayi* as positive controls were included in the analysis.

Species confirmation in the 10 previous *D. immitis*- and *B. pahangi*-positive samples was achieved by DNA sequencing of the 5.8s-ITS2-28S amplicons using pan-filarial primers given the clear and distinctive amplicon. The PCR product (50 μL) was purified using a NucleoSpin® PCR Clean-up Kit (Macherey-Nagel GmbH, Duren, Germany) and submitted for direct fluorescent dye-terminator sequencing in the sense and antisense directions by BioBasic Inc. (The Elitist, Singapore).

Table 1. Primer pairs used for PCR amplification.

Gene Target	Primer Pairs	Primer Sequence (5′-3′)	Filarial Species	Product Size (bp)
ITS1	ITS1-F ITS1-R	GGT GAA CCT GCG GAA GGA TC GAG TTA CGC AGA CGT TAA GCG	*W. bancrofti* *B. malayi* *B. pahangi* *D. immitis* *D. repens*	482 504 510 595 602
5.8S-ITS2-28S	DIDR-F1 DIDR-R1	AGT GCG AAT TGC AGA CGC ATT GAG AGC GGG TAA TCA CGA CTG AGT TGA	*D. immitis* *A. reconditum* *D. repens* *A. dracunculoides* *B. pahangi* *B. malayi*	542 578 484 584 664 615
COI	DI COI -F1 DI COI-R1	AGT GTA GAG GGT CAG CCT GAG TTA ACA GGC ACT GAC AAT ACC AAT	*D. immitis*	203

2.3. Data and Statistical Analysis

2.3.1. Determination of the Presence and Prevalence of Filariasis

Diagnostic steps for identification and determination of the prevalence of filariasis are presented in Figure 2. The determination of the presence of filariasis was based on three diagnostic techniques: (1) *D. immitis* infection when positive with CHW Ag kit and/or *D. immitis*-DNA-positive with any molecular technique; (2) occult *D. immitis* infection when positive with CHW Ag kit but negative with any PCR; (3) *B. pahangi* infection when *B. pahangi*-DNA-positive with any molecular technique. Descriptive statistics summarizing the prevalence of filarial worm infection are presented as a percentage of each species. Differences in prevalence between groups or categories were analyzed using proportion tests. All statistical analyses were performed using R software [22] and *p*-values < 0.05 were considered statistically significant.

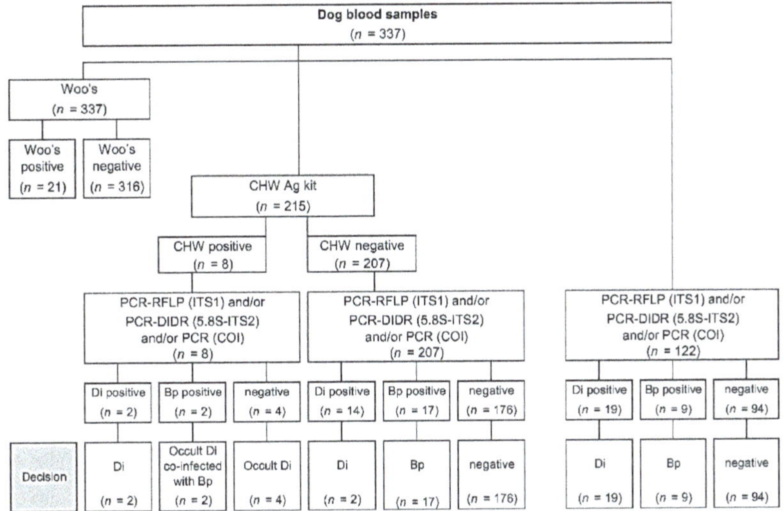

Figure 2. Diagnostic diagram of canine filariasis using CHW Ag test kit and three molecular techniques: PCR-RFLP of ITS1 region, PCR of 5.8S-ITS2-28S region, and PCR of *cox*1 gene. Di: *D. immitis*; Bp: *Brugia pahangi*.

2.3.2. Sequencing and Phylogenetic Analyses

The obtained sequences were compared with the previously deposited ones in GenBank using the BLASTn program [23] for specific diagnosis. Seven successful 5.8S-ITS2 nucleotide sequences originating from two *D. immitis* isolates (Di CM329 and CM331) and five isolates of *B. pahangi* (Bp-CM22, CM188, CM189, CM328, and CM337) (Table S1) were used to analyze the phylogenetic relationship among the regions. In addition, the available sequences of *D. immitis* and *B. pahangi* from other countries (Brazil, Bulgaria, China, India, Iran, Lithuania, Portugal, Taiwan, Thailand, Turkey, Tunisia, and the U.S.; Table S1) were compared with the sequences obtained in this study for the construction of a phylogenetic tree. The sequences of *B. malayi*, *Onchocerca volvulus*, *A. reconditum* (previously *Dipetalonema reconditum*) and *D. repens* (GenBank accession: EU373624, EU272179, AF217801, and MK942385, respectively), and the sequence of *Setaria digitata* (EF196091) as the outgroup were included for evolution analysis. All sequences obtained in this study were submitted to the DDBJ/EMBL/GenBank database (Table S1).

Evolutionary analyses were conducted using MEGA X [24]. Multiple sequences were aligned using ClustalW and a phylogenetic tree using a maximum likelihood (ML) method based on the Tamura–Nei model [25]. The Neighbor-Join and BioNJ algorithms were applied to a matrix of pairwise distances estimated using the Tamura-Nei model, and then the topology was selected with superior log likelihood value. A consensus tree was obtained after bootstrap analysis of 100 replications. The tree is drawn to scale, with branch lengths indicating the number of substitutions per site. Thirty nucleotide sequences were included in this analysis. Codon positions included were first, second, third, and noncoding.

2.3.3. Geographical Information System Mapping and Distribution by Altitude

Geographical Information System (GIS) locations (latitude, longitude, and altitude) from a mobile phone application were input to a computer. The digital elevation model (DEM) was obtained from the NASA Earth data website and used to create a raster image and to convert the model into contour lines representing altitude in the GIS application (QGIS software version 3.6, GNU General Public License). The altitudes of the infected sites were measured using a GPS. Altitudes of the positive prevalence spots were divided into three classes modified from those used by Devi and Jauhari [26]: below 400, 400–800, and over 800 m. A proportional test was used to determine differences in prevalence of *B. pahangi* and *D. immitis* among altitude classes using R software [22] and p-values < 0.05 were considered statistically significant. Geographic distribution by altitude was generated from geographic information systems and GPS tracking data using QGIS software version 3.6. Notably, NASA Earth data indicates that all NASA data are available without restrictions.

2.3.4. Geographic Spatial Cluster Analysis

Spatial cluster analyses of the positive *D. immitis* and *B. pahangi* cases were attained using a Bernoulli model and SaTScan™ v9.6 software [27]. Generally, this model requires data from cases and controls. The definition of a case was a positive filarial nematode using any of the three diagnostic techniques (Figure 2), and a negative individual was the control. In the study area, 50% of the total population was set up in the spatial scanning window. The Monte Carlo hypothesis testing technique (number of replications = 999) was used to determine the statistical significance of the cluster. The primary clusters of *D. immitis* and *B. pahangi* were classified based on the highest log-likelihood ratio (LLR), in which the likelihood function of the Bernoulli model is:

$$\left(\frac{c}{n}\right)^{c}\left(\frac{n-c}{c}\right)^{n-c}\left(\frac{C-c}{N-n}\right)^{C-c}\left(\frac{(N-n)-(C-c)}{N-n}\right)^{(N-n)-(C-c)}I(),\qquad(1)$$

where C is the total number of cases, N is the combined total number of cases, the control c is the observed number of cases within the window, n is the total number of cases and controls within the window, and $I()$ is the indicator of function I. As the analysis was only focused on detecting clusters with higher than expected rates, $I()$ was set to 1 [27]. The illustrated spatial cluster of each species combined with the altitude and the geographic functional land use was generated by QGIS software version 3.6.

3. Results

3.1. Prevalence of Filariasis Using a Combination of Conventional and Molecular Techniques

Various diagnostic techniques were employed to identify filarial species infection and molecular techniques were applied to identify the species of filarial nematodes. Of the total of 337 blood samples, microfilaria was detected in only 21 dogs (6.23%) using Woo's method and most of those were from the central zone (Table 2). Of the 215 blood samples examined using a CHW Ag kit, only eight dogs (3.72%) were *D. immitis* antigen positive (Figure 2). Only three CHW Ag-positive dogs had microfilaremia, of which molecular techniques could discriminate one positive with *D. immitis* and two positives with *B. pahangi*. The overall prevalence of *D. immitis* infection was 12.17% (41/337 dogs), higher than the prevalence of 8.31% for *B. pahangi* (28/337 dogs). However, the difference in infection rate between the two species was not significant ($p > 0.05$). The distributions of *D. immitis* and *B. pahangi* infection in the three zones of Chiang Mai province are shown in Table 2, Table S2, and Figure 1B. The observed *D. immitis* and *B. pahangi* infection incidence was highest in the central zone at 22.96% (31/135), followed by the south zone at 22.45% (22/98). However, the prevalence rates of *D. immitis* and *B. pahangi* infection among the three zones were not significantly different ($p > 0.05$). Two dogs in the central zone had dual infections with *D. immitis* and *B. pahangi*. No adult *D. immitis* antigens were detected in the two *D. immitis* DNA-positive samples. Finally, an important proportion of occult heartworm infections (6/8; 75%) was distinguished (Figure 2).

Table 2. Prevalence of microfilariae and filarial nematode species in the community dogs among three zones of Chiang Mai, Thailand.

Zone	No. of Examined	Microfilariae (Woo's Method)	Filarial DNA (Molecular Techniques)			
		% Positive (no.)	*B. pahangi* Infection (*n1*)	*D. immitis* Infection (*n2*)	Dual Infection (*n3*)	Total (*n1 + n2 − n3*)
North	104	2.88 (3) [a,b]	3.85 (4)	9.62 (10)	0	13.46 (14)
Central	135	11.11 (15) [b]	10.37 (14)	14.07 (19)	2	22.96 (31)
South	98	3.06 (3) [a]	10.20 (10)	12.24 (12)	0	22.45 (22)
Total	337	6.23 (21)	8.31 (28)	12.17 (41)	2	19.88 (67)

[a,b] values in the same column with different superscripts are statistically different ($p < 0.05$); *n1*, *n2*, *n3* are the number of positive cases.

The range of the percent PCV (%PCV) of dogs with positive microfilariae was 15–46%, with a mean $\pm$ SD of 32.48 $\pm$ 8.38 and a mode of 31%. Fourteen dogs with positive microfilariae (71.43%; 15/21) had a %PCV lower than 37%, which is considered anemic. Anemic status was categorized as mild (30–37% PCV; 9/21, 42.86%), moderate (20–29% PCV; 4/21, 19.05%), and severe (13–19% PCV; 2/21, 9.52%). No icteric plasma was observed in any of the samples.

3.2. Phylogenetic Relationship of D. immitis and B. pahangi

The nucleotide sequences of the partial 5.8S rRNA and ITS2 region contained approximately 300–350 bp that allowed the filarial nematodes to be classified as *D. immitis* or *B. pahangi*. The nucleotide sequences of five isolates of *B. pahangi* were identical and grouped in a haplotype that showed 100% identity with *B. pahangi* (AY988600; [3]) and 95.67% with isolates from cats from the Narathiwat province in Southern Thailand (EU373655; [28]). However, two nucleotide sequences of *D. immitis* had some nucleotide differences, which showed 95–100% identity with the *D. immitis* references (Table S1). The nucleotide sequences of *D. immitis* and *B. pahangi* in this study were deposited in the

DDBJ/EMBL/GenBank database, including *D. immitis* (accession No. LC554219-554220) and *B. pahangi* (accession No. LC554214-554218 (Table S1)).

A comparative genomic analysis of *D. immitis* and *B. pahangi* was performed to investigate the phylogenetic relationship among the different geographic regions (Figure 3). The phylogenetic tree based on the partial 5.8S and ITS2 sequence showed that all isolated samples from *D. immitis* clustered together in one group and were similar to *D. immitis* from China (EU182331), Brazil (KX932106), Iran (JX889636), Bulgaria (MN596213), and Turkey (KF273906 and HM126606). Five isolates from *B. pahangi* were clustered together in the same clade with *B. pahangi* published by Rishniw et al. [3], in which the worm was from an unknown region. Comparison of nucleotide sequences with other species of filarial nematode found *D. immitis* and *B. pahangi* could be distinguished from *B. malayi, D. repens, O. volvulus,* and *A. reconditum* (*Dipetalonema reconditum* in Figure 3).

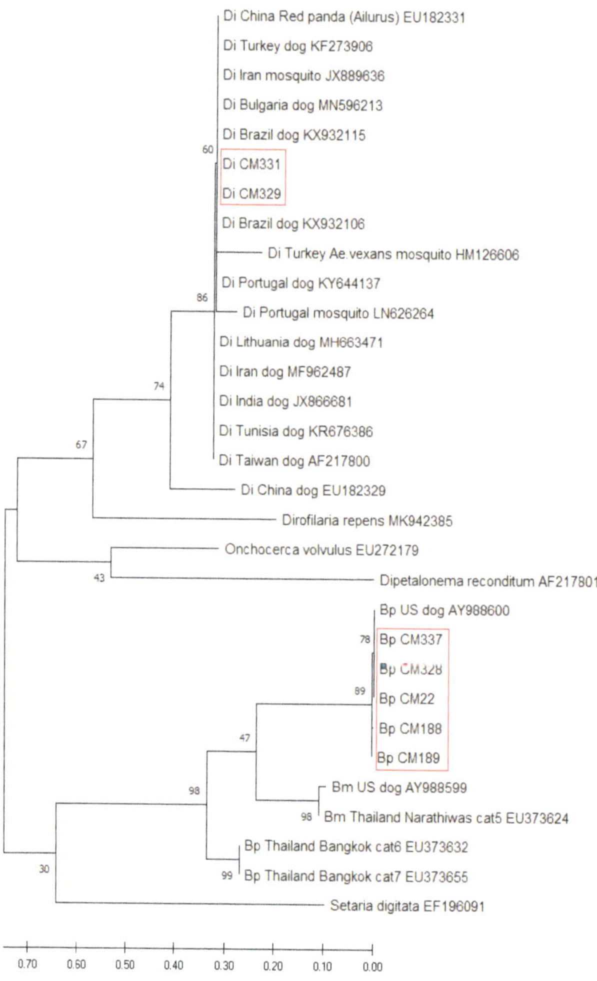

Figure 3. Evolutionary analysis of *D. immitis* and *B. pahangi* 5.8S-ITS2 region. A phylogenetic tree based on the 5.8S-ITS2 region of *D. immitis* (Di) and *B. pahangi* (Bp) from Thailand and other countries was inferred using the maximum likelihood method and the Tamura–Nei model. The tree with the highest log likelihood (−2536.04) is shown. The percentage of trees in which the associated taxa clustered together is shown next to the branches. There was a total of 344 positions in the final dataset. Sequences in the red outline boxes originated from the present study.

3.3. Geographical Distribution of Filarial Infection

3.3.1. Altitudinal Distribution of Filarial Infection

There was no difference in the filarial positivity of either *B. pahangi* or *D. immitis* among the different altitude classes (Table 3). The range of *B. pahangi* infection was 4.08–10.27%, with the highest occurrence observed in the low areas <400 m altitude (10.27%; 23 cases). The range of *D. immitis* infection was 12.05–14.29%, with the highest occurrence observed at 400–800 m (14.29%; 14 cases). The prevalence of *D. immitis* was significantly higher than *B. pahangi* in the middle altitude class of 400–800 m, at 14.29% and 4.08%, respectively ($p < 0.05$; Table 3). The altitudinal distribution of filarial positive cases is illustrated on a GIS map of Chiang Mai, Thailand (Figures 4C and 5C).

Table 3. The altitudinal distribution and prevalence of canine filariasis in Chiang Mai, Thailand.

Level	Altitude Level (m.)	No. of Examined	*B. pahangi* Infection (n)	*D. immitis* Infection (n)	Total Canine Filariasis (n)
1	<400	224	10.27% (23)	12.05% (27)	21.43% (48)
2	400–800	98	4.08% (4) [a]	14.29% (14) [b]	18.37% (18)
3	>800	15	6.67% (1)	0	6.67% (1)
Total		337	8.31% (28)	12.17% (41)	19.88% (67)

[a,b] Values in the same column with different superscripts are statistically different ($p < 0.05$).

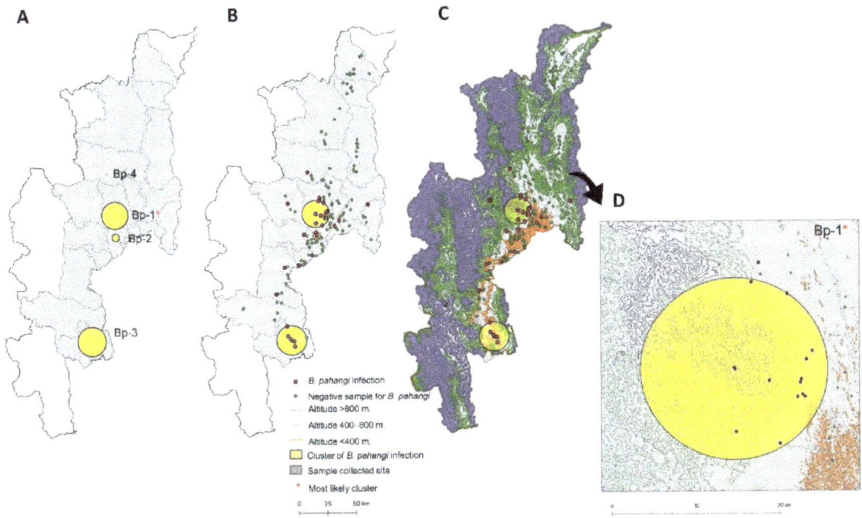

Figure 4. Geographic spatial clusters of *B. pahangi* infection obtained using the Bernoulli scan statistic model. (**A**). The distribution of four clusters of *B. pahangi* infection; (**B**). the distribution of *B. pahangi* cases in each cluster; (**C**). the altitudinal distribution of *B. pahangi* cases in each cluster; (**D**). the distribution of *B. pahangi* cases in the most likely cluster.

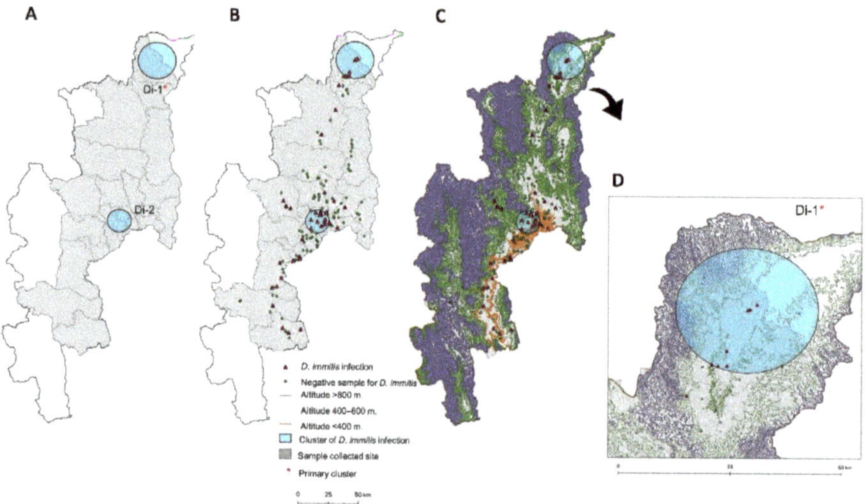

Figure 5. Geographic spatial clusters of *D. immitis* infection obtained using the Bernoulli scan statistic model. (**A**). the distribution of four clusters of *D. immitis* infection; (**B**). the distribution of *D. immitis* cases in each cluster; (**C**). the altitudinal distribution of *D. immitis* cases in each cluster; (**D**). the distribution of *D. immitis* cases in the most likely cluster.

3.3.2. Spatial Cluster Distribution of Filarial Infection

The canine filariasis spatial clusters were analyzed using the Bernoulli special scanning statistic method. Tthe relative risk (RR) and size of each cluster are outlined in Table 4. Four and two spatial clusters were identified for *B. pahangi* and *D. immitis*, respectively. Regarding *B. pahangi* distribution, a significant cluster (cluster Bp-1; $p < 0.001$) was identified with RR at 6.69 and a clustering radius of 10.55 km. This most likely cluster was distributed in three districts: Hang Dong, Mae Rim, and Mueang. The primary cluster of *B. pahangi* infection was found to occur in various altitudinal areas (0 to >800 m Figure 4C,D). Replacement of forests by agricultural activities and irrigated crops occurred in scattered locations in this cluster area. However, no significant spatial patterns were identified for *D. immitis* infection (Table 4), although the number of cases observed was higher than with *B. pahangi*. The most likely cluster of *D. immitis* infection (Di-1) was located in the Fang district (18.737609° N, 98.931619° E) where the clustering radius was 9.32 km, and the RR was 4.03. Sites of positive identification of *D. immitis* in this cluster were distributed in the low areas at altitudes of 400 to 800 m (Figure 5C,D) and were found mainly in areas with irrigated agricultural crops.

Table 4. Geographic spatial clusters obtained by Bernoulli scan statistic model on canine filariasis in Chiang Mai, Thailand.

Cluster Number	Cluster Type	Centroid (X,Y)/Radius (km)	Cases	Expected Cases	Observed to Expected Cases	RR [a]	LLR [b]	*p*-Value	Districts
				B. pahangi infection					
Bp-1	Most likely	18.815706 N, 98.883035 E/10.55	8	1.58	5.07	6.69	8.848	0.010	Hang Dong, Mae Rim, Mueang, Samoeng
Bp-2	Secondary	18.660241 N, 98.889764 E/2.91	2	0.17	12.04	12.88	5.043	0.439	Hang Dong, San Pa Tong
Bp-3	Secondary	17.924869 N, 98.704168 E/11.42	5	1.41	3.54	4.09	3.457	0.754	Doi Tao
Bp-4	Secondary	19.108108 N, 98.824182 E/0	2	0.25	8.02	8.56	3.214	0.883	Mae Taeng
				D. immitis infection					
Di-1	Most likely	18.737609 N, 98.931619 E/9.32	9	2.68	3.36	4.03	6.386	0.143	Fang
Di-2	Secondary	19.922648 N, 99.208706 E/15.01	6	1.70	3.52	3.96	4.392	0.393	Hang Dong, Mueang, Saraphi

[a] RR = relative risk, [b] LLR = log likelihood ratio.

4. Discussion

The current study updates our understanding of the situation of *D. immitis* infection and proven *B. pahangi* infection in community dogs in Northern Thailand. *B. malayi* was not found in Chiang Mai, although the province does have a large free-roaming dog population.

Heartworm disease caused by *D. immitis* infection is widespread in tropical and subtropical regions and is a primary life-threatening disease of dogs and cats worldwide. The prevalence of *D. immitis* infection in dogs worldwide was 10.91% (95% CI = 10.18–11.65) in 2020. The prevalence of *D. immitis* in dogs varies across countries and continents, e.g., 22.68% in Australia, 12.07% in Asia, 11.60% in the Americas, 10.45% in Europe, and 7.57% in Africa [1]. Over the past few decades, the regional prevalence of *D. immitis* infection in dogs in Thailand has been reported as 10–58% in Bangkok [16,29–31], 23–25% in the southern region [32], and 6–25% in the northern region [19,33–35]. In Chiang Mai, the prevalence gradually decreased from 45.76% in 1987 to 24.71% in 1992, to 18.20% in 2008, and then to 12.17% in 2020. The reduction of dirofilariasis is possibly due to effective regular preventive chemotherapy of dogs and to vector control by dog owners in collaboration with the active support of District Administration Organizations, which should be extended to other high prevalence districts such as Fang.

B. pahangi, a lymphatic filarial worm that infests mammals, is commonly found in cats, dogs, and wild carnivores [36], and is closely related to *B. malayi*. Recently, five cases of clinically typical LF caused by *B. pahangi* were reported in suburban areas in the city of Kuala Lumpur, Malaysia [9]. Additionally, there was a report of a human subconjunctival infection with *B. pahangi* in Malaysia, which was identified by cyclooxygenase-1 (COX-1) PCR [37]. To the best of our knowledge, that study was the first to confirm *B. pahangi* infection in an animal in Chiang Mai. Elsewhere in Thailand, the *B. pahangi* infection rate range of 4–25% in dogs and cats in Bangkok has been reported [16,31,38]. In addition, in Western and Southern Thailand, accidental zoonotic filariasis of *B. pahangi* infections in children were reported, and *Armigeres subalbatus* was found to be a common mosquito in the infected areas [39]. The recently discovered parasite *B. pahangi* exhibits novel aspects and adaptations. For that reason, *Brugia* infections in humans in this region should be closely monitored, especially the environment and geographical distribution in areas with a high vector density and densely crowded human populations.

Natural infections of canine dirofilariasis enhance the risk of transmission to humans [40]. The relative risk of canine dirofilariasis has been found to be related to various factors, e.g., outdoor living, lack of heartworm prevention, and infected vector exposure [19,40]. The present study observed *D. immitis* cases in all altitude classes (0 to >800 m), probably due to the broad range of mosquito species, which are potential transmission vectors. Although *Mansonia uniformis* is an important natural mosquito vector of *D. immitis* in Chiang Mai [33], *Culex pipiens quinquefasciatus*, *Aedes albopictus*, and *Ae. aegypti* are additional vector species for *D. immitis* transmission. The majority of *Aedes*, *Anopheles*, *Armigeres*, and *Culex* mosquitoes have been found at altitudes between 300 and 900 m [26]. *Ae. albopictus* is distributed in a wide range of elevations, between 300 and 1300 m, whereas *Cx. quinquefasciatus* has been found at elevations of 500 m and above [26] and *Mansonia* species are spread across different altitudes [41]. The number of *Mansonia* species is likely to be higher in areas with floating aquatic plants.

B. pahangi is considered to be a new causative agent of urban *Brugia* filariasis in Thailand [42]. Infective-stage larvae of *B. pahangi* microfilariae can develop in *An. quadrimaculatus* and *Ae. aegypti* [43]. Development and transmission of *B. pahangi* microfilariae to a vertebrate host can occur via natural vectors such as *Ar. subalbatus* [44,45]. *Ar. subalbatus* has also been found to be a common mosquito in infected areas and is suspected of being the main vector for zoonotic *B. pahangi* infection of children in Thailand [39]. Chaves et al. [46] demonstrated that density of adult *Ar. subalbatus* decreases with altitude. Its preferred range is 109–330 m, but it tends to increase in areas with abundant leaf litter on the ground. However, other mosquitoes, including the genera *Culex*, *Aedes*, *Mansonia*,

Anopheles, and *Armigeres*, are potential mosquito vectors of *Brugia* spp. [47] which might explain the distribution of *B. pahangi* cases at all altitudes from 0 to >800 m. Climatic and geographic conditions, as well as environmental factors and changes in land use, may support the high prevalence of *B. pahangi* at low altitudes. The occurrence of more hotspots of *D. immitis* and *B. pahangi* in suburban areas of Chiang Mai, such as the Fang and Mae Rim districts (Figure 4), could be influenced by factors driving transmission such as urbanization, rural–urban migration, and expansion of irrigated agricultural land, which is common in this area and provides suitable humidity for mosquitoes to thrive. Spatial mapping of the mosquito vector distribution and breeding habitats in canine filarial-infected areas on finer scales should be conducted to complete our current understanding of transmission dynamics and to help with the establishment of appropriate control strategies.

Although results of analyses of the correlation among diagnostic techniques for filariasis have so far been inconclusive [48,49], a combination of various diagnostic methods remains crucial for precise canine filariasis mapping of the region. Although the sensitivity of microhematocrit centrifugation for microfilaria detection was reported to be only 30% [50], this concentration technique is commonly applied in clinical practice to screen for microfilaria prior to confirmation using a heartworm antigen kit. We did not differentiate the microfilaria species by microscopic examination because this is labor-intensive and because of the difficulty in resolving the identification of species using blood smears. However, molecular techniques such as PCR can successfully detect microfilaria from blood samples. Additionally, PCR can confirm species identification using specific primers or by sequencing, making it more precise than microscopic procedures, especially when many filarial species and coinfection are involved. In the present study, two *D. immitis* DNA-positive dogs did not present with adult *D. immitis* antigen in their blood. This may be linked to the existence of an immune complex that blocks the antigen detection process or to the low antigen secretion by adult worms [51]. However, one occult infection was apparently observed in the current study. This detection problem might be due to many factors, including detection in dogs with prepatent infection, immature female worms, unisexual worm infection, drug-induced sterility of adult heartworms, senile infertility of female worms, host immune responses, and treatment with macrocyclic lactones [6,52,53].

This study has some limitations. First, although we covered a wide range of geographical features in Northern Thailand, the results should be interpreted with caution as they cannot be generalized to all endemic settings. More variations in geographical characteristics should be included in future studies. Second, as sampling was conducted mainly among free-roaming dogs in the studied communities, information about the dogs, e.g., housing, routine drugs, and vaccination, was limited; thus, an analysis of risk factors associated with the risk of *B. pahangi* could not be determined in this study. Last, the sampling strategy in this study was from an unknown prevalence; thus, it may pose a challenge, especially for *B. pahangi*, possibly contributing to an underestimation of number of samplings. However, for a follow-up study, the prevalence obtained from this study might be used and modified for a sample size determination.

5. Conclusions

In this study, we covered the detection and confirmation of species of canine filarial infection in free-roaming community dogs found in Chiang Mai, Northern Thailand, using a combination of diagnostic methods. The infection rates of *B. pahangi* were found to be comparable to those of *D. immitis*. Both types of filarial nematodes had a high prevalence in the low altitude suburban agricultural areas. The residents of the region should be made aware that lymphatic filariasis is caused by *B. pahangi*. For practicing veterinarians, filarial species identification is necessary for determination of the proper heartworm treatment as well as for public health efforts related to zoonotic transmission. Spatial distribution of filariasis in consecutive areas as well as mosquito vector distribution and breeding habitats of *D. immitis* and *B. pahangi* in infected areas should be determined at finer scales to identify and implement appropriate control strategies.

Supplementary Materials: The following are available online at https://www.mdpi.com/2076-2615/11/1/33/s1, Table S1: The 5.8S-ITS2 sequence list of *D. immitis, B. pahangi,* and other filarial nematodes used in the phylogenetic analysis, Table S2: Prevalence of canine filariasis in different districts in Chiang Mai Province, Thailand.

Author Contributions: Conceptualization, S.T.; methodology, M.K., D.K., S.F. and S.T.; software, O.A., V.P. and S.T.; validation, S.T.; formal analysis, M.K., O.A., V.P. and S.T.; investigation, M.K., D.K. and S.T.; resources, M.K., K.T., C.B. and S.T.; data curation, M.K., V.P. and S.T.; writing—original draft preparation, M.K., D.K. and S.T.; writing—review and editing, M.K., S.F., V.P. and S.T.; visualization, M.K., V.P. and S.F.; supervision, K.T., C.B., V.P. and S.T.; project administration, S.T.; funding acquisition, V.P. All authors have read and agreed to the published version of the manuscript.

Funding: This work was partially supported by Chiang Mai University (CMU), the research fund of Faculty of Veterinary Medicine, CMU (Grant Number R000016307 to V.P.), Research Grants for Young Researchers, CMU Y2014 (Grant Number H5701 to S.T.), and the partially supported by JSPS KAKENHI (Grant Numbers 19KK0175, 19H03121 to S.F.).

Institutional Review Board Statement: The study was conducted according to the guidelines of the Institutional Animal Care and Use Committee (IACUC), and approved by the Faculty of Veterinary Medicine, Chiang Mai University Animal Care and Use Committee (FVM, ACUC; protocol code R15/2562, approved on 12 June 2019).

Informed Consent Statement: Informed consent was obtained from all subjects involved in the study.

Data Availability Statement: Data available on request due to privacy for sequential projects.

Acknowledgments: The authors appreciate the cooperation of the monks and animal keepers at all temples and community locations where blood samples and data were collected. We would also like to thank Piyanan Taweethavonsawat (Faculty of Veterinary Science, Chulalongkorn University, Thailand) and Atiporn Saeung (Faculty of Medicine, Chiang Mai University, Thailand) for the positive control of *Brugia* spp.

Conflicts of Interest: The authors declare no conflict of interest.

References

1. Ghasemi, E.; Shamsinia, S.; Taghipour, A.; Anvari, D.; Bahadory, S.; Shariatzadeh, S.A.; Kordi, B.; Majidiani, H.; Borji, H.; Chaechi Nosrati, M.; et al. Filarial worms: A systematic review and meta-analysis of diversity in animals from Iran with emphasis on human cases. *Parasitology* **2020**, *147*, 909–921. [CrossRef] [PubMed]
2. Irwin, P.J. Companion animal parasitology: A clinical perspective. *Int. J. Parasitol.* **2002**, *32*, 581–593. [CrossRef]
3. Rishniw, M.; Barr, S.C.; Simpson, K.W.; Frongillo, M.F.; Franz, M.; Dominguez Alpizar, J.L. Discrimination between six species of canine microfilariae by a single polymerase chain reaction. *Vet. Parasitol.* **2006**, *135*, 303–314. [CrossRef] [PubMed]
4. Simón, F.; Kramer, L.H.; Román, A.; Blasini, W.; Morchón, R.; Marcos-Atxutegi, C.; Grandi, G.; Genchi, C. Immunopathology of *Dirofilaria immitis* infection. *Vet. Res. Commun.* **2007**, *31*, 161–171. [CrossRef] [PubMed]
5. Bowman, D.D. *Georgis' Parasitology for Veterinarians*; Saunders, Elsevier: St. Louis, MO, USA, 2009.
6. Simón, F.; Siles-Lucas, M.; Morchón, R.; González-Miguel, J.; Mellado, I.; Carretón, E.; Montoya-Alonso, J.A. Human and animal dirofilariasis: The emergence of a zoonotic mosaic. *Clin. Microbiol. Rev.* **2012**, *25*, 507–544. [CrossRef] [PubMed]
7. Anderson, R.C. *Nematode Parasites of Vertebrates: Their Development and Transmission*, 2nd ed.; CABI: London, UK, 2000.
8. Edeson, J.F.; Wilson, T.; Wharton, R.H.; Laing, A.B. Experimental transmission of *Brugia malayi* and *B. pahangi* to man. *Trans. R. Soc. Trop. Med. Hyg.* **1960**, *54*, 229–234. [CrossRef]
9. Tan, L.H.; Fong, M.Y.; Mahmud, R.; Muslim, A.; Lau, Y.L.; Kamarulzaman, A. Zoonotic *Brugia pahangi* filariasis in a suburbia of Kuala Lumpur City, Malaysia. *Parasitol. Int.* **2011**, *60*, 111–113. [CrossRef]
10. Melrose, W.D. Lymphatic filariasis: New insights into an old disease. *Int. J. Parasitol.* **2002**, *32*, 947–960. [CrossRef]
11. Woo, P.T. Evaluation of the haematocrit centrifuge and other techniques for the field diagnosis of human trypanosomiasis and filariasis. *Acta Trop.* **1971**, *28*, 298–303.
12. Kawthalkar, S.M. *Essentials of Clinical Pathology*; Jaypee Brothers Medical Publishers: New Delhi, India, 2018.
13. Nuchprayoon, S.; Junpee, A.; Poovorawan, Y.; Scott, A.L. Detection and differentiation of filarial parasites by universal primers and polymerase chain reaction-restriction fragment length polymorphism analysis. *Am. J. Trop. Med. Hyg.* **2005**, *73*, 895–900. [CrossRef]
14. Dickson, B.F.R.; Graves, P.M.; McBride, W.J. Lymphatic filariasis in mainland Southeast Asia: A systematic review and meta-analysis of prevalence and disease burden. *Trop. Med. Infect. Dis.* **2017**, *2*, 32. [CrossRef] [PubMed]
15. Kaikuntod, M.; Thongkorn, K.; Tiwananthagorn, S.; Boonyapakorn, C. Filarial worms in dogs in Southeast Asia. *Vet. Integr. Sci.* **2008**, *16*, 1–17.

16. Nithiuthai, S.; Chungpivat, S. Lymphatic filaria (*Brugia pahangi*) in dogs. *Thai J. Vet. Med.* **1992**, *4*, 123–133.
17. Iyengar, M.O. Filariasis in Thailand. *Bull World Health Organ* **1953**, *9*, 731–766.
18. Yokmek, S.; Warunyuwong, W.; Rojanapanus, S.; Jiraamornimit, C.; Boitano, J.J.; Wongkamchai, S. A case report of Brugian filariasis outside an endemic area in Thailand. *J Helminthol.* **2013**, *87*, 510–514. [CrossRef]
19. Boonyapakorn, C.; Srikitjakarn, L.; Morakote, N.; Hoerchner, F. The epidemiology of *Dirofilaria immitis* infection in outpatient dogs at Chiang Mai university small animal hospital, Thailand. *Southeast Asian J. Trop. Med. Public. Health* **2008**, *39*, 33–38.
20. World Health Organization. *UNDP-World Bank-WHO Special Programme for Research and Training in Tropical Diseases WHO-UNICEF Joint Programme for Health Mapping World Health Organization: Division of Tropical Diseases. Research on Rapid Geographical Assessment of Bancroftian Filariasis*; World Health Organization: Geneva, Switzerland, 1998.
21. Tvedten, H. Basic approach to anemia diagnosis. In Proceedings of the World Small Animal Veterinary Association World Congress Proceedings, Mexico city, Mexico, 11–14 May 2005; Available online: https://www.vin.com/apputil/content/defaultadv1.aspx?id=3854233&pid=11196#:~{}:text=One%20documents%20the%20presence%20and,and%20how%20severe%20it%20is. (accessed on 15 May 2019).
22. R Core Team. R: A Language and Environment for Statistical Computing. Available online: http://www.R-project.org/ (accessed on 15 May 2019).
23. Altschul, S.F.; Gish, W.; Miller, W.; Myers, E.W.; Lipman, D.J. Basic local alignment search tool. *J. Mol. Biol.* **1990**, *215*, 403–410. [CrossRef]
24. Kumar, S.; Stecher, G.; Li, M.; Knyaz, C.; Tamura, K. MEGA X: Molecular evolutionary genetics analysis across computing platforms. *Mol. Biol. Evol.* **2018**, *35*, 1547–1549. [CrossRef]
25. Tamura, K.; Nei, M. Estimation of the number of nucleotide substitutions in the control region of mitochondrial DNA in humans and chimpanzees. *Mol. Biol. Evol.* **1993**, *10*, 512–526. [CrossRef]
26. Devi, N.P.; Jauhari, R.K. Altitudinal distribution of mosquitoes in mountainous area of Garhwal region: Part-I. *J Vector Borne Dis.* **2004**, *41*, 17–26.
27. Kulldorff, M. SaTScan™ User Guide for Version 9.6. Available online: https://www.satscan.org/cgi-bin/satscan/register.pl/SaTScan_Users_Guide.pdf?todo=process_userguide_download (accessed on 15 May 2019).
28. Areekit, S.; Singhaphan, P.; Khuchareontaworn, S.; Kanjanavas, P.; Sriyaphai, T.; Pakpitchareon, A.; Khawsak, P.; Chansiri, K. Intraspecies variation of *Brugia* spp. in cat reservoirs using complete ITS sequences. *Parasitol. Res.* **2009**, *104*, 1465–1469. [CrossRef] [PubMed]
29. Jittapalapong, S.; Pinyopanuwat, N.; Chimnoi, W.; Nimsupan, B.; Saengow, S.; Simking, P.; Kaewmongkol, G. Prevalence of heartworm infection of stray dogs and cats in Bangkok metropolitan areas. *Kasetsart J. Nat. Sci.* **2005**, *39*, 30–34.
30. Tiawsirisup, S.; Thanapaisarnkit, T.; Varatorn, E.; Apichonpongsa, T.; Bumpenkiattikun, N.; Rattanapuchpong, S.; Chungpiwat, S.; Sanprasert, V.; Nuchprayoon, S. Canine heartworm (*Dirofilaria immitis*) infection and immunoglobulin G antibodies against *Wolbachia* (Rickettsiales: Rickettsiaceae) in stray dogs in Bangkok, Thailand. *Thai J. Vet. Med.* **2015**, *40*, 165–170.
31. Satjawongvanit, H.; Phumee, A.; Tiawsirisup, S.; Sungpradit, S.; Brownell, N.; Siriyasatien, P.; Preativatanyou, K. Molecular analysis of canine filaria and its *Wolbachia endosymbionts* in domestic dogs collected from two animal university hospitals in Bangkok metropolitan region, Thailand. *Pathogens* **2019**, *8*, 114. [CrossRef]
32. Kamyingkird, K.; Junsiri, W.; Chimnoi, W.; Kengradomkij, C.; Saengow, S.; Sangchuto, K.; Ka jeerum, W.; Pangjai, D.; Nimsuphan, B.; Inpankeaw, T.; et al. Prevalence and risk factors associated with *Dirofilaria immitis* infection in dogs and cats in Songkhla and Satun provinces, Thailand. *Agric. Nat. Resour.* **2017**, *51*, 299–302. [CrossRef]
33. Choochote, W.; Sukhavat, K.; Keha, P.; Somboon, P.; Khamboonruang, C.; Suwanpanit, P. The prevalence of *Dirofilaria immitis* in stray dog and its potential vector in Amphur Muang Chiang Mai, Northern Thailand. *Southeast Asian J. Trop. Med. Public Health* **1987**, *18*, 131–134.
34. Choochote, W.; Suttajit, P.; Rongsriyam, Y.; Likitvong, K.; Tookyang, B.; Pakdicharoen, A. The prevalence of *Dirofilaria immitis* in domestic dogs and their natural vectors in amphur Muang Chiang Mai, Northern Thailand. *J. Trop. Med. Parasitol.* **1992**, *15*, 11–17.
35. Colella, V.; Nguyen, V.L.; Tan, D.Y.; Lu, N.; Fang, F.; Zhijuan, Y.; Wang, J.; Liu, X.; Chen, X.; Dong, J.; et al. Zoonotic vectorborne pathogens and ectoparasites of dogs and cats in eastern and southeast Asia. *Emerg. Infect. Dis.* **2020**, *26*, 1221–1233. [CrossRef]
36. Denham, D.A.; McGreevy, P.B. Brugian filariasis: Epidemiological and experimental studies. *Adv. Parasitol.* **1977**, *15*, 243–309. [CrossRef]
37. Chang, K.M.; Shamala, R.; Mangalam, S.; Rohela, M.; Fong, M.Y.; Sivanandam, S.; Azdayanti, M.; Jamaiah, I.; Noraishah, M.A.A. Live *Brugia pahangi* immature female worm recovered from the subconjunctiva of a patient in Malaysia. In Proceedings of the 46th MSPTM Annual Scientific Conference, Kuala Lumpur, Malaysia, 11–12 March 2020; p. 101.
38. Chungpivat, S.; Sucharit, S. Microfilariae in cats in Bangkok. *Thai J. Vet. Med.* **1993**, *23*, 75–87.
39. Klinklom, A.; Santarattiwong, P. Update on emerging infectious disease: Zoonotic filariasis. In *Proceedings of PIDST Gazette*; Royal Cliff Hotel Pattaya: Chon Buri, Thailand; pp. 18–20.
40. Capelli, G.; Frangipane di Regalbono, A.; Simonato, G.; Cassini, R.; Cazzin, S.; Cancrini, G.; Otranto, D.; Pietrobelli, M. Risk of canine and human exposure to *Dirofilaria immitis* infected mosquitoes in endemic areas of Italy. *Parasit Vectors* **2013**, *6*, 60. [CrossRef] [PubMed]

41. Anwar, C.; Ghiffari, A.; Kuch, U.; Taviv, Y. The Abundance of Mosquitoes (Family: Culicidae) Collected in an altitudinal gradient In South Sumatra, Indonesia. In Proceedings of the International Conference on Agricultural, Ecological and Medical Sciences (AEMS-2015), Phuket, Thailand, 10 February 2015; pp. 28–30.

42. Zielke, E.; Hinz, E.; Sucharit, S. Lymphatic filariasis in Thailand a review on distribution and transmission. *Trop. Med. Parasitol.* **1993**, *15*, 141–148.

43. Nayar, J.K.; Knight, J.W.; Vickery, A.C. Susceptibility of *Anopheles quadrimaculatus* (Diptera: Culicidae) to subperiodic *Brugia malayi* and *Brugia pahangi* (Nematoda: Filarioidea) adapted to nude mice and jirds. *J. Med. Entomol.* **1990**, *27*, 409–411. [CrossRef] [PubMed]

44. Beerntsen, B.T.; Luckhart, S.; Christensen, B.M. *Brugia malayi* and *Brugia pahangi*: Inherent difference in immune activation in the mosquitoes *Armigeres subalbatus* and *Aedes aegypti*. *J. Parasitol.* **1989**, *75*, 76–81. [CrossRef]

45. Muslim, A.; Fong, M.Y.; Mahmud, R.; Lau, Y.L.; Sivanandam, S. *Armigeres subalbatus* incriminated as a vector of zoonotic *Brugia pahangi* filariasis in suburban Kuala Lumpur, Peninsular Malaysia. *Parasit Vectors* **2013**, *6*, 219. [CrossRef]

46. Chaves, L.F.; Imanishi, N.; Hoshi, T. Population dynamics of *Armigeres subalbatus* (Diptera: Culicidae) across a temperate altitudinal gradient. *Bull. Entomol. Res.* **2015**, *105*, 589–597. [CrossRef]

47. Carton, Y.; Nappi, A.J.; Poirie, M. Genetics of anti-parasite resistance in invertebrates. *Dev. Comp. Immunol.* **2005**, *29*, 9–32. [CrossRef]

48. Cano, J.; Moraga, P.; Nikolay, B.; Rebollo, M.P.; Okorie, P.N.; Davies, E.; Njenga, S.M.; Bockarie, M.J.; Brooker, S.J. An investigation of the disparity in estimates of microfilaraemia and antigenaemia in lymphatic filariasis surveys. *Trans. R. Soc. Trop. Med. Hyg.* **2015**, *109*, 529–531. [CrossRef]

49. Irvine, M.A.; Njenga, S.M.; Gunawardena, S.; Wamae, C.N.; Cano, J.; Brooker, S.J.; Hollingsworth, T.D. Understanding the relationship between prevalence of microfilariae and antigenaemia using a model of lymphatic filariasis infection. *Trans. R. Soc. Trop. Med. Hyg.* **2016**, *110*, 317. [CrossRef]

50. Gadahi, J.A.; Arijo, A.G.; Abubakar, M.; Javaid, S.B.; Arshed, M.J. Prevalence of blood parasites in stray and pet dogs in Hyderabad Area: Comparative sensitivity of different diagnostic techniqes for the detection of microfilaria. *Vet. World* **2008**, *1*, 229–232.

51. Little, S.; Saleh, M.; Wohltjen, M.; Nagamori, Y. Prime detection of *Dirofilaria immitis*: Understanding the influence of blocked antigen on heartworm test performance. *Parasit Vectors* **2018**, *11*, 186. [CrossRef] [PubMed]

52. American Heartworm Society, Current Canine Guidelines for the Prevention, Diagnosis, and Management of Heartworm (Dirofilaria Immitis) Infection in Dogs. Available online: https://www.heartwormsociety.org/veterinary-550resources/american-heartworm-society-guidelines (accessed on 15 May 2019).

53. Prieto, G.; Cancrini, G.; Muro, A.; Genchi, C.; Simon, F. Seroepidemiology of *Dirofilaria immitis* and *Dirofilaria repens* in humans from three areas of Southern Europe. *Res. Rev. Parasitol.* **2000**, *60*, 95–98.

Article

A Comparison of the Colonic Microbiome and Volatile Organic Compound Metabolome of *Anoplocephala perfoliata* Infected and Non-Infected Horses: A Pilot Study

Rachael Slater [1,*] , Alessandra Frau [1] , Jane Hodgkinson [2], Debra Archer [2] and Chris Probert [1]

[1] Institute of Systems, Molecular and Integrative Biology, University of Liverpool, Crown Street, Liverpool L69 3GE, UK; afrau@liverpool.ac.uk (A.F.); mdcsjp@liverpool.ac.uk (C.P.)
[2] Institute of Infection, Veterinary and Ecological Science, University of Liverpool, Leahurst Campus, Chester High Road, Wirral CH64 7TE, UK; jhodgkin@liverpool.ac.uk (J.H.); darcher@liverpool.ac.uk (D.A.)
* Correspondence: rsh14@liverpool.ac.uk

Simple Summary: In horses, tapeworm infection is associated with specific forms of colic (abdominal pain) that can be life-threatening without surgical treatment. There is growing evidence that intestinal parasites interact with the gut bacteria, and the consequences of these interactions may influence the ability of the host to resist infection and parasite-associated disease. We aimed to compare the intestinal bacteria and the gases produced by metabolic processes in the gut between horses that had varying levels of tapeworms and those with no tapeworm present. Overall, the diversity of gut bacteria was similar in horses with and without tapeworms. There were some decreases in beneficial bacteria in horses with tapeworms, indicating a possible negative consequence of infection. Intestinal gases correlated with some bacteria indicating their functionality and use as potential markers of active bacteria. Our study validates further research investigating tapeworm and gut bacteria interactions in the horse.

Abstract: *Anoplocephala perfoliata* is a common equine tapeworm associated with an increased risk of colic (abdominal pain) in horses. Identification of parasite and intestinal microbiota interactions have consequences for understanding the mechanisms behind parasite-associated colic and potential new methods for parasite control. *A. perfoliata* was diagnosed by counting of worms in the caecum post-mortem. Bacterial DNA was extracted from colonic contents and sequenced targeting of the 16S rRNA gene (V4 region). The volatile organic compound (VOC) metabolome of colonic contents was characterised using gas chromatography mass spectrometry. Bacterial diversity (alpha and beta) was similar between tapeworm infected and non-infected controls. Some compositional differences were apparent with down-regulation of operational taxonomic units (OTUs) belonging to the symbiotic families of Ruminococcaceae and Lachnospiraceae in the tapeworm-infected group. Overall tapeworm burden accounted for 7–8% of variation in the VOC profile (permutational multivariate analysis of variance). Integration of bacterial OTUs and VOCs demonstrated moderate to strong correlations indicating the potential of VOCs as markers for bacterial OTUs in equine colonic contents. This study has shown potential differences in the intestinal microbiome and metabolome of *A. perfoliata* infected and non-infected horses. This pilot study did not control for extrinsic factors including diet, disease history and stage of infection.

Keywords: anoplocephala perfoliata; equine; gut microbiome; volatile organic compounds (VOCs); omics integration

Citation: Slater, R.; Frau, A.; Hodgkinson, J.; Archer, D.; Probert, C. A Comparison of the Colonic Microbiome and Volatile Organic Compound Metabolome of *Anoplocephala perfoliata* Infected and Non-Infected Horses: A Pilot Study. *Animals* **2021**, *11*, 755. https://doi.org/10.3390/ani11030755

Academic Editors: Robert Li and Gary Muscatello

Received: 19 January 2021
Accepted: 5 March 2021
Published: 9 March 2021

Publisher's Note: MDPI stays neutral with regard to jurisdictional claims in published maps and institutional affiliations.

1. Introduction

The gastrointestinal tapeworm *Anoplocephala perfoliata* is a common inhabitant of the equine gut [1,2]. High levels of A. perfoliata infection are known to be risk factors for specific forms of colic (abdominal pain) in the horse including spasmodic colic, ileal impaction, caecal intussusception and caecal rupture [3–7] and some of these forms of colic may be fatal without surgical treatment. The exact mechanisms by which tapeworm-associated colic develops are not known; such knowledge would enhance our understanding of the pathophysiology of these and other forms of colic. Currently licenced anti-cestode drugs (pyrantel and praziquantel) appear effective at reducing burdens of *A. perfoliata*. However, anthelmintic resistance is a growing concern regarding the effective control of other equine parasites, such as strongyles and *Parascaris equorum* and may limit our ability to treat tapeworm infection in the future [8,9]. This is a particular concern given the small number of drug classes that are available to treat cestode burdens in horses [9,10]. This places greater emphasis on the need to better detect the presence of tapeworms and target treatments accordingly. Current diagnosis of tapeworm infection in live horses is based on serological or saliva antibody assays [11,12]. Studies have shown the antibodies may remain raised long-beyond clearance of infection, making them inaccurate for determining real-time infection [11,13,14]. Alternatively a faecal egg count (FEC) can be conducted, but this has only 61% sensitivity [15].

The horse is a hindgut fermenter and relies on volatile fatty acids (VFAs) produced by the microbiota to provide a large proportion of daily energy requirements [16,17]. Murine studies have shown evidence of the role of the commensal intestinal microbiota in the establishment and maintenance of parasites within the gut and this interaction has potential implications for host immunity [18–20]. Correlations between helminths and the faecal microbiota in live horses have been investigated using strongyle FEC as an indicator of infection [21–23]. In ponies predicted to be resistant or susceptible to strongyle infection, modest differences in microbiome composition were observed post-exposure to natural infection [21]. Another study demonstrated significant clustering in the microbial composition associated with FEC rather than anthelmintic treatment, but there were no differences in faecal bacterial diversity between horses with high and low FECs [22]. Both studies reported an increase in members of the phylum Proteobacteria in horses with high FECs. Three studies in horses have shown a shift in microbial composition following anthelmintic treatment, irrespective of whether FECs were high or low [23–25]. Differences in microbial composition following treatment may also be more evident in young animals [23,24]. These results indicate a possible effect of anthelmintic medication, or clearance of strongyles (even at low levels) from the gastrointestinal tract, on the faecal microbiome in some populations of horses. Collectively, these initial studies imply that intestinal helminth-microbiota interactions occur in the horse and warrant further study in other equine species of parasite, including cestodes. The relationship of *A. perfoliata* with the gut microbiome and metabolome (functional microbiome) of the horse has not previously been investigated. Volatile organic compounds (referred to as the VOC metabolome) may be generated by physiological processes from the host or the microbiota [26,27]. In the horse, VOC patterns appear to mirror the faecal microbiota [28]. Studies incorporating both microbiome and metabolome interactions with parasites are important in understanding the functional effect parasites have on the gut microbiota and intercommunications within the host [20]. The knowledge of such interactions may have beneficial consequences for equine health through alternatives to parasite control [29].

We hypothesised that the colonic microbiome and VOC metabolome would differ between *A. perfoliata*-positive and -negative horses. To enable accurate classification of tapeworm-positive and -negative horses, we collected samples post-mortem from abattoir material (gold-standard diagnosis of *A. perfoliata*). Collection of abattoir material enabled colonic contents to be taken for microbiome and metabolome analysis rather than using a faecal proxy. Faeces has been shown to be representative of the distal regions of the hindgut but is less representative of more proximal regions [30,31]. The aims of this study were to compare the microbiome and VOC metabolome of the distal colonic contents of *A. perfoliata* infected and non-infected horses and those with high and low strongyle FECs, to rule this out as a confounding factor. We also aimed to investigate correlations between 16S rRNA sequence data and VOCs to infer active species and functionality of the microbiome that may be associated with tapeworm infection.

2. Materials and Methods

2.1. Sample Collection

Samples were collected from 51 horses slaughtered for non-experimental purposes at an abattoir in the South of the UK in November 2015. A power calculation for this pilot study was not performed as, to the authors' knowledge, this was the first study to compare the gut microbiome of tapeworm infected and non-infected horses. However, studies investigating strongyle species and microbiome interactions sampled similar numbers of horses to those used here [21,22,24]. Metadata including previous diet, breed, anthelmintic treatment and medications history were not available to be recorded. Specific horse age was also not available, but no pre-weaned foals were included. Post-mortem (immediately after slaughter and removal of the viscera—within 15 min) the caecum was opened to expose the caecal mucosa and visual assessment was used to estimate numbers of *A. perfoliata* present (gold standard). Samples were categorised as low (1–20 tapeworms, n = 7), medium (21–49, n = 5), high ($\geq$50, n = 5), tapeworm positive but numbers unrecorded (n = 4) and tapeworm negative (n = 30). The low threshold was chosen based on evidence that small numbers of tapeworms (1–20 worms) cause little intestinal pathology [32]. The thresholds set for the medium and high categories were arbitrary; they were chosen to demonstrate patterns with increased worm burden as there is a positive correlation between infection intensity and the risk of tapeworm associated colic [4]. For microbiome and metabolome analysis luminal contents (~25 mL) were collected from the large colon (pelvic flexure) for consistency in sampling. The regions of the hindgut where the tapeworms were located (caecum) were empty in many horses, possibly because of withholding of feed prior to slaughter. Colonic contents were transferred on dry ice to the laboratory and stored at -80 °C. Samples of rectal contents (~10 g) were also collected into universal containers and transported in a cool box and stored at 4 °C upon reaching the laboratory. Strongyle and ascarid worm faecal egg counts (FEC) (recorded in Table S1) were performed on rectal contents using a centrifugal floatation technique with a detection limit of one egg per gram (e.p.g.) within 10 days of collection [33]. Frozen colonic contents (~5 g) were freeze-dried for 48 h (Edwards High Vacuum, West Sussex, UK). Two aliquots of freeze-dried colonic-contents were created, one for VOC profiling (100 mg in a glass headspace vial) and one for DNA extraction (100 mg). For DNA extraction two glass beads (4 mm, undrilled, Fisher Scientific, Loughborough, UK) were added before the vials were placed in a frozen rack (-80 °C) and bead-beaten (TissueLyser II, Qiagen, Manchester, UK) for 2 min at full power (30 Hz). Freeze-drying of faecal and colonic contents prior to bead-beating has previously been applied in equine studies to disrupt the fibrous material [31,34].

2.2. Microbiome Profiling: 16S rRNA Sequencing

DNA extraction was performed on freeze-dried samples using Qiagen QIamp DNA stool mini kits (Qiagen, Manchester, UK) and steps were followed according to the manufacturer's instructions. After the addition of ASL buffer (Qiagen, Manchester, UK) to the sample and briefly vortexing, it was incubated in a heat block at 95 °C for 5 min, as suggested by the manufacturer. To amplify the bacterial 16S rRNA gene (V4 region), a universal tail tag dual indexing barcode approach was used [35]. For the first round of PCR the primers used were F515 (NNN NNG TGC CAG CMG CCG CGG TAA) and R806 (GG ACT ACH VGG GTW TCT AAT) (HPLC grade, Integrated DNA Technologies, Coralville, IA, USA) [36]. A PCR master mix was made containing (per total 20 µL reaction): 1× Q5 reaction buffer, 0.02 U/µL Q5 High-Fidelity DNA polymerase, 200 µM dNTPs (NEB), 0.125 µM of each forward and reverse primers made up to 15 µL with molecular-grade water. The master mix was divided between wells, with 5 µL (1 ng/µL) of template DNA. The plate was loaded into a pre-heated thermocycler (Multigene™ and Multigene™ OptiMax, Labnet International, Edison, NJ, USA) with an initial denaturation of 2 min at

98 °C, followed by 14 cycles of 20 s at 95 °C, 15 s at 65 °C, 30 s at 72 °C. There was a final extension step of 5 min at 72 °C and then samples were held at 4 °C until purification (within 24 h). The first round of PCR was carried out in duplicate to reduce PCR bias. Purification was performed using the AxyPrep Mag PCR Clean-up kit (Axygen, Corning, NY, USA) and the entire eluate (10 μL) was used as template for index PCR. Index PCR was carried out in a volume of 20 μL with the same reagents and method used in the first round of PCR. An exception was the use of a set of barcoded index primers (listed in Table S2, TruGrade, Integrated DNA Technologies, Coralville, IA, USA) used at 0.25 μM per sample and 15 PCR cycles. Index PCR was carried out in duplicate, and purification was repeated on the pooled duplicates. The final elute was 25 μL and the concentration of DNA was recorded using a Qubit (Qubit dsDNA HS assay kit, Life Technologies, Carlsbad, CA, USA). Sample pooling was performed based on sample concentration and fragment size determined by a bioanalyzer (2100 Bioanalyzer, Agilent Technologies, Santa Clara, CA, USA) by the Centre for Genomic Research (CGR), Liverpool, UK. Paired-end (2 × 250 bp) sequencing was performed on an Illumina MiSeq platform by the CGR, Liverpool, UK.

2.3. Volatile Organic Compound (VOC) Metabolome Profiling

Aliquots of freeze-dried colonic contents were analysed by headspace solid-phase microextraction gas chromatography mass spectrometry (HS-SPME-GCMS). Freeze-dried colon contents were taken forward for analysis rather than wet samples because greater numbers of VOCs were detected (results in Supplementary: Figure S1). The SPME fibre coating used was divinylbenzene-carboxen-polydimethylsiloxane (DVB-CAR-PDMS) 50/30μm (1 cm). Samples were heated at 60 °C for 30 min prior to SPME fibre exposure to the sample headspace for 20 min. GCMS conditions were as described previously [37].

2.4. Data Analysis

2.4.1. Microbiome

Illumina adapter sequences were trimmed from raw FastQ files using Cutadapt (1.2.1) [38]. The reads were further trimmed using Sickle (1.2) [39] with a minimum window quality score of 20. Reads shorter than 200 bp after trimming were removed. Samples with fewer than 1000 reads were also removed at this stage. Sequence clustering (99% similarity) was performed by SWARM 2.0 (d = 3) [40] and chimeras were filtered out using UCHIME [41]. The following steps were then carried out using Quantitative Insights Into Microbial Ecology (QIIME) (1.9.1) [42]. The classifier tool BLAST [43] was applied together with the SILVA (SILVA_123) database [44] to assign OTUs at a 99% threshold. The final steps were alignment of OTUs with the database using PyNAST [45] and construction of a phylogenetic tree using FastTree [46]. Prior to statistical analysis three samples were excluded. One sample (C25) failed to freeze-dry and two samples (T6 and C11) contained higher relative abundances of Fusobacteria and Proteobacteria than Bacteroidetes (the second most dominant phyla in all other samples) (Figure 1). The demographics of the animals in this study were unknown, and therefore these samples were removed to prevent bias. Removal of outliers using these criteria has been applied in equine faecal microbiome studies previously [22]. In R (version 3.5.1) the following statistical analysis was performed using methods described previously [47]. Specifically, a taxonomy plot was constructed at phylum level. The data were normalised to relative abundance for these visualisation plots. Numbers of horses with low (n = 7), medium (n = 4) and high (n = 5) tapeworm burdens were considered too small to compare to controls (CO, n = 28) by statistical analysis. Therefore, groups to be compared by statistical analysis were divided as follows: all tapeworm positive horses (TP, n = 20) were compared to CO and horses with clinically important burdens (≥21 tapeworms) referred to as TP_21 (n = 9) were also compared to CO. Diversity indices were calculated at OTU level using the R package vegan [48]. The alpha diversity (Richness, Shannon and Fisher indices) was calculated for each group and differences between groups were compared using the aov() function in R. The beta diversity was plotted on non-metric multidimensional scaling (NMDS) ordination plots (distances: Bray-Curtis, Unifrac and Weighted Unifrac). Statistical differences in the beta diversity between groups was assessed by permutational multivariate analysis of variance (PERMANOVA) using the adonis() function. Differential abundance of taxa (phylum, order, class, family, genus and OTU level) between CO and tapeworm groups (TP and TP_21) was evaluated using the DESeq2 R package [49]. To investigate the association of strongyle FEC in this cohort of horses alpha and beta diversity indices and the DESeq R package were applied to assess differences in the microbiota of groups of horses with low ≤10 e.p.g (n = 10, Low_FEC) or high ≥100 e.p.g (n = 24, High_FEC) strongyle FEC, as investigated previously [22]. A PERMANOVA (gower distance) model to access the variation described by tapeworm burden and FEC in the gut microbiome was performed using the vegan R package, function adonis2().

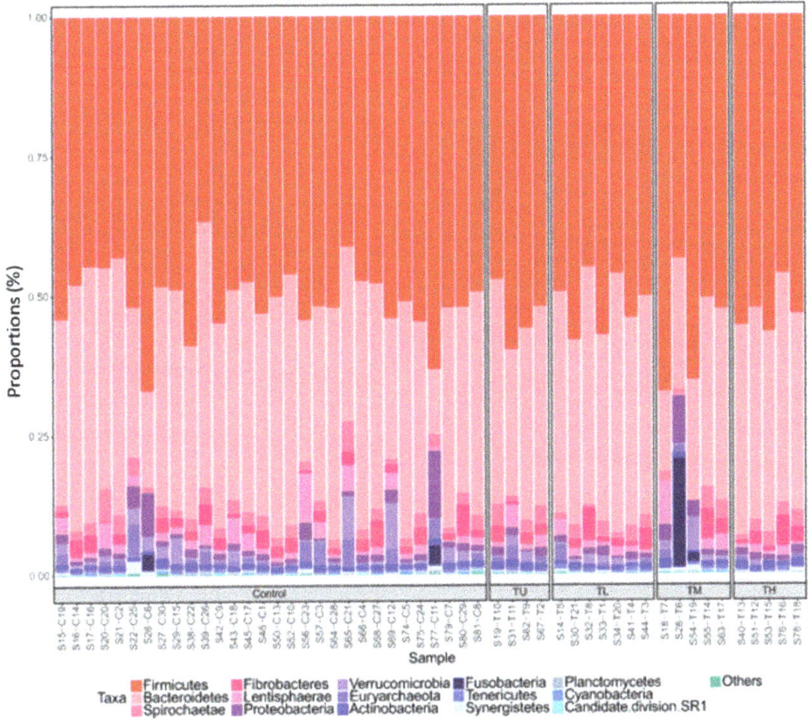

Figure 1. Taxonomic summary at phylum level for all samples. Key: TU = tapeworm positive with burden unrecorded, TL = low tapeworm (1–20 worms), TM = medium tapeworm (21–49 worms), TH = high tapeworm (≥50 worms).

2.4.2. Volatile Organic Compound (VOC) Metabolome

VOC data were processed using Automated Mass Spectral Deconvolution System (AMDIS-version 2.71, 2012, www.amdis.net, (accessed on 29 November 2020)) coupled to the National Institute of Standards and Technology (NIST) mass spectral library (version 2.0, 2011, purchased from PerkinElmer, Beaconsfield, UK) to putatively identify VOCs. The R package Metab [50] was used to align data. Low quality data points were removed using the criteria that a VOC must be present in at least 50% of samples within at least one condition to remain present. Following this, half-minimum values were imputed to replace missing data as performed previously [28]. Statistical analysis of VOC metabolome data was performed on the same groups as the microbiome data. The number of VOCs and differences in their abundance between tapeworm groups (TP and TP_21) and CO was evaluated by an independent *t*-test. To evaluate VOC abundance associated with high and low FECs an independent *t*-test was performed on the same groups outlined in the microbiome analysis. The same PERMANOVA model in the microbiome analysis including FEC and tapeworm burden was applied to the VOC metabolome data.

2.5. Integration of Omics Data

The R package mixOmics using the Data Integration Analysis for Biomarker discovery (DIABLO) framework was used to identify correlations between OTUs and VOCs. Specifically, the N-integrative supervised analysis was performed with DIABLO [51]. Prior to input of data into the DIABLO model, 16S rRNA data were normalised by Total Sum Scaling normalisation (TSS) followed by centred log-ratio (CLR). VOC data were normalised as described in Section 2.4. The first step involved fitting the model prior to variable selection to assess performance (based on correct assignment of groups) using the function perf(). A 10-fold cross validation of perf() was undertaken. From the cross-validation, a performance plot was constructed to calculate the balanced error rate (BER). The balanced error rate calculates the average proportion of wrongly classified samples in each class, weighted by the number of samples in each class. Based in the performance of the model the optimal number of components was chosen using the function choice.ncomp(). Next, the tune.block.splsda function was used to choose the optimum number of variables for each type of omics (OTUs and VOCs). Using these optimal variables, the final DIABLO model was constructed. A scatterplot (Pearson's correlation) was generated for OTUs and VOCs separately to evaluate which omics was the most important in distinguishing between time points using the function plotIndiv(). The cor.test() function (for paired data) was used to determine significant pair-wise correlations identified between OTUs and VOCs. P values were adjusted by FDR. A plot generated by the plotDIABLO() function was used to show overall correlation between selected OTUs and VOCs. For visualisation of correlations between specific OTUs and VOCs a heatmap was constructed using the plotVar() function. Integrated analysis was not performed on the TP_21 sub-set or FEC groupings because of the small sample sizes.

3. Results

3.1. Microbiome Profiling

The total number of sequences was 6,186,806 after filtering. The sequences clustered to a total number of 1,458. There was a minimum of 75,244 sequences per sample and a maximum of 167,089; there were 116,212 sequences per sample on average. In total, 19 phyla were identified, and the most abundant phyla were Firmicutes (51.6%), Bacteroidetes (36.1%), Spirochaetae (2.3%) and Fibrobacteres (2.0%) (Figure 1).

Box plots of the alpha diversity indices (Richness, Shannon and Fisher alpha) between tapeworm and control groups are shown in Figure 2A,B. Pair-wise comparisons between groups were not significant ($p > 0.05$). NMDS plots of the beta diversity (Bray-Curtis) of tapeworm and control groups are shown in Figure 2C,D and demonstrated no distinct clustering. PERMANOVA of beta diversity values (Bray-Curtis, Weighted Unifrac, Unifrac) revealed there were no significant differences in beta diversity between groups for both TP vs. CO and TP_21 vs. CO. Box plots of the alpha diversity indices (Richness, Shannon and Fisher alpha) and NMDS plots for beta diversity (Bray-Curtis) for comparisons between Low_FEC and High_FEC are shown in Figure 3. There were no significant differences in alpha (pair-wise ANOVA, $p > 0.05$) or beta (PERMANOVA, $p > 0.05$) diversity between Low_FEC and High_FEC groups.

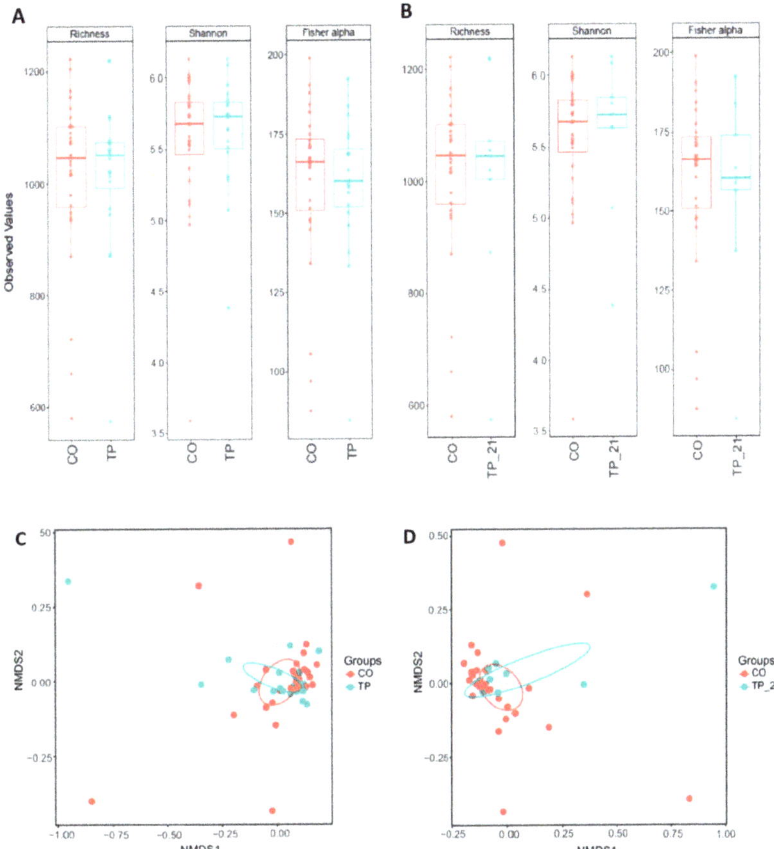

Figure 2. Alpha (**A,B**) and beta diversity (Bray-Curtis distance is shown) (**C,D**) indices (OTU level) of the colonic contents of horses infected with *Anoplocephala perfoliata* and non-infected controls. Alpha diversity was not significantly different between groups, pair-wise ANOVA ($p > 0.05$). The beta diversity between groups was not significant (PERMANOVA, $p > 0.05$). In (**A,C**) CO (n = 28) vs. TP (n = 20), (**B,D**) CO vs. TP_21 (n = 9). Key: TP = tapeworm positive (n = 20), TP_21 = tapeworm samples with >21 worms (n = 9), CO = control (tapeworm negative), NMDS = non-metric multidimensional scaling.

Using the DESeq2 R package, there were no significant differences in taxa abundance at phylum, order or class levels between tapeworm and control samples. Some differences at family and genus level were observed and are listed in Table 1. At OTU level, 69 OTUs (Figure S2A) were found to be significantly different between TP and CO; of these, 17 (24.6%) were more abundant in the TP group. When TP_21 and CO were compared, significant differences in taxa between groups were observed at order, family, genus (Table 1) and OTU level. One hundred and eighteen OTUs (Figure S2B) were significantly different with 28 (23.7%) more abundant in the TP_21 group. Taxa differential analysis revealed significant differences at order, family, genus (Table 1) and OTU level between Low_FEC and High_FEC samples. Ninety-eight OTUs were significantly different with 76 (77.6%) more abundant in the High_FEC group (Figure S3). A PERMANOVA model, which included TP and CO samples, revealed tapeworm burden accounted for 3% (p = 0.03) of variation and strongyle FEC accounted for the same, 3% (p = 0.04). When the PERMANOVA model included TP_21 and CO samples, 4% (p = 0.04) of variation in the VOC profile could be described by tapeworm burden but no significant variation could be described by strongyle FEC (4%, $p = 0.10$).

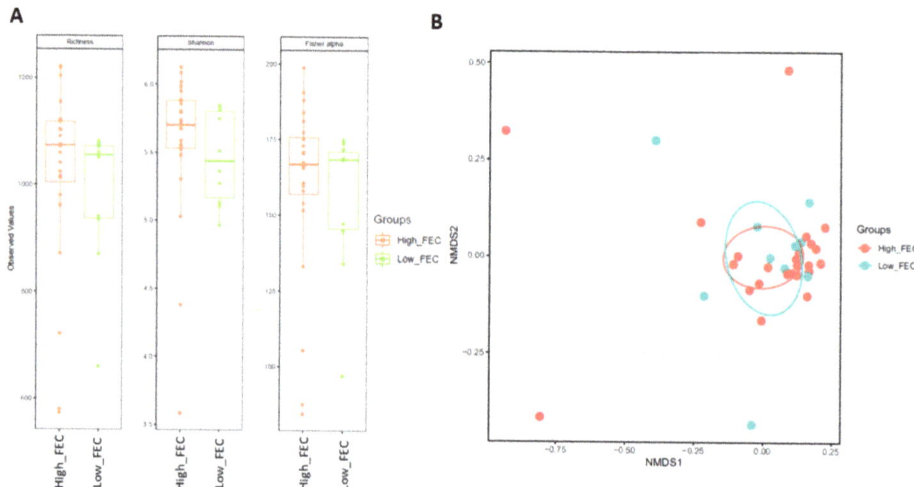

Figure 3. Alpha (**A**) and beta (**B**) (Bray-Curtis distance is shown) diversity indices for horses with low (<10 e.p.g) and high (>100 e.p.g) strongyle FECs. Alpha diversity was not significantly different between groups, pair-wise ANOVA ($p > 0.05$). The beta diversity between groups was not significant (permutational multivariate analysis of variance, $p > 0.05$) Key: High_FEC = high strongyles (>100 e.p.g) n = 24, Low_FEC = low strongyles (<10 e.p.g) n = 10, FEC = faecal egg count, e.p.g = eggs per gram, NMDS = non-metric multidimensional scaling.

3.2. Volatile Organic Compound (VOC) Profiling

Across all samples, 85 VOCs were identified. A complete list with the VOCs identified within TP and CO samples is shown in Table S3. A significantly higher mean number of VOCs was detected in the TP group (mean 72 SE $\pm$ 1.44) compared to CO (67 $\pm$ 1.21) ($p = 0.03$, *t*-test). For the TP_21 group the mean number of VOCs (67 $\pm$ 2.43) was not significantly different from CO (72 $\pm$ 1.21, $p = 0.12$, *t*-test). Principal component analysis did not reveal any distinct clustering of groups Figure S4A,B.

In total, 24 VOCs were significantly different in abundance between TP and CO ($p < 0.05$, *t*-test, pre-correction for multiple comparisons) (Table 2). For TP_21 and CO, 17 VOCs were significantly different in abundance between groups ($p < 0.05$, *t*-test, pre-correction for multiple comparisons) (Table 2). For visualisation purposes, the 3 VOCs with the smallest *p*-values from both tapeworm groups were plotted as box and whisker plots in Figure 4. For strongyle FEC, there were two VOCs in significantly greater abundance in the High _FEC group ($p < 0.05$, *t*-test, pre-correction for multiple comparisons) (Table 2). However, after correction for multiple comparisons, none of the VOCs were significantly different in abundance between horses with tapeworm and controls or between those with high and low strongyle FECs. When TP and CO were included in a PERMANOVA model, tapeworm burden explained 7% ($p = 0.01$) of variation and strongyle FEC was not able to explain any variation (2%, $p = 0.23$). The inclusion of TP_21 and CO in a PERMANOVA model resulted in 8% ($p = 0.02$) of variation in the VOC profile explained by tapeworm burden and no effect of strongyle FEC (2%, $p = 0.55$).

Table 1. Taxa that were significantly different in abundance between tapeworm groups and controls and those with high and low FEC, irrespective of tapeworm infection.

Taxon	*p*-Value (FDR Corrected)	More Abundant in
TP vs. CO		
Family		
Bacteroidales UCG-001	0.016	CO
Genus		
Ruminococcaceae UCG-004	0.006	CO
Jeotgalicoccus	0.012	TP
Candidatus Soleaferrea	0.012	CO
Romboutsia	0.028	CO
TP_21 vs. CO	*p*-value (FDR corrected)	More abundant in
Order		
Rickettsiales	0.004	CO
Family		
Rickettsiales Incertae Sedis	0.005	CO
Genus		
Candidatus Hepatincola	0.01	CO
Selenomonas 3	0.02	TP_21
High_FEC vs. Low_FEC	*p*-value (FDR corrected)	More abundant in
Order		
Rhodospirillales	0.004	High_FEC
Family		
Fusobacteriaceae	0.003	High_FEC
Rhodospirillaceae	0.003	High_FEC
Bacteroidaceae	0.003	High_FEC
Genus		
Bacteroides	0.005	High_FEC
Fusobacterium	0.006	High_FEC
Thalassospira	0.007	High_FEC
Gordonibacter	0.021	High_FEC
Prevotellaceae Ga6A1 group	0.031	High_FEC

Differential abundance was calculated by the DESeq2 R package. Key: TP = tapeworm positive (n = 20), TP_21 = tapeworm samples with $\geq$21 worms (n = 9), CO = control (tapeworm negative, n = 28), Low_FEC = low strongyles ($\leq$10 e.p.g), High_FEC = high strongyles ($\geq$100 e.p.g). Key: FEC = faecal egg count, e.p.g = eggs per gram.

Table 2. Comparison of volatile organic compound (VOC) abundance from the colonic contents of horses positive for *Anoplocephala perfoliata* and non-infected controls.

TP vs. CO	*p*-Value	Adjusted *p*-Value	More Abundant in
VOCs			
Undecane, 2,6-dimethyl-	0.006	0.147	TP
Acetic acid	0.006	0.147	TP
Nonane	0.008	0.147	TP

Table 2. *Cont.*

TP vs. CO	*p*-Value	Adjusted *p*-Value	More Abundant in
Furan, 2-methyl-	0.009	0.147	TP
2-Undecanone	0.013	0.147	TP
1H-Pyrrole-2,5-dione, 3-ethyl-4-methyl-	0.013	0.147	TP
Dodecane, 2,6,10-trimethyl-	0.014	0.147	TP
3-Pentanone, 2-methyl-	0.015	0.147	TP
Heptane, 2-methyl-	0.017	0.147	TP
2-Heptanone, 6-methyl-	0.019	0.147	TP
2-Nonanone	0.020	0.147	TP
2-Octene, (E)-	0.025	0.147	TP
2-Decanone	0.026	0.147	TP
2-Octanone	0.026	0.147	TP
1-Penten-3-one	0.026	0.147	TP
Cyclohexanone, 2,2,6-trimethyl-	0.029	0.147	TP
3-Octanone	0.030	0.147	TP
D-Limonene	0.033	0.147	TP
2-Butanone	0.033	0.147	TP
Propanal	0.035	0.147	TP
Furan, 2-pentyl-	0.036	0.147	TP
2-Heptanone	0.038	0.147	TP
2-Butanone, 4-(2,6,6-trimethyl-1-cyclohexen-1-yl)-	0.042	0.156	TP
1-Cyclohexene-1-carboxaldehyde, 2,6,6-trimethyl-	0.044	0.156	TP
TP_21 vs. CO	*p*-value	Adjusted *p*-value	More abundant in
VOCs			
Undecane, 2,6-dimethyl-	0.001	0.127	TP_21
1H-Pyrrole-2,5-dione, 3-ethyl-4-methyl-	0.005	0.128	TP_21
Heptane, 2-methyl-	0.006	0.128	TP_21
2-Heptanone, 6-methyl-	0.006	0.128	TP_21
1-Cyclohexene-1-carboxaldehyde, 2,6,6-trimethyl-	0.012	0.174	TP_21
2-Undecanone	0.015	0.174	TP_21
Cyclohexanone, 2,2,6-trimethyl-	0.015	0.174	TP_21
2-Octene, (E)-	0.017	0.174	TP_21
Furan, 2-pentyl-	0.020	0.174	TP_21
Acetic acid	0.021	0.174	TP_21
Octane	0.023	0.178	TP_21
Nonane	0.027	0.189	TP_21
Decane	0.031	0.192	TP_21
Furan, 2-methyl-	0.034	0.192	TP_21
TP_21 vs. CO	*p*-value	Adjusted *p*-value	More abundant in
5-Hepten-2-one, 6-methyl-	0.036	0.192	TP_21
2-Decanone	0.036	0.192	TP_21
1-Octen-3-ol	0.041	0.203	TP_21
High vs. Low FEC			
VOCs			
1-Propanol, 2-methyl	0.0219	0.903	High_FEC
1-Hexanol	0.0381	0.903	High_FEC

Results are shown for VOCs which were significantly different in abundance ($p < 0.05$, *t*-test) prior to FDR and their corrected *p*-values. Key: TP = tapeworm positive (n = 20), TP_21 = tapeworm samples with $\geq$21 worms (n = 9), CO = control (tapeworm negative, n = 28), FEC = faecal egg count, High_FEC = $\geq$100 strongyle eggs per gram (n = 24), Low_FEC = $\leq$10 strongyle eggs per gram (n = 10).

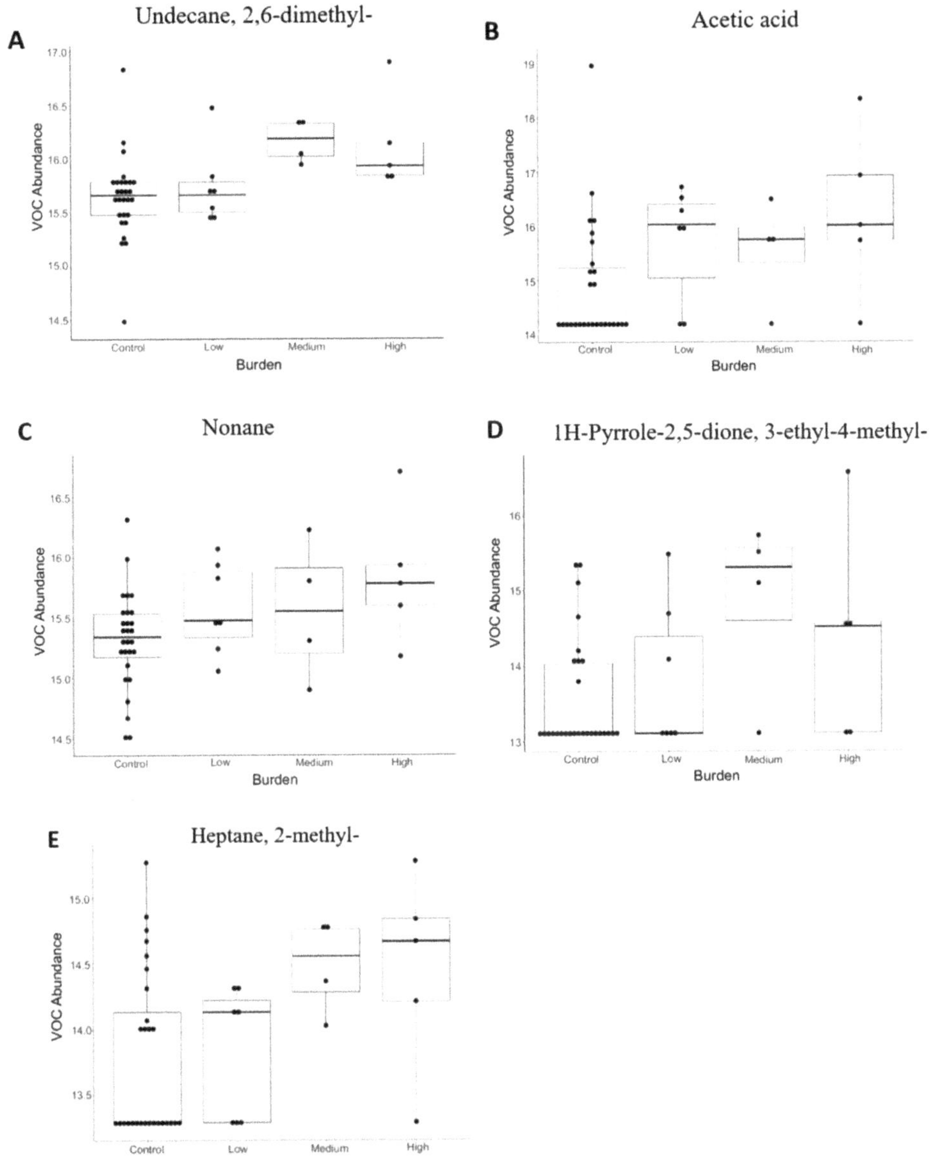

Figure 4. Box and whisker plots of VOC abundance of the colonic contents of horses infected with *Anoplocephala perfoliata* and non-infected controls. The three volatile organic compounds (VOCs) with the smallest *p*-values, identified by *t*-test for both TP vs. CO (**A–C**) and TP_21 vs. CO (**A,D,E**) are shown. Plots were constructed to show VOC abundance change in individual groups: control, low (1–20 worms), medium (21–49 worms) and high (≥50 worms) to show gradient change of compounds with level of tapeworm burden.

3.3. Integrated-Omics

The DIABLO model generated a balanced error rate of 35–44% (Figure S5). A Pearson's correlation plot demonstrated a strong correlation between bacterial OTUs and VOCs, but the combination of these omics resulted in only subtle clustering of TP and

CO (Figure 5A,B). Overall, bacterial OTUs were better at separating TP and CO groups than VOCs (Figure 5C,D). A total of 1257 correlations (0.3 and above) were identified between OTUs and VOCs. Correlations identified using the plotVar() function are shown in a heatmap in Figure 6. To assess the use of VOCs as markers for OTUs in higher or lower abundance in control or tapeworm samples identified in single omics, a table of VOCs and OTUs which were significant in the single omics analysis (pre-FDR corrected) and had significant correlations in the integrated analysis was constructed (Table 3). Of interest, OTU2331 and OTU147 (both Prevotellaceae) were in higher abundance in TP in single omics analysis and each positively correlated (0.36–0.6) with 17 and 19 VOCs respectively, which were also in significantly greater abundance in TP in single omics analysis. OTU2051 (Rikenellaceae), identified in higher abundance in the CO group, was found to be negatively correlated (−0.6) with furan,-2-pentyl and 5-hepten-2-one-6-methyl.

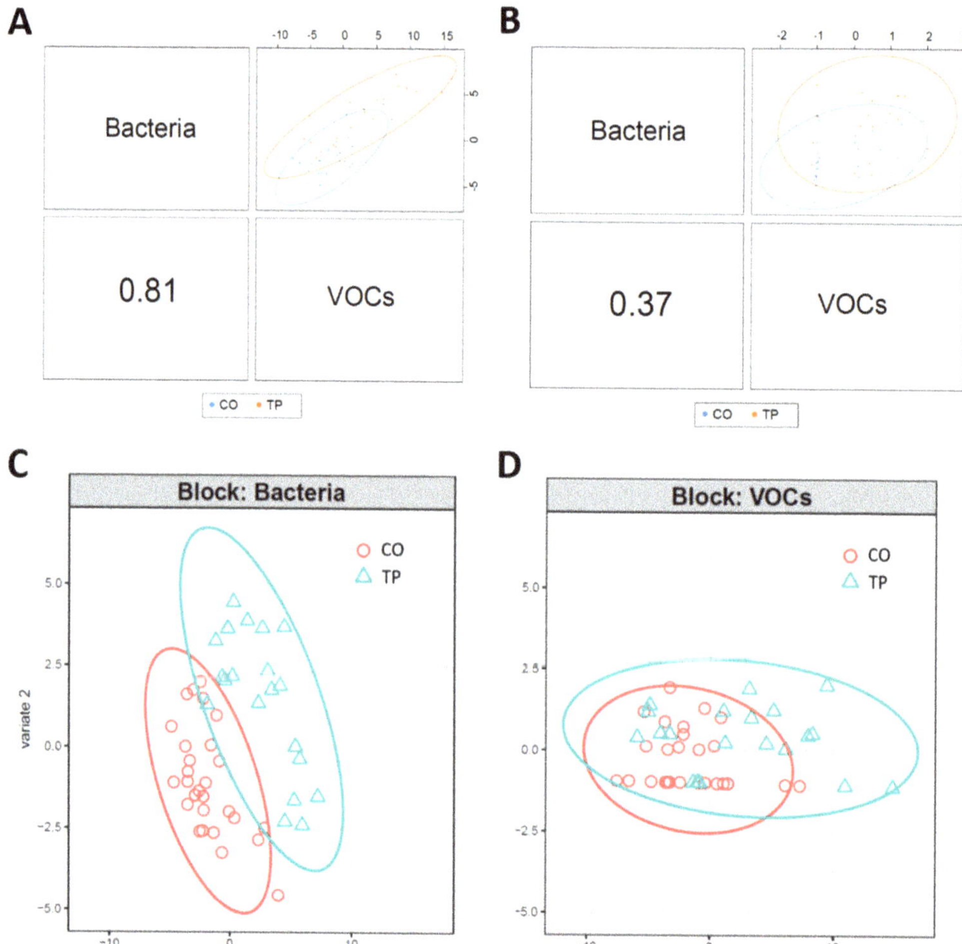

Figure 5. A Pearson's correlation plot of bacteria and volatile organic compound (VOC) data of the colonic contents of horses infected with *Anoplocephala perfoliata* and non-infected controls. Component 1 is shown in (**A**) and component 2 is shown in (**B**). The ability of the model to separate CO and TP by bacteria alone is demonstrated by plot (**C**), and VOCs alone by plot (**D**). Key: TP = tapeworm positive (n = 20), CO = control (tapeworm negative, n = 28).

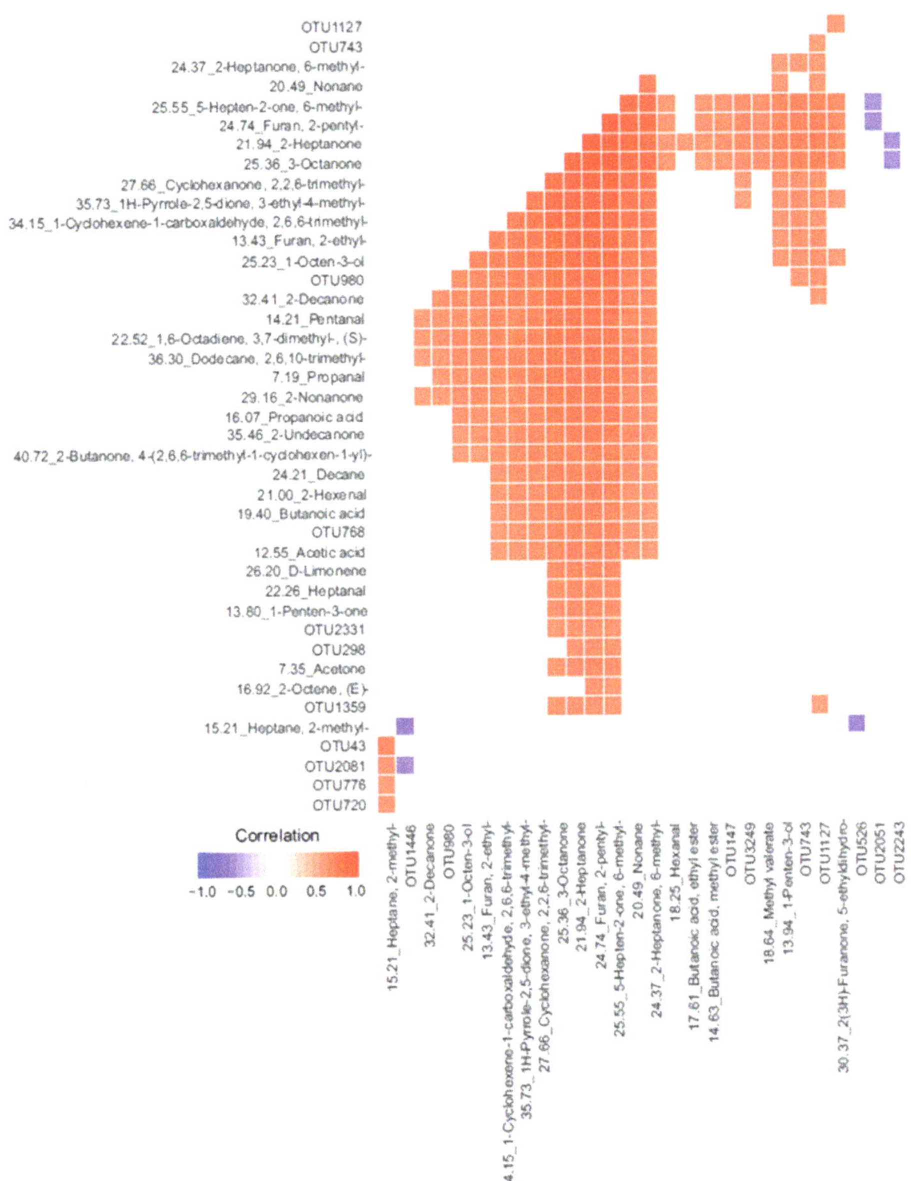

Figure 6. A heatmap of correlations (sorted by correlation values) between bacterial operational taxonomic units (OTUs) and volatile organic compounds (VOCs) found in the colonic contents of horses infected with *Anoplocephala perfoliata* and non-infected controls. Details of OTU taxonomic identification are in Table S4.

Table 3. Operational taxonomic units (OTUs) and volatile organic compounds (VOCs) identified in single omics analysis that were significantly correlated with each other when integrated.

OTU	*p*-Value	OTU More Abundant in	VOCs Significantly Correlated	Correlation	VOC More Abundant in
OTU5074	<0.001	CO	Acetic acid	−0.42	TP
OTU3878	<0.001	TP	2-Nonanone	0.46	TP
			2-Decanone	0.43	TP
			1-Octen-3-ol	0.43	TP
OTU705	<0.001	CO	2-Octanone	−0.46	TP
			1-Octen-3-ol	−0.47	TP
OTU314	<0.001	TP	1H-Pyrrole-2,5-dione, 3-ethyl-4-methyl-	0.43	TP
			Cyclohexanone, 2,2,6-trimethyl-	0.41	TP
			Undecane, 2,6-dimethyl-	0.41	TP
			2-Heptanone, 6-methyl-	0.4	TP
			1-Cyclohexene-1-carboxaldehyde, 2,6,6-trimethyl-	0.4	TP
			Dodecane, 2,6,10-trimethyl-	0.37	TP
			Acetic acid	0.37	TP
OTU673	<0.001	CO	Heptane, 2-methyl-	−0.37	TP
OTU2331	<0.001	TP	2-Heptanone, 6-methyl-	0.6	TP
			Decane	0.58	TP
			Propanal	0.56	TP
			1-Cyclohexene-1-carboxaldehyde, 2,6,6-trimethyl-	0.55	TP
			Furan, 2-pentyl-	0.55	TP
			Cyclohexanone, 2,2,6-trimethyl-	0.53	TP
			Undecane, 2,6-dimethyl-	0.53	TP
			1-Octen-3-ol	0.51	TP
			1H-Pyrrole-2,5-dione, 3-ethyl-4-methyl-	0.48	TP
			2-Heptanone	0.45	TP
			Pentanal	0.45	TP
			2-Decanone	0.43	TP
			Furan, 2-methyl-	0.43	TP
			3-Octanone	0.42	TP
			2-Nonanone	0.4	TP
			2-Undecanone	0.4	TP
			Nonane	0.37	TP
OTU2051	<0.001	CO	Acetic acid	−0.37	TP
OTU	*p*-value	OTU more abundant in	VOCs significantly correlated	Correlation	VOC more abundant in
			1H-Pyrrole-2,5-dione, 3-ethyl-4-methyl-	−0.38	TP
			2-Heptanone	−0.39	TP
			1-Octen-3-ol	−0.41	TP
			2-Decanone	−0.45	TP
			2-Octanone	−0.47	TP
OTU142	0.015	CO	2-Nonanone	−0.37	TP
			Furan, 2-pentyl-	−0.38	TP
			2-Butanone, 4-(2,6,6-trimethyl-1-cyclohexen-1-yl)-	−0.4	TP

Table 3. Cont.

OTU	*p*-Value	OTU More Abundant in	VOCs Significantly Correlated	Correlation	VOC More Abundant in
			2-Decanone	−0.43	TP
			Propanal	−0.43	TP
			2-Heptanone	−0.43	TP
			1-Octen-3-ol	−0.44	TP
OTU147	0.015	TP	2-Heptanone	0.56	TP
			3-Octanone	0.53	TP
			1H-Pyrrole-2,5-dione, 3-ethyl-4-methyl-	0.53	TP
			Furan, 2-pentyl-	0.53	TP
			2-Decanone	0.51	TP
			2-Nonanone	0.51	TP
			1-Penten-3-one	0.49	TP
			Nonane	0.48	TP
			2-Heptanone, 6-methyl-	0.47	TP
			2-Undecanone	0.47	TP
			1-Octen-3-ol	0.45	TP
			Cyclohexanone, 2,2,6-trimethyl-	0.43	TP
			2-Octanone	0.43	TP
			1-Cyclohexene-1-carboxaldehyde, 2,6,6-trimethyl-	0.42	TP
			Propanal	0.42	TP
			Furan, 2-methyl-	0.41	TP
			Dodecane, 2,6,10-trimethyl-	0.4	TP
			Acetic acid	0.39	TP
			D-limonene	0.36	TP
OTU239	0.020	CO	Decane	0.41	TP
			2-Heptanone, 6-methyl-	0.39	TP

OTUs were identified as significantly different between CO and TP by DeSeq2 R package. VOC abundances were identified as significantly different (pre-FDR) between CO and TP by *t*-test. Correlations between OTUs and VOCs were recorded from 0.3 to 1 (−0.3 to −1) to indicate positive (negative) linear relationships. The correlations were calculated from a Pearson's correlation plot and the cor.test() function was used to determine if correlations were statistically significant. OTU classifications: OTU5074 Treponema 2, OTU3878 Alloprevotella, OTU705 Bacteroidales S24-7 group, OTU314 (Eubacterium) oxidoreducens group, OTU673 Lachnospiraceae NK4A136 group, OTU2331 Prevotellaceae UCG-003, OTU2051 Rikenellaceae RC9 gut group, OTU142 Ruminococcaceae UCG-005, OTU147 Prevotellaceae UCG-001, OTU239 Prevotella 1. CO = control (tapeworm negative, n = 28), TP = tapeworm positive (n = 20).

4. Discussion

This is the first study to investigate the intestinal microbiome and metabolome of horses with and without infection of the important equine parasite *A. perfoliata*, based on the gold standard of physical identification of parasites at post-mortem. Knowledge about whether tapeworm infection is associated with changes in the microbiome and metabolome of the horse may further our understanding of the parasite-microbiota interactions in the equine gut and help us to better understand the pathophysiology of tapeworm-associated colic.

The present study found that Firmicutes (51.6%) and Bacteroidetes (36.1%) were the dominant phyla in the samples obtained, which is consistent with other equine studies [21,52,53]. The bacterial diversity in samples from horses that were either tapeworm positive or negative (CO group) were similar, a finding consistent with many other studies comparing the gut microbiome of parasite infected and non-infected animals [21,22,54–57]. Significant differences in microbiota diversity in parasite positive and negative animals have been more widely reported in rodent and rabbit studies in which the level of infection and diet can be closely controlled and larger sample sizes can be used [58–61], and which is something that is difficult to practically achieve in studies of horses, which are usually client-owned rather than research horses.

Here, a decreased abundance of some fibrolytic bacteria considered symbiotic, including an unidentified genus belonging to the family Ruminococcaceae UCG-004, was observed in the tapeworm groups when compared to the controls [62,63]. Fewer Ruminococcaceae were also seen in rabbits and pigs (genus level, *Ruminococcus*) infected with intestinal parasites [60,64]. A decrease in *Ruminococcus* was previously identified as a consequence of strongyle infection in ponies identified as being susceptible [21]. The authors of the latter study suggested that a reduction in butanoic acid (because of a reduction in *Ruminococcus*) could influence the level of inflammation caused by infection. However, the authors did not measure butanoic acid, and in the present study, a correlation between butanoic acid and symbiotic bacteria was not observed. The role of symbiotic bacteria during parasite infection requires further investigation to determine if interactions occur as a result of, or as a pre-cursor to infection, and what consequences this may have for the host.

The genus *Selenomonas 3* was more abundant in the tapeworm positive (TP_21) group compared to non-tapeworm infected controls (CO). Little is known about *Selenomonas*. Previous studies have found that *Selenomonas ruminatum* was more abundant in goats infected with the gastrointestinal parasite *Haemonchus contortus* [55]. Some *Selenomonas* species have also been associated with inflammation in human patients with periodontal disease [65]. Species of *Selenomonas* are involved in the fermentation pathway of starch and sugars in the hindgut so an increased presence may be attributable to the diet or to an adaption of the host or gut microbiota to compete for nutrients with the parasite [66]. In the current study, just two out of nine horses in the TP_21 group had high levels of *Selenomonas 3*.

In general, as tapeworm burdens increase, severity of lesions, inflammation of the mucosa and the risk of colic increases [4,32]. However, fewer tapeworms are required to produce severe lesions at the ileo-caecal junction than the caecal wall [67]. Therefore, numbers of tapeworms may not have been fully representative of mucosal inflammation in this work, and the specific site and extent of inflammation was not recorded. Future studies should record sites of parasite attachment as well as taking mucosal samples from inflamed and non-inflamed regions, furthermore mucosal samples from controls would also help to determine parasite related and non-parasite related inflammation. A decrease in bacterial diversity, as well as an increase in inflammatory biomarkers, has been reported in horses post-treatment for cyathostomins [24]. Post-dosing colic has been linked to high ELISA optical densities for tapeworm, suggesting colic may be a result of the removal of significant tapeworm burdens [68]. The role of the gut microbiota and tapeworm in the development of or reduction in likelihood of tapeworm-associated equine colic remains unclear, but the subtle changes in the gut microbiota observed in this work warrant further study in this area. Sampling from gut regions where tapeworm attachment and associated gastrointestinal pathology (e.g., ileo-caecal junction and caecal wall) may provide further insight into whether parasite-microbiota interactions have a role in specific types of equine colic.

A PERMANOVA analysis revealed that tapeworm burden explained a small amount (7–8%) of variation in the VOC profile, but evidently, this was not enough to demonstrate clear clustering for tapeworm in PCA analysis. The VOC metabolome changed little with the FEC, and only two VOCs were significantly different in samples with high FEC (prior to FDR). There was no significant variation described in the PERMANOVA analysis. These findings were similar to a previous study, using a different metabolomics platform, where only moderate differences in faecal metabolite profiles were observed between horses with high and low FECs [23]. Other factors, which have been described previously to influence the equine hindgut microbiome and metabolome, including horse age, breed, diet, and history of gastrointestinal disease, as well as antibiotic and anthelmintic treatment [25,62,69–71], were not recorded in the current study. These factors may have been responsible for other variations observed in the VOC profile. The inability to record this metadata was the main limitation of this post-mortem study. For future investigations, to maximise the ability to determine variation influenced by parasites alone on the microbiota, extrinsic factors should be controlled for. However, this information may be difficult to obtain for horses submitted to an abattoir which, at present is the most practical method for sampling large numbers of horses using the gold standard for equine tapeworm diagnosis (direct evaluation of parasite status in the gastrointestinal tract post-mortem).

Integrated analysis demonstrated a strong correlation (0.81) between bacterial OTUs and VOCs. In the single omics analysis 69 OTUs were found to be significantly different between TP and CO. Of these, 10 were correlated with VOCs. A lack of correlation for the other 59 OTUs with VOCs could imply the organisms were not active, as DNA-based studies are not representative of active species [72]. Furthermore, the use of one metabolomics platform and extraction technique will not encompass all metabolites within a biological matrix so there is a possibility that metabolites directly correlating with the most abundant organisms were not detected here [73].

There were no significantly differences in abundance of VOCs between tapeworm infected and non-infected controls after p-values were adjusted for multiple comparisons. Despite not reaching statistical significance in the single omics analysis many of these VOCs were responsible for the variation shown in the Pearson's correlation plot (supervised analysis) and were correlated with several OTUs. This demonstrates that, although statistical significance was not reached after correction for multiple comparisons, these VOCs were still important in discriminating between tapeworm infected and non-infected controls when combined with other analysis and the study may have simply been under-powered when evaluated as a single omics. Furthermore, the use of supervised analysis in noisy datasets with confounding factors such as this, may be advantageous [74]. However, when supervised analysis is employed, it is important to check that the data are not being over fitted. In the present study the BER was used to evaluate the supervised model for over-fitting. We achieved a BER of 35–44%, an indication that the group separations observed between TP and CO should be interpreted with caution [75]. The high error rate may be because of the small sample size (more features than samples were included in the model) or because the differences between the two groups were not enough to accurately classify the samples.

Several VOCs in higher abundance in TP samples were correlated with two OTUs (OTU2331 and OTU147) belonging to the family Prevotellaceae. Members of the Prevotellaceae family have been associated with driving chronic intestinal inflammation in mice with inflammasome-mediated dysbiosis [76,77]. It may be reasonable to speculate that the more highly abundant OTUs were associated with parasite-driven inflammation. On the other hand, some of the correlating VOCs, including 2-heptanone, 1-octen-3-ol, 2-octanone and furan compounds, have previously been identified as being of possible fungal origin [78,79]. Furthermore, in healthy human subjects consuming a high-carbohydrate diet, a correlation was observed between the fungal genus *Candida* and members of Prevotellaceae (genus *Prevotella*) [80]. The authors speculated that *Candida* were able to break down complex carbohydrates into simple sugars to be fermented by *Prevotella*, to produce acetic acid. In the current work acetic acid was also more abundant in the tapeworm group, supporting this hypothesis. Fungal populations were not characterised in this study, so it can only be speculated that these VOCs were related to fungal metabolism. In addition, the OTUs of interest could not specifically be identified as *Prevotella*, but as members of the family Prevotellaceae, in which there are three other possible genera to which the OTUs may have belonged. In horses, this is one of the few studies to correlate findings of a 16S rRNA study with the gut metabolome in a statistical model. The integration of omics to address the same hypothesis was able to provide stronger evidence for the conclusions made. A larger sample size is required to clarify whether the combination of bacterial OTUs and VOCs can accurately distinguish between horses with and without tapeworm infection.

A comparison between horses with high and low strongyle FECs was performed to evaluate the association of FEC with the gut microbiome and metabolome as a potential confounding factor in this study. Here, we categorised horses into low and high FEC groups, as performed previously by others [22,23]. There was no significant association between the strongyle FEC and the overall gut microbiota diversity, as observed previously in faeces [21,22]. Differential analysis demonstrated some significant differences in microbial taxa at order, family and genus level between horses with high and low FECs that were not shared with tapeworm and control comparisons. An increase in members of the phylum Proteobacteria was observed in horses with high or a susceptibility to a high strongyle FEC [21,22], but this was not observed in tapeworm infected horses or in those in the High_FEC group in the present study, despite a larger sample size. However, the order Rhodospirillales, family and genus *Thalassospira* (all members of phylum Proteobacteria) were more abundant in those with High_FEC compared to Low_FEC. Possibly the levels of infection were not high enough to enable differences to be observed at phylum level between groups or other confounding factors of the current study, including mixed infections, a lack of control for diet, age, breed and anthelmintic and antibiotic treatment history. To date, differences in the microbiota between horses with high levels of mixed parasite burdens with those with low mixed burdens has not been compared and was not possible in the current study because of small sample sizes. However, investigating the impact of general high parasite burdens on the equine microbiota and overall health would be interesting and may our further knowledge of parasite-microbiota interactions in the horse. It should be noted that FEC and numbers of parasites present in the gastrointestinal tract have a weak correlation [81]. A validated serum ELISA for encysted cyathostomins has become available since the samples were collected for this study [82,83]. Future studies should aim to use this alongside other diagnostic methods.

Understanding the mechanisms responsible for microbiota, parasite and host interactions were beyond the scope of this study. Whether *A. perfoliata* infection and the gut microbiota interact to produce positive (stimulation of the immune system) or negative (a role in the development of colic) impacts on the host remains unknown. Others have proposed the idea that the use of probiotics and prebiotics to manipulate the gut microbiota may help control gastrointestinal parasites as an alternative to drugs [84]. However, before further exploration is performed, the wider impacts of such therapies on the host should be considered.

5. Conclusions

This is the first study to integrate microbiome and metabolome data in relation with the tapeworm status of the horse. Furthermore, the microbiome and metabolome of gut contents have not previously been reported in previous equine parasitology studies. We found statistically significant correlations between specific OTUs and the presence of tapeworm. Some VOCs and bacterial OTUs were also correlated. The relationship between VOCs and tapeworm infection could not be demonstrated after adjustment for multiple comparisons. Further investigation of the equine gut microbiome and metabolome using larger studies is therefore warranted.

Supplementary Materials: The following are available online at https://www.mdpi.com/2076-261 5/11/3/755/s1, Table S1: Parasitology results of horses included in the study, Table S2: Barcoded index primers, Table S3: A list of VOCs identified and frequencies of occurrence in TP and CO samples, Table S4: Taxonomic identification of OTUs in Figure 6., Figure S1: Comparison of VOC numbers and chromatogram of colon contents freeze-dried (FD) and not freeze-dried (NFD), Figure S2: (**A**) Significant OTUs comparing TP and CO. (**B**) Significant OTUs comparing TP_21 and CO., Figure S3 Significant OTUs between Low_FEC and High_FEC groups, Figure S4: (**A**) PCA of VOC profile (tapeworm positive = TP and tapeworm negative = CO), (**B**) PCA of VOC profile ($\geq$21 tapeworms = TP_21 and tapeworm negative = CO), Figure S5: Balanced error rate for mixOmics model.

Author Contributions: Conceptualization, C.P., D.A., J.H., R.S.; investigation, D.A., R.S.; resources, C.P., D.A., J.H.; formal analysis, A.F., R.S.; draft preparation, R.S.; writing—review and editing, R.S., C.P., D.A., J.H., A.F.; supervision, C.P., D.A. All authors have read and agreed to the published version of the manuscript.

Funding: This research received no external funding.

Institutional Review Board Statement: Ethical review and approval were waived for this study, as samples were abattoir-sourced.

Data Availability Statement: The raw microbiome and metabolome data are available at https://data.mendeley.com/datasets/44sn7kbn6x/1 (accessed on 29 November 2020).

Acknowledgments: We would like to thank Umer Zeeshan Ijaz (University of Glasgow) for writing the microbiome and mixOmics R scripts, Shebl Salem (University of Liverpool) for writing some of the metabolomics R scripts, The Centre for Genomic Research (University of Liverpool) for performing the sequencing, demultiplexing, quality filtering and adaptors trimmings of reads. We would also like to thank Cara Hallowell-Evans (University of Liverpool) for assistance with strongyle faecal egg counting. We acknowledge Ibrandify/Freepik for the design of some of the icons used in the graphical abstract.

Conflicts of Interest: The authors declare no conflict of interest.

References

1. Tomczuk, K.; Kostro, K.; Szczepaniak, K.O.; Grzybek, M.; Studzińska, M.; Demkowska-Kutrzepa, M.; Roczeń-Karczmarz, M. Comparison of the sensitivity of coprological methods in detecting Anoplocephala perfoliata invasions. *Parasitol. Res.* **2014**, *113*, 2401–2406. [CrossRef] [PubMed]
2. Sallé, G.; Guillot, J.; Tapprest, J.; Foucher, N.; Sevin, C.; Laugier, C. Compilation of 29 years of postmortem examinations identifies major shifts in equine parasite prevalence from 2000 onwards. *Int. J. Parasitol.* **2020**, *50*, 125–132. [CrossRef]
3. Owen, R.A.; Jagger, D.W.; Quan-Taylor, R. Caecal intussusceptions in horses and the significance of Anoplocephala perfoliata. *Vet. Rec.* **1989**, *124*, 34–37. [CrossRef] [PubMed]
4. Proudman, C.J.; French, N.P.; Trees, A.J. Tapeworm infection is a significant risk factor for spasmodic colic and ileal impaction colic in the horse. *Equine Vet. J.* **1998**, *30*, 194–199. [CrossRef] [PubMed]
5. Proudman, C.J.; Holdstock, N.B. Investigation of an outbreak of tapeworm-associated colic in a training yard. *Equine Vet. J.* **2000**, *32*, 37–41. [CrossRef]
6. Ryu, S.H.; Bak, U.B.; Kim, J.G.; Yoon, H.J.; Seo, H.S.; Kim, J.T.; Park, J.Y.; Lee, C.W. Cecal rupture by Anoplocephala perfoliata infection in a thoroughbred horse in Seoul Race Park, South Korea. *J. Vet. Sci.* **2001**, *2*, 189–193. [CrossRef] [PubMed]
7. Boswinkel, M.; Sloet van Oldruitenborgh-Oosterbann, M. Correlation between colic and antibody levels against Anoplocephala perfoliata in horses in The Netherlands. *Tijdschr. Diergeneeskd.* **2007**, *132*, 508–512. [PubMed]
8. Nielsen, M.; Reinemeyer, C.; Donecker, J.; Leathwick, D.; Marchiondo, A.; Kaplan, R. Anthelmintic resistance in equine parasites—Current evidence and knowledge gaps. *Vet. Parasitol.* **2014**, *204*, 55–63. [CrossRef] [PubMed]
9. Nielsen, M.K. Equine tapeworm infections: Disease, diagnosis and control. *Equine Vet. Educ.* **2015**, *28*, 388–395. [CrossRef]
10. Lyons, E.; Bellaw, J.; Dorton, A.; Tolliver, S. Efficacy of moxidectin and an ivermectin-praziquantel combination against ascarids, strongyles, and tapeworms in Thoroughbred yearlings in field tests on a farm in Central Kentucky in 2016. *Vet. Parasitol. Reg. Stud. Rep.* **2017**, *8*, 123–126. [CrossRef]
11. Lightbody, K.L.; Davis, P.J.; Austin, C.J. Validation of a novel saliva-based ELISA test for diagnosing tapeworm burden in horses. *Vet. Clin. Pathol.* **2016**, *45*, 335–346. [CrossRef] [PubMed]
12. Proudman, C.; Trees, A. Use of excretory/secretory antigens for the serodiagnosis of Anoplocephala perfoliata cestodosis. *Vet. Parasitol.* **1996**, *61*, 239–247. [CrossRef]
13. Abbott, J.B.; Mellor, D.J.; Barrett, E.J.; Proudman, C.J.; Love, S. Serological changes observed in horses infected with Anoplocephala perfoliata after treatment with praziquantel and natural reinfection. *Vet. Rec.* **2008**, *162*, 50–53. [CrossRef] [PubMed]
14. Proudman, C.; Trees, A. Correlation of antigen specific IgG and IgG(T) responses with Anoplocephala perfoliata infection intensity in the horse. *Parasite Immunol.* **1996**, *18*, 499–506. [CrossRef] [PubMed]
15. Proudman, C.; Edwards, G. Validation of a centrifugation/floatation technique for the diagnosis of equine cestodiasis. *Vet. Rec.* **1992**, *131*, 71–72. [CrossRef]
16. Bergman, E.N. Energy contributions of volatile fatty acids from the gastrointestinal tract in various species. *Physiol. Rev.* **1990**, *70*, 567–590. [CrossRef] [PubMed]
17. Daly, K.; Stewart, C.S.; Flint, H.J.; Shirazi-Beechey, S.P. Bacterial diversity within the equine large intestine as revealed by molecular analysis of cloned 16S rRNA genes. *FEMS Microbiol. Ecol.* **2001**, *38*, 141–151. [CrossRef]

18. Reynolds, L.A.; Smith, K.A.; Filbey, K.J.; Harcus, Y.; Hewitson, J.P.; Redpath, S.A.; Valdez, Y.; Yebra, M.J., Finlay, B.B.; Maizels, R.M. Commensal-pathogen interactions in the intestinal tract: Lactobacilli promote infection with, and are promoted by, helminth parasites. *Gut Microbes* **2014**, *5*, 522–532. [CrossRef] [PubMed]

19. Zaiss, M.M.; Harris, N.L. Interactions between the intestinal microbiome and helminth parasites. *Parasite Immunol.* **2016**, *38*, 5–11. [CrossRef]

20. Brosschot, T.P.; Reynolds, L.A. The impact of a helminth-modified microbiome on host immunity. *Mucosal Immunol.* **2018**, *11*, 1039–1046. [CrossRef]

21. Clark, A.; Sallé, G.; Ballan, V.; Reigner, F.; Meynadier, A.; Cortet, J.; Koch, C.; Riou, M.; Blanchard, A.; Mach, N. Strongyle Infection and Gut Microbiota: Profiling of Resistant and Susceptible Horses over a grazing season. *Front. Physiol.* **2018**, *9*, 272. [CrossRef] [PubMed]

22. Peachey, L.; Molena, R.; Jenkins, T.; Di Cesare, A.; Traversa, D.; Hodgkinson, J.; Cantacessi, C. The relationships between faecal egg counts and gut microbial composition in UK Thoroughbreds infected by cyathostomins. *Int. J. Parasitol.* **2018**, *48*, 403–412. [CrossRef] [PubMed]

23. Peachey, L.E.; Castro, C.; Molena, R.A.; Jenkins, T.P.; Griffin, J.L.; Cantacessi, C. Dysbiosis associated with acute helminth infections in herbivorous youngstock—observations and implications. *Sci. Rep.* **2019**, *9*, 11121. [CrossRef]

24. Walshe, N.; Duggan, V.; Cabrera-Rubio, R.; Crispie, F.; Cotter, P.; Feehan, O.; Mulcahy, G. Removal of adult cyathostomins alters faecal microbiota and promotes an inflammatory phenotype in horses. *Int. J. Parasitol.* **2019**, *49*, 489–500. [CrossRef]

25. Daniels, S.P.; Leng, J.; Swann, J.R.; Proudman, C.J. Bugs and drugs: A systems biology approach to characterising the effect of moxidectin on the horse's faecal microbiome. *Anim. Microbiome* **2020**, *2*, 1–14. [CrossRef]

26. Korpi, A.; Järnberg, J.; Pasanen, A.-L. Microbial volatile organic compounds. *Crit. Rev. Toxicol.* **2009**, *39*, 139–193. [CrossRef]

27. Proudman, C.; Hunter, J.O.; Darby, A.C.; Escalona, E.E.; Batty, C.; Turner, C. Characterisation of the faecal metabolome and microbiome of Thoroughbred racehorses. *Equine Vet. J.* **2014**, *47*, 580–586. [CrossRef] [PubMed]

28. Salem, S.E.; Hough, R.; Probert, C.; Maddox, T.W.; Antczak, P.; Ketley, J.M.; Williams, N.J.; Stoneham, S.J.; Archer, D.C. A longitudinal study of the faecal microbiome and metabolome of periparturient mares. *PeerJ* **2019**, *7*, e6687. [CrossRef] [PubMed]

29. Walshe, N.; Mulcahy, G.; Hodgkinson, J.; Peachey, L. No Worm is an island; the influence of commensal gut microbiota on cyathostomin infections. *Animals* **2020**, *10*, 2309. [CrossRef] [PubMed]

30. Costa, M.; Silva, G.; Ramos, R.; Staempfli, H.; Arroyo, L.; Kim, P.; Weese, J. Characterization and comparison of the bacterial microbiota in different gastrointestinal tract compartments in horses. *Vet. J.* **2015**, *205*, 74–80. [CrossRef]

31. Dougal, K.; Harris, P.A.; Edwards, A.; Pachebat, J.A.; Blackmore, T.M.; Worgan, H.J.; Newbold, C.J. A comparison of the microbiome and the metabolome of different regions of the equine hindgut. *FEMS Microbiol. Ecol.* **2012**, *82*, 642–652. [CrossRef]

32. Pearson, G.R.; Davies, L.W.; White, A.L.; O'Brien, J.K. Pathological lesions associated with Anoplocephala perfoliata at the ileo-caecal junction of horses. *Vet. Rec.* **1993**, *132*, 179–182. [CrossRef]

33. Christie, M.; Jackson, F. Specific identification of strongyle eggs in small samples of sheep faeces. *Res. Vet. Sci.* **1982**, *32*, 113–117. [CrossRef]

34. Blackmore, T.M.; Dugdale, A.; Argo, C.M.; Curtis, G.; Pinloche, E.; Harris, P.A.; Worgan, H.J.; Girdwood, S.E.; Dougal, K.; Newbold, C.J.; et al. Strong stability and host specific bacterial community in faeces of ponies. *PLoS ONE* **2013**, *8*, e75079. [CrossRef]

35. D'Amore, R.; Ijaz, U.Z.; Schirmer, M.; Kenny, J.G.; Gregory, R.; Darby, A.C.; Shakya, M.; Podar, M.; Quince, C.; Hall, N. A comprehensive benchmarking study of protocols and sequencing platforms for 16S rRNA community profiling. *BMC Genom.* **2016**, *17*, 55. [CrossRef] [PubMed]

36. Bates, S.T.; Berg-Lyons, D.; Caporaso, J.G.; Walters, W.A.; Knight, R.; Fierer, N. Examining the global distribution of dominant archaeal populations in soil. *ISME J.* **2010**, *5*, 908–917. [CrossRef] [PubMed]

37. Hough, R.; Archer, D.; Probert, C. A comparison of sample preparation methods for extracting volatile organic compounds (VOCs) from equine faeces using HS-SPME. *Metabolomics* **2018**, *14*, 19. [CrossRef] [PubMed]

38. Martin, M. Cutadapt removes adapter sequences from high-throughput sequencing reads. *EMBnet J.* **2011**, *17*, 10–12. [CrossRef]

39. Joshi, N.; Fass, J. Sickle: A Sliding-Window, Adaptive, Quality-Based Trimming Tool for FastQ Files. Available online: https://github.com/najoshi/sickle/releases/tag/v1.2 (accessed on 29 November 2020).

40. Mahé, F.; Rognes, T.; Quince, C.; De Vargas, C.; Dunthorn, M. Swarm v2: Highly-scalable and high-resolution amplicon clustering. *PeerJ* **2015**, *3*, e1420. [CrossRef] [PubMed]

41. Edgar, R.C.; Haas, B.J.; Clemente, J.C.; Quince, C.; Knight, R. Uchime improves sensitivity and speed of chimera detection. *Bioinformatics* **2011**, *27*, 2194–2200. [CrossRef]

42. Caporaso, J.G.; Kuczynski, J.; Stombaugh, J.; Bittinger, K.; Bushman, F.D.; Costello, E.K.; Fierer, N.; Peña, A.G.; Goodrich, J.K.; Gordon, J.I.; et al. QIIME allows analysis of high-throughput community sequencing data. *Nat. Methods* **2010**, *7*, 335–336. [CrossRef] [PubMed]

43. Edgar, R.C. Search and clustering orders of magnitude faster than BLAST. *Bioinformatics* **2010**, *26*, 2460–2461. [CrossRef] [PubMed]

44. Quast, C.; Pruesse, E.; Yilmaz, P.; Gerken, J.; Schweer, T.; Yarza, P.; Peplies, J.; Glöckner, F.O. The SILVA ribosomal RNA gene database project: Improved data processing and web-based tools. *Nucleic Acids Res.* **2012**, *41*, 590–596. [CrossRef] [PubMed]

45. Caporaso, J.G.; Bittinger, K.; Bushman, F.D.; DeSantis, T.Z.; Andersen, G.L.; Knight, R. PyNAST: A fexible tool for aligning sequences to a template alignment. *Bioinformatics* **2010**, *26*, 266–267. [CrossRef] [PubMed]

46. Price, M.N.; Dehal, P.S.; Arkin, A.P. FastTree 2—Approximately maximum-likelihood trees for large alignments. *PLoS ONE* **2010**, *5*, e9490. [CrossRef]

47. Ijaz, U.Z.; Sivaloganathan, L.; McKenna, A.; Richmond, A.; Kelly, C.; Linton, M.; Stratakos, A.C.; Lavery, U.; Elmi, A.; Wren, B.W.; et al. Comprehensive longitudinal microbiome analysis of the chicken cecum reveals a shift from competitive to environmental drivers and a window of opportunity for campylobacter. *Front. Microbiol.* **2018**, *9*, 2452. [CrossRef]

48. Oksanen, J.; Blanchet, F.; Kindt, R.; Legendre, P.; Michen, P.; O'Hara, R.; Simpson, G.; Solymos, P.; Stevens, M.; Wagner, H. *Vegan*: Ordination Methods, Diversity Analysis and Other Functions for Community and Vegetation Ecologists. Available online: https://cran.r-project.org/web/packages/vegan/index.html (accessed on 29 November 2020).

49. Love, M.I.; Huber, W.; Anders, S. Moderated estimation of fold change and dispersion for RNA-seq data with DESeq2. *Genome Biol.* **2014**, *15*, 550. [CrossRef]

50. Aggio, R.; Villas-Bôas, S.G.; Ruggiero, K. Metab: An R package for high-throughput analysis of metabolomics data generated by GC-MS. *Bioinformatics* **2011**, *27*, 2316–2318. [CrossRef] [PubMed]

51. Rohart, F.; Gautier, B.; Singh, A.; Lê Cao, K.-A. mixOmics: An R package for 'omics feature selection and multiple data integration. *PLoS Comput. Biol.* **2017**, *13*, e1005752. [CrossRef]

52. Steelman, S.M.; Chowdhary, B.P.; Dowd, S.; Suchodolski, J.; Janečka, J.E. Pyrosequencing of 16S rRNA genes in fecal samples reveals high diversity of hindgut microflora in horses and potential links to chronic laminitis. *BMC Vet. Res.* **2012**, *8*, 231. [CrossRef]

53. Weese, J.S.; Holcombe, S.J.; Embertson, R.M.; Kurtz, K.A.; Roessner, H.A.; Jalali, M.; Wismer, S.E. Changes in the faecal microbiota of mares precede the development of post partum colic. *Equine Vet. J.* **2014**, *47*, 641–649. [CrossRef] [PubMed]

54. Duarte, A.M.; Jenkins, T.P.; Latrofa, M.S.; Giannelli, A.; Papadopoulos, E.; De Carvalho, L.M.; Nolan, M.J.; Otranto, D.; Cantacessi, C. Helminth infections and gut microbiota—A feline perspective. *Parasites Vectors* **2016**, *9*, 625. [CrossRef] [PubMed]

55. Li, R.W.; Li, W.; Sun, J.; Yu, P.; Baldwin, R.L.; Urban, J.F. The effect of helminth infection on the microbial composition and structure of the caprine abomasal microbiome. *Sci. Rep.* **2016**, *6*, 20606. [CrossRef] [PubMed]

56. Šlapeta, J.; Dowd, S.E.; Alanazi, A.D.; Westman, M.E.; Brown, G.K. Differences in the faecal microbiome of non-diarrhoeic clinically healthy dogs and cats associated with Giardia duodenalis infection: Impact of hookworms and coccidia. *Int. J. Parasitol.* **2015**, *45*, 585–594. [CrossRef]

57. Li, R.W.; Wu, S.; Li, W.; Huang, Y.; Gasbarre, L.C. Metagenome plasticity of the bovine abomasal microbiota in immune animals in response to Ostertagia Ostertagi infection. *PLoS ONE* **2011**, *6*, e24417. [CrossRef]

58. Houlden, A.; Hayes, K.S.; Bancroft, A.J.; Worthington, J.J.; Wang, P.; Grencis, R.K.; Roberts, I.S. Chronic trichuris muris infection in C57BL/6 mice causes significant changes in host microbiota and metabolome: Effects reversed by pathogen clearance. *PLoS ONE* **2015**, *10*, e0125945. [CrossRef]

59. Jenkins, T.P.; Peachey, L.E.; Ajami, N.J.; Macdonald, A.S.; Hsieh, M.H.; Brindley, P.J.; Cantacessi, C.; Rinaldi, G. Schistosoma mansoni infection is associated with quantitative and qualitative modifications of the mammalian intestinal microbiota. *Sci. Rep.* **2018**, *8*, 12072. [CrossRef]

60. Cattadori, I.M.; Sebastian, A.; Hao, H.; Katani, R.; Albert, I.; Eilertson, K.E.; Kapur, V.; Pathak, A.; Mitchell, S. Impact of helminth infections and nutritional constraints on the small intestine microbiota. *PLoS ONE* **2016**, *11*, e0159770. [CrossRef]

61. Arrazuria, R.; Elguezabal, N.; Juste, R.A.; Derakhshani, H.; Khafipour, E. Mycobacterium avium subspecies paratuberculosis infection modifies gut microbiota under different dietary conditions in a rabbit model. *Front. Microbiol.* **2016**, *7*, 446. [CrossRef]

62. Daly, K.; Proudman, C.J.; Duncan, S.H.; Flint, H.J.; Dyer, J.; Shirazi-Beechey, S.P. Alterations in microbiota and fermentation products in equine large intestine in response to dietary variation and intestinal disease. *Br. J. Nutr.* **2011**, *107*, 989–995. [CrossRef]

63. Biddle, A.S.; Stewart, L.; Blanchard, J.L.; Leschine, S. Untangling the genetic basis of fibrolytic specialization by Lachnospiraceae and Ruminococcaceae in diverse gut communities. *Diversity* **2013**, *5*, 627–640. [CrossRef]

64. Wu, S.; Li, R.W.; Li, W.; Beshah, E.; Dawson, H.D.; Urban, J.F. Worm burden-dependent disruption of the porcine colon microbiota by Trichuris suis infection. *PLoS ONE* **2012**, *7*, e35470. [CrossRef]

65. Nagpal, D.; Prakash, S.; Bhat, K.; Singh, G. Detection and comparison of Selenomonas sputigena in subgingival biofilms in chronic and aggresive periodontitis patients. *Indian Soc. Periodontol.* **2016**, *20*, 286–291.

66. Shirazi-Beechey, S.P. Molecular insights into dietary induced colic in the horse. *Equine Vet. J.* **2008**, *40*, 414–421. [CrossRef]

67. Williamson, R.; Gasser, R.; Middleton, D.; Beveridge, I. The distribution of Anoplocephala perfoliata in the intestine of the horse and associated pathological changes. *Vet. Parasitol.* **1997**, *73*, 225–241. [CrossRef]

68. Barrett, E.J.; Blair, C.W.; Farlam, J.; Proudman, C.J. Postdosing colic and diarrhoea in horses with serological evidence of tapeworm infection. *Vet. Rec.* **2005**, *156*, 252–253. [CrossRef] [PubMed]

69. Dougal, K.; De La Fuente, G.; Harris, P.A.; Girdwood, S.E.; Pinloche, E.; Geor, R.J.; Nielsen, B.D.; Schott, H.C.; Elzinga, S.; Newbold, C.J. Characterisation of the faecal bacterial community in adult and elderly horses fed a high fibre, high oil or high starch diet using 454 pyrosequencing. *PLoS ONE* **2014**, *9*, e87424. [CrossRef]

70. Yamano, H.; Koike, S.; Kobayashi, Y.; Hata, H. Phylogenetic analysis of hindgut microbiota in Hokkaido native horses compared to light horses. *Anim. Sci. J.* **2008**, *79*, 234–242. [CrossRef]

71. Costa, M.C.; Stämpfli, H.R.; Arroyo, L.G.; Allen-Vercoe, E.; Gomes, R.G.; Weese, J.S. Changes in the equine fecal microbiota associated with the use of systemic antimicrobial drugs. *BMC Vet. Res.* **2015**, *11*, 19. [CrossRef]

72. Schirmer, M.; Franzosa, E.A.; Lloyd-Price, J.; McIver, L.J.; Schwager, R.; Poon, T.W.; Ananthakrishnan, A.N.; Andrews, E.; Barron, G.; Lake, K.; et al. Dynamics of metatranscription in the inflammatory bowel disease gut microbiome. *Nat. Microbiol.* **2018**, *3*, 337–346. [CrossRef]

73. De Raad, M.; Fischer, C.R.; Northen, T.R. High-throughput platforms for metabolomics. *Curr. Opin. Chem. Biol.* **2016**, *30*, 7–13. [CrossRef]

74. Gromski, P.S.; Muhamadali, H.; Ellis, D.I.; Xu, Y.; Correa, E.; Turner, M.L.; Goodacre, R. A tutorial review: Metabolomics and partial least squares-discriminant analysis—a marriage of convenience or a shotgun wedding. *Anal. Chim. Acta* **2015**, *879*, 10–23. [CrossRef] [PubMed]

75. Bijlsma, S.; Bobeldijk, I.; Verheij, E.R.; Ramaker, R.; Kochhar, S.; Macdonald, I.A.; Van Ommen, B.; Smilde, A.K. Large-scale human metabolomics studies: A strategy for data (pre-) processing and validation. *Anal. Chem.* **2006**, *78*, 567–574. [CrossRef]

76. Scher, J.U.; Sczesnak, A.; Longman, R.S.; Segata, N.; Ubeda, C.; Bielski, C.; Rostron, T.; Cerundolo, V.; Pamer, E.G.; Abramson, S.B.; et al. Expansion of intestinal Prevotella copri correlates with enhanced susceptibility to arthritis. *eLife* **2013**, *2*, e01202. [CrossRef]

77. Elinav, E.; Strowig, T.; Kau, A.L.; Henao-Mejia, J.; Thaiss, C.A.; Booth, C.J.; Peaper, D.R.; Bertin, J.; Eisenbarth, S.C.; Gordon, J.I.; et al. NLRP6 inflammasome regulates colonic microbial ecology and risk for colitis. *Cell* **2011**, *145*, 745–757. [CrossRef] [PubMed]

78. Effmert, U.; Kalderás, J.; Warnke, R.; Piechulla, B. Volatile mediated interactions between bacteria and fungi in the soil. *J. Chem. Ecol.* **2012**, *38*, 665–703. [CrossRef] [PubMed]

79. Hertel, M.; Hartwig, S.; Schütte, E.; Gillissen, B.; Preissner, R.; Schmidt-Westhausen, A.M.; Paris, S.; Kastner, I.; Preissner, S. Identification of signature volatiles to discriminate Candida albicans, glabrata, krusei and tropicalis using gas chromatography and mass spectrometry. *Mycoses* **2016**, *59*, 117–126. [CrossRef] [PubMed]

80. Hoffmann, C.; Dollive, S.; Grunberg, S.; Chen, J.; Li, H.; Wu, G.D.; Lewis, J.D.; Bushman, F.D. Archaea and fungi of the human gut microbiome: Correlations with diet and bacterial residents. *PLoS ONE* **2013**, *8*, e66019. [CrossRef]

81. Nielsen, M.; Baptiste, K.; Tolliver, S.; Collins, S.; Lyons, E. Analysis of multiyear studies in horses in Kentucky to ascertain whether counts of eggs and larvae per gram of feces are reliable indicators of numbers of strongyles and ascarids present. *Vet. Parasitol.* **2010**, *174*, 77–84. [CrossRef] [PubMed]

82. Dowdall, S.M.J.; Matthews, J.B.; Mair, T.; Murphy, D.; Love, S.; Proudman, C.J. Antigen-specific IgG(T) responses in natural and experimental cyathostominae infection in horses. *Vet. Parasitol.* **2002**, *106*, 225–242. [CrossRef]

83. Tzelos, T.; Geyer, K.K.; Mitchell, M.C.; McWilliam, H.E.G.; Kharchenko, V.O.; Burgess, S.T.G.; Matthews, J.B. Characterisation of serum IgG(T) responses to potential diagnostic antigens for equine cyathostominosis. *Int. J. Parasitol.* **2020**, *50*, 289–298. [CrossRef] [PubMed]

84. Peachey, L.E.; Jenkins, T.P.; Cantacessi, C. This gut ain't big enough for both of us. Or is it? Helminth–microbiota interactions in veterinary species. *Trends Parasitol.* **2017**, *33*, 619–632. [CrossRef] [PubMed]

Article

Parasite Occurrence and Parasite Management in Swedish Horses Presenting with Gastrointestinal Disease—A Case–Control Study

Ylva Hedberg-Alm [1,*], Johanna Penell [2], Miia Riihimäki [3], Eva Osterman-Lind [4], Martin K. Nielsen [5] and Eva Tydén [6]

1 Horse Clinic, University Animal Hospital, Swedish University of Agricultural Sciences, 750 07 Uppsala, Sweden
2 Division of Veterinary Nursing, Department of Clinical Sciences, Swedish University of Agricultural Sciences, 750 07 Uppsala, Sweden; Johanna.Penell@slu.se
3 Equine Medicine Unit, Department of Clinical Sciences, Swedish University of Agricultural Sciences, 750 07 Uppsala, Sweden; Miia.Riihimaki@slu.se
4 National Veterinary Institute, Department of Microbiology, Section for Parasitology diagnostics, 751 89 Uppsala, Sweden; eva.osterman-lind@sva.se
5 Maxwell H. Gluck Equine Research Center, Department of Veterinary Science, University of Kentucky, Lexington, KY 40546, USA; martin.nielsen@uky.edu
6 Parasitology Unit, Department of Biomedical Science and Veterinary Public Health, Swedish University of Agricultural Sciences, 750 07 Uppsala, Sweden; eva.tyden@slu.se
* Correspondence: ylva.hedberg.alm@uds.slu.se; Tel.: +46-18-672935

Received: 3 March 2020; Accepted: 31 March 2020; Published: 7 April 2020

Simple Summary: Abdominal pain, colic, is a common clinical sign in horses, sometimes reflecting life-threatening disease. One cause of colic is parasitic infection of the gut. Various drugs, anthelmintics, can be used to reduce or eliminate such parasites. However, frequent use has led to problems of drug resistance, whereby many countries now allow anthelmintics to be used on a prescription-only basis. In Sweden, this has led to a concern that parasitic-related colic in horses is increasing. This study aimed to investigate whether horses with colic differed in parasitological status compared to horses without colic. A secondary aim was to collect information regarding current parasite control measures used by horse owners. Exposure to *S. vulgaris*, a parasite with the potential to cause life-threatening disease, appeared high as determined by the presence of antibodies in the blood. Horses with inflammation in the abdominal cavity had higher antibody levels than other causes of colic. Despite new legislation, 29% of owners did not use fecal analyses for parasites and the use of extended methods to diagnose specific parasites was low. Also, owners rarely used alternative methods to reduce the pasture parasite burden. The study suggests a need for education in the use of both fecal analyses and pasture management.

Abstract: All grazing horses are exposed to intestinal parasites, which have the potential to cause gastrointestinal disease. In Sweden, there is a concern about an increase in parasite-related equine gastrointestinal disease, in particular *Strongylus vulgaris*, since the implementation of prescription-only anthelmintics approximately 10 years ago. In a prospective case–control study, parasitological status, using fecal analyses for strongyle egg counts, the presence of *Anoplocephala perfoliata* eggs and *S. vulgaris* Polymerase chain reaction (PCR) as well as serology for *S. vulgaris*, were compared between horses presenting with or without gastrointestinal disease at a University hospital during a one-year period. Information regarding anthelmintic routines and pasture management was gathered with an owner-filled questionnaire. Although the prevalence of *S. vulgaris* PCR was 5.5%, 62% of horses were positive in the enzyme-linked immunosorbent assay (ELISA) test and horses with peritonitis showed higher antibody levels for *S. vulgaris*, as compared to other diagnoses or controls. Overall, 36% of the horse owners used only fecal egg counts (FEC), 32% used FEC combined with specific diagnostics for *S. vulgaris* or *A. perfoliata*, and 29% dewormed routinely without prior parasite diagnostics. Effective management methods to reduce the parasitic burden on pastures were rare and considering exposure to *S. vulgaris* appears high; the study indicates a need for education in specific fecal diagnostics and pasture management.

Keywords: horse; colic; gastrointestinal disease; *Strongylus vulgaris*; *Anoplocephala perfoliata*; cyathostominae

1. Introduction

Colic in horses is a significant cause of mortality and morbidity and thus, is a major contributing factor to economic loss within the equine industry [1]. A vast number of potential risk factors have been implicated as a cause of gastrointestinal disease in horses, including intestinal parasites [2,3]. As grazing animals, horses are hosts to a large number of internal parasites that, to a varying degree, may result in pathology of the gastrointestinal tract [2].

Of all the equine internal parasites, *Strongylus vulgaris* is regarded as the most pathogenic and was therefore the main target of parasitic control programs in the 1960s, when modern anthelmintic drugs were first introduced [4]. This parasite has a long lifecycle of about six months, of which four months involve migration of larvae in the mesenteric arteries, in particular the cranial mesenteric artery [5]. Due to its larval migration within the mesenteric vasculature, *S. vulgaris* may cause severe arterial inflammation and damage to the otherwise smooth endothelial surface, with subsequent thrombus formation [5]. This potentially leads to the occlusion of arteries and arterioles supplying the intestinal wall, causing intestinal infarction and septic peritonitis. This condition is often fatal and requires surgical resection of the infarcted intestinal segment for a chance of survival [6].

The cestode *Anoplocephala perfoliata* has also been implicated as a cause of colic of varying degrees of severity [7]. This parasite's predilection for the ileocecal junction may cause lesions in this area, such as ileocecal invaginations and ileal impactions, most likely due to parasite-induced intestinal pathology at the site of attachment, but perhaps also secondary to physical obstruction and/or modifications in intestinal motility [8,9]. In addition, studies have demonstrated an association between *A. perfoliata* infection and spasmodic colic [10] as well as colic in general [11].

In contrast to other parasites, small strongyles, cyathostomins, have a direct life cycle and the early third larval stage can be arrested in the gut wall as encysted larvae for prolonged periods of time. The emergence of large numbers of such arrested larval stages can cause a severe inflammatory reaction in the colon and cecum resulting in diarrhea, sometimes with fatal outcomes, predominately in young animals; a condition known as larval cyathostominosis [12,13]. Moreover, large burdens of cyathostomins have been related to ill-thrift and weight-loss [14,15]. However, it is still under debate if cyathostomins are associated with colic in general, and a recent case–control study was unable to demonstrate a connection between cyathostomins and colic [16].

In the 1960s, regular anthelmintic treatment of horses was introduced, and as a result, some previously common parasite-induced gastrointestinal diseases, such as non-strangulating intestinal infarction caused by *S. vulgaris*, appeared to be largely eradicated [2]. However, the increasing reports of anthelmintic resistance in both cyathostomin parasites and *Parascaris* spp. [17,18], have led several countries, including Denmark, Finland, the Netherlands, Italy and Sweden, to implement a prescription-only sale of equine and livestock anthelmintic drugs [19]. In Sweden, the legislation advocates the prescription of anthelmintics to adult horses only after fecal analyses. By treating horses exceeding a chosen cut-off value, often 200 strongyle eggs per gram of feces (EPG), fewer anthelmintic treatments are administered, thus reducing the risk of developing of anthelmintic resistance [19,20]. However, since the implementation of more selective anthelmintic treatment strategies, Denmark and Sweden have recently documented a clear increase in the prevalence of *S. vulgaris* in the equine population in these countries [21,22]. Furthermore, post-mortem studies conducted in Italy indicate that pathological lesions caused by *S. vulgaris* are still widespread and in Denmark, several clinical cases of *S. vulgaris*-associated non-strangulating infarctions with subsequent septic peritonitis with a high mortality rate have been reported [6,23,24].

In Sweden, the legislation restricting anthelmintic products to prescription-only came into effect in October 2007. Just prior to this, only a minority (1%) of Swedish horse owners based anthelmintic treatment on fecal analyses of their horses [25]. As yet, there are few studies investigating how the prescription-only legislation has affected anthelmintic routines in Sweden and its possible effect on parasite-related disease in Swedish horses. Despite similar legislation in several European countries, there appears to be substantial national differences in how the law is implemented in practice [19]. Recently, an *S. vulgaris* prevalence of 61% was demonstrated in Swedish horse farms [22]. Whether this has resulted in an increase in clinical disease in Swedish horses is, however, unknown.

The aim of the present study was to compare the parasitological status between horses presenting for gastrointestinal disease (colic, colitis, peritonitis, weight loss) and horses presenting for non-intestinal disease at a university equine referral hospital during a one-year period, using coprological and serological assays. In addition, a secondary aim was to gather information regarding current anthelmintic treatment routines and pasture management practices using an owner-filled questionnaire at admission.

2. Materials and Methods

The study was approved by the Uppsala animal welfare ethics committee (license number: Dnr 68/16). Owner consent was required for study participation and participating owners were asked to fill in a consent form at admission.

2.1. Project Outline

The study was designed as a prospective case–control study at the Horse Clinic, University Animal Hospital, Swedish University of Agricultural Sciences, over a one-year period (February 2017–February 2018). The Horse Clinic receives both first opinion as well as referred equine cases with an annual case load of approximately 6000 cases. A horse presenting to the clinic and diagnosed with a disease related to the gastrointestinal canal was classified as a case. Each case was matched with a control horse presenting during the same week and, when possible, of same or similar age, and diagnosed with either a disease unrelated to the gastrointestinal canal (e.g., respiratory, skin, eye, cardiac, and orthopedic conditions) or presenting for prophylactic reasons (e.g., general health check, oral examination). Fecal and blood samples were collected by veterinarians from each case and control during the horse´s visit at the clinic. In addition, each owner was asked to respond to a questionnaire for collecting information regarding previous colic history, previous anthelmintic treatment, anthelmintic routines, and pasture management (Table 1). Regarding the question about deworming routines, there were five alternatives for the horse owner with the highest number of treatments being 2–4 times/year (Table 1). The reason for this is that in Sweden the recommendation for adult horses is to perform parasite diagnostics in April/May and only treat individuals exceeding 200 EPG and positive for *S. vulgaris* and/or *A. perfoliata* [26]. It is very rare that adult horses are dewormed more than 4 times/year in Sweden.

Table 1. Questionnaire data collected from participating cases and controls.

Information	Descriptor
Age of horse	Years
Time in ownership	Years
Any previous history of colic [a]	Yes/No
If yes, specify number of colic episodes	Number
Treated for colic by a veterinarian during last 24 months	Yes/No
Signs of diarrhea [b] during last 2 months	Yes/No
Lost weight [c] during last 2 months	Yes/No
Last anthelmintic treatment (months)	0–3; 3–6; 6–12; >12
Drug used during last treatment	FBZ [d]; PYR [e]; IVM [f]; MOX [g]; COMB [h]
Deworming routines applied on the farm	(i) only after FEC [i] (ii) after FEC and *S. vulgaris* diagnostics (iii) diagnostics for *A. perfoliata* at least 1/year (iv) routine deworming once/year (v) routine deworming 2–4 times/year
Presence of parasites in fecal sample during last 24 months	Small strongyles; *S. vulgaris*; *A. perfoliata*; *Parascasris spp*; do not know
Access to separate winter and summer pasture	Yes/No
Grazing with other animals species	Yes/No
Size of pasture or paddock, winter	Hectare
Size of pasture or paddock, summer	Hectare
Number of horses in pasture/paddock, winter	Number
Number of horses in pasture/paddock, summer	Number
Use of fecal removal	Yes/No
If yes, how often are feces removed	Number/week
Hours spent outdoors (paddock), winter	1–24
Hours spent outdoors (pasture), summer	1–24

[a] restlessness and pawing at the ground, irritated kicking to the stomach, rolling or attempting to roll; [b] loose consistency of feces; [c] as subjectively appreciated by owner; [d] fenbendazole; [e] pyrantel; [f] ivermectin; [g] moxidectin; [h] ivermectin/moxidectin and praziquantel; [i] fecal egg count.

2.2. Processing of Samples

Each horse included in the study had a fecal sample collected either at the time of rectal exam (majority of case horses), or from feces passed in the loose box (cases and controls) or transport vehicle in which the horse was brought to the clinic (controls). Fecal samples, approximately 200 g, were stored at 4 °C until parasite analyses the following day. In addition, two serum blood collection tubes (10 mL BD Vacutainer®(BD Technologies, Durham, NC, USA)) were obtained, either by vacutainer technique (20 G × 25 mm) from the jugular vein, or, in some case horses, from a permanent jugular catheter (Extended Use MILACATH® (MILA International Inc, Florence, KY, USA) placed in the jugular vein. In the latter case, at least 10 mL of blood was discarded before obtaining the sample. Blood samples were refrigerated at 4 °C for 12–24 h and then centrifuged at 2000× g for ten minutes. Serum was pipetted off and stored in 1.8 mL cryo tubes at −80 °C until the enzyme-linked immunosorbent assay (ELISA) was performed.

2.3. Parasite Analyses

Strongyle fecal egg counts (FECs) were carried out for each horse using a modified McMaster technique with a minimum detection limit of 50 EPG [27]. Parasite eggs in fecal samples (3 g) were floated using a saturated NaCl solution (SG = 1.18) [27]. Irrespective of FEC, larval cultures for detection of *S. vulgaris* were performed on 50 g feces from each horse according to Bellaw and Nielsen [28]. In brief, feces were mixed with an equal volume of vermiculite (Weibulls, Sweden), tap water was added to obtain a moist condition and samples were cultured at 20 °C for 14 ds. Third stage larvae were harvested after sedimentation for 12–16 h at 20 °C by the inverted Petri dish method [29]. Pellet of harvested L3 was performed by collecting approximately 20 mL of the fluid into a 50 mL Falcon tube, centrifuging at 248× g for three minutes and discarding the supernatant. Larval DNA was extracted with NuceloSpin®Tissue (Macherey-Nagel, Düren, Germany) according to the manufacturer's instruction. An ITS-2 specific PCR was used for identification of *S. vulgaris* according to Nielsen et al. (2008) and Tydén et al. (2019) [22,30].

Anoplocephala perfoliata were examined on 30 g feces using the modified flotation technique described by Berozoa et al. [31]. Feces were mixed with 60 mL tap water, sieved and collected into four 15 mL test tubes and centrifuged 1000× g for ten minutes. The supernatant was removed and pellet suspended in sugar salt solution (saturated sodium chloride solution with 50% glucose with a density of 1.280) to form a convex meniscus. Cover glasses (18 mm × 18 mm) were put on the top and centrifuged for 5 min at 214× g (no brakes) in a swing out centrifuge. Samples were left 5 min after centrifugation and thereafter transferred to microscope slides and microscopically examined at 40–100× magnification.

2.4. Strongylus vulgaris Specific ELISA

Serum samples were packed with dry ice and shipped to the Nielsen Parasitology Laboratory at the University of Kentucky. Here, an indirect ELISA using recombinant SvSXP protein as antigen was performed as previously described [32]. Samples were diluted 1:50 and horseradish peroxidase-conjugated goat anti-horse IgG(T) (Bethyl Laboratories, Inc., Montgomery, TX, USA) was used as a secondary antibody at a dilution of 1:40,000. Results were reported as the normalised value, percentage of a positive control to reduce inter-assay variability [32]. The positive control sample was obtained from a horse known to be infected with *S. vulgaris*.

2.5. Statistical Analyses

All statistical analyses were interpreted as statistically significant up to the p-value < 0.05 with the application of Bonferroni correction of the p-value, i.e., the division of the p-value for significance with the number of tests applied, to account for multiple testing, when appropriate. All statistical analyses were performed using Stata SE (version 14.2, StataCorp, College Station, TX, USA).

2.5.1. Data on Horses, Colic Diagnoses and Season

Statistical analyses on data on horses, colic diagnosis and season were performed as follows. Horses were categorised into six age-groups (1–2; 3–5; 6–10; 11–15; 16–20 and >20 years) and gender (mare, stallion, gelding). Gastrointestinal disease was classified into nine categories: impaction (large colon, small colon, and cecal), large intestinal displacement, acute colitis, peritonitis, equine eosinophilic gastroenteritis/enterocolitis, gas distention colic, chronic colitis, undiagnosed, and other. Season was categorised into four seasons (December–February, March–May, June–August, September–November).

Comparisons between diagnostic categories and season was performed using Fisher's exact test. Comparisons between cases and controls regarding demographic profile was done using univariate logistic regression generating crude ORs (for age, as continuous variable) and chi 2 test (for gender), respectively. For cases, comparisons between gastrointestinal diagnoses and age categories were done using univariate logistic regression generating crude ORs.

2.5.2. Questionnaire Data Including Previous Medical History, Anthelmintic Routines and Pasture Management

Comparisons between cases and controls regarding previous medical history (colic, veterinary treated colic, diarrhea, weight loss), anthelmintic routines (last drug used, time since last deworming, farm deworming routine, prevalence of previous parasitic infection) and management factors (use of co-grazing, fecal removal routines in pasture/paddock, time at pasture/in paddock, stocking density) were made using univariate logistic regression generating crude ORs.

2.5.3. Parasitological Status

The parasitological status of each horse was determined for cyathostomins as FECs, and for *S. vulgaris* PCR, *S. vulgaris* ELISA and *A. perfoliata* as positive/not positive. Strongyle egg counts were clustered into the following EPG groups: low egg shedders (EPG ≤ 200), medium egg shedders (250–500 EPG) and high egg shedders (>500 EPG) [33]. The difference in distribution of the EPG categories for each determinant, respectively, was investigated using univariate ordered logistic regression generating crude ORs. The *S. vulgaris* ELISA was identified as positive using a cut-off value of 13.47% of positive control according to Andersen et al. [32]. The number of positive samples (i.e., above the positive threshold) were presented as total number and proportion of the total. For *S. vulgaris* PCR, *S. vulgaris* ELISA, and *A. perfoliata*, the association between each parasitological outcome and various determinants (season, case/control, age, previous medical history, anthelmintic routines, gastrointestinal diagnosis and pasture management factors) was evaluated in separate logistic regression models (one per each specific combination of exposure and outcome). Additional analysis using the Wilcoxon matched-pairs signed-rank test for *S. vulgaris* ELISA was performed as distribution plots (using the original continuous data) for cases and controls, respectively, for each gastrointestinal disease category as well as for all cases compared with all controls.

2.5.4. Multivariable Logistic Regression Analysis

Variables with a univariable p value < 0.2 were considered for subsequent inclusion in a multivariable model. To avoid problems associated with collinearity, where variables were considered to be measuring the same exposure or were shown to be highly correlated (Pearson correlation coefficient > 0.8), the most statistically significant or biologically plausible variable was selected for inclusion in the model. Five submodels were initially created for the following dependent variables: cases/controls, strongyle fecal egg counts, *S. vulgaris* PCR, *S. vulgaris* ELISA, *A. perfoliata*. For each model, all variables with $p < 0.20$ from the univariate analysis with the specific dependent variable were added to create the full model. From each model, backward selection of variables was performed: the variable with the highest p-value was removed, one at a time, until all variables remaining in the model had $p < 0.05$.

3. Results

3.1. Data on Horses

A total of 137 cases and 137 controls were included in the study. A total of 66.7% of all presenting gastrointestinal patients at the clinic during the period February 2017 to February 2018 were included in the study. The mean and median ages of the cases and control horses were 10.8 (± SD 5.8) and 11 years and 10.2 (± SD 6.0) and 9 years, respectively, with a range of one to 25 years in both groups. The control population consisted of 74 geldings (54.0%), 58 mares (42.3%) and five stallions (3.6%) and the case population of 62 geldings (45.3%), 65 mares (47.4%) and 10 stallions (7.3%). There were no statistical differences in either age ($p = 0.53$) or gender distribution ($p = 0.83$) between cases and controls. The control horses were represented by 23 different breeds, the majority of which were Warmbloods (54.7%), followed by pony breeds (16.8%), Icelandic Horses (7.3%), Thoroughbreds (5.8%), draft breeds (4.4%), and Standardbreds (1.5%), with 9.5% consisting of various other breeds. Similarly, cases, consisting of 26 different breeds, were predominantly Warmbloods (40.1%), followed by Icelandic Horses (15.3%), pony breeds (13.7%), Standardbreds (7.2%), draft breeds (6.6%) and Thoroughbreds (4.4%), with 10.9% consisting of various other breeds.

3.2. Colic Diagnoses

The diagnoses of cases included impaction colic (large colon, small colon and cecal impactions) (38.0%), acute colitis (10.2%), large intestinal displacement (9.5%), peritonitis (8.0%), gas distention colic (5.8%), eosinophilic enteritis/colitis (3.6%), chronic colitis (2.2%), and other (7.3%). No specific diagnosis was reached in 15.3% of the cases (Table 2). Age had no association with type of diagnosis, with impaction colic common in all ages ($p = 0.32$). Of the horses diagnosed with peritonitis, the underlying etiology was known in three cases; esophageal rupture (n = 1) and non-strangulating intestinal infarction (n = 2). The remainder were classified as idiopathic. Abdominal surgery was performed in 12/137 cases (8.0%). Of the horses that had surgery performed, the final diagnoses were as follows: large colon displacement (4), ileocecal invagination (2), equine eosinophilic gastroenteritis/enterocolitis (2), non-strangulating intestinal infarction (2), gas distention colic (1) and undiagnosed (1). A total of four (2.9%) cases were euthanised, all of which had had abdominal surgery. Of the horses that had abdominal surgery and/or were euthanised, parasitic injuries were found in two cases, both of which were diagnosed with *S. vulgaris*-associated non-strangulating intestinal infarction of the colon. Both cases of ileocecal invagination had abdominal surgery without parasitic findings, e.g., *A. perfoliata*. The final diagnoses and number of cases in each group are presented in Table 2.

Table 2. Distribution of gastrointestinal diagnoses in cases.

Diagnosis	Number of Cases (% of Total Case Load)
Impaction (small colon, large colon, cecal)	52 (38.0%)
Acute colitis	14 (10.2%)
Large colon displacement	13 (9.5%)
Peritonitis	11 (8.0%)
Gas distention colic	8 (5.8%)
Equine eosinophilic gastroenteritis/enterocolitis (EEG)	5 (3.6%)
Chronic colitis	3 (2.2%)
Undiagnosed	21 (15.3%)
Other	10 (7.3%)

3.3. Association with Season

A total of 47 horses presented with colic during the winter months (December–February), equivalent to a third (34.3%) of the total number of cases. The most common diagnoses made during this season included different types of impaction colic (31.9%), intestinal displacement (17%), acute colitis (12.8%), peritonitis (10.6%), and equine eosinophilic gastroenteritis/enterocolitis (6.4%). In spring (March–May), 37 cases were included (27% of total caseload), the majority of which were impaction colic (37.8%), peritonitis (10.8%) and intestinal displacement (8.1%). Fewer horses with gastrointestinal disease presented during the summer, with 21 cases (15.3% of total caseload) during the months June–August. Furthermore, during the summer, impaction colic (large intestinal) was most commonly diagnosed (38%). Peritonitis, gas distention colic and acute colitis were each diagnosed in 9.5% of the cases during these three months. Twenty-three percent of the total number of cases presented during the autumn (September–November). Again, impactions were common and were diagnosed in almost half of the colic cases during this time period (46.9%). The second and third most common diagnoses during these months were acute colitis (15.6%) and gas distention colic (12.5%), respectively. No peritonitis cases were diagnosed in the autumn. There was no statistical association with season and the type of colic diagnosed ($p = 0.32$).

3.4. Questionnaire Data

The owner of each horse participating in the study completed a questionnaire on previous colic history, deworming routines and pasture management (Table 1). The overall questionnaire response rate for cases was 92% (126/137) and for control horses, 98% (134/137); however, within the questionnaire, the response rate varied depending on the questions asked (74.5%–98%). The results from the questionnaire are presented in Tables 3–5.

3.5. Previous Medical History

Case horses had been owned by the present owner for a mean of 4.3 years (range of one month to 25 years), and control horses had been in the present ownership for a mean of 4.9 years (range of one month to 25 years). Of the cases, 50.4% (60/119) had previously reported episodes of colic, with an average of 2.2 colic episodes prior to the study (range 1–12). The reported prevalence of previous colic was 23.9% for the controls (32/134), with an average of 1.8 previous colic episodes (range 1–10). Cases were significantly more likely to have had previous episodes of colic, compared to controls ($p < 0.01$) (Table 3). A third of the cases (33.3%) had been attended by a veterinarian due to colic within the previous 24 months, which was significantly more often compared with control horses (8.0%) ($p < 0.01$). In addition, cases were more likely to have had diarrhea within the previous two months compared to controls, although with a conservative Bonferroni correction approach, this difference did not reach statistical significance ($p = 0.03$).

Table 3. Results of questionnaire data: owner-reported previous medical history of cases and controls as outlined in a questionnaire presented after agreement of study participation. The *p*-value for statistical significance was corrected for the number of analyses (n = 5) according to Bonferroni, i.e., $p < 0.01$ (0.05/5).

Information	Cases	Controls	OR (95% C.I.)	*p*-Value
Prevalence with previous history of colic [a] (%)	50.4%	23.9%	3.13 (1.83–5.35)	<0.01
Number of previous colic episodes (mean ± SD)	2.2 (± 2.3)	1.8 (± 1.9)	1.09 (0.86–1.39)	0.47
Prevalence treated for colic during last 24 months (%)	33.3%	8%	5.73 (2.78–11.84)	<0.01
Prevalence with diarrhea [b] during last 2 months	15.8%	7.5%	0.42 (0.18–0.94)	0.03
Prevalence with weight loss [c] during last 2 months	9.2%	6.0%	0.45 (0.18–1.10)	0.08

[a] restlessness and pawing at the ground, irritated kicking to the stomach, rolling or attempting to roll; [b] loose consistency of feces; [c] as subjectively appreciated by owner.

3.6. Anthelmintic Routines

Deworming history was known in 88.9% (112/126) of cases and 98.5% (132/134) of controls that responded to the questionnaire, although the overall response rate to individual anthelmintic routine questions again varied (74.5%–98.5%). Anthelmintic treatment was performed at various times prior to presentation: 0–3 months (18.9%), 3–6 months (25.4%), 6–12 months (31.1%) and over 12 months (24.6%). There was no significant difference in time from deworming between cases and controls ($p = 0.46$) (Table 4). Macrocyclic lactones (ML) was the anthelmintic group most commonly used (59%) at the last treatment occasion. In detail, 41.7% had treated with ivermectin (IVM), 3.3% with moxidectin (MOX), 14.2% with a combination of IVM/moxidectin and praziquantel (COMB), 5.4% with pyrantel (PYR) and 3.3% had used fenbendazole (FBZ). Cases and controls showed a tendency to differ in terms of which drug was used at the last treatment, with control horses more often treated with IVM or COMB, and owners of cases, more often unable to recall the drug used ($p = 0.02$) (Table 4). Overall, 32.1% of the owners did not remember which drug they had used on the most recent treatment occasion (cases 43.8%, controls 24.2%).

In this data set, 28.8% of the horses were dewormed routinely without parasitic diagnostics being performed, where 12.3% of the horses were dewormed once per year and 16.5% of the horses 2–4 times per year. Selective treatment based on FECs was used by 35.8% of the owners, and 31.9% of owners used selective treatment based on both FECs and larval culture. Overall, 16% performed routine diagnostics for *A. perfoliata*, but this was more common within the control group (18.7%) compared to cases (13.5%). In addition, routine deworming 2–4 times per year was more often employed by owners of control horses (20.2%) compared to cases (12.7%). However, there were no significant differences between cases and controls in fecal diagnostics and deworming routines (p= 0.11–0.99) (Table 4).

One third of all horses were reported to have been infected with cyathostomins within the previous 24 months (cases 29.4%, controls 36.8%). In contrast, only 4.6% of horses were known to have been positive for *S. vulgaris* during the same time period. Case horses showed a higher prevalence of previous *S. vulgaris* infection (6.9% versus 2.6%), although this difference did not reach statistical significance (p = 0.10). Infection with *A. perfoliata* and *Parascaris* spp. were rarely reported to have occurred within the previous 24 months, and the data showed same frequency for both parasites (1.4% (1.0% cases, 1.7% controls)), see Table 4. However, previous parasitic infection history was unknown in the majority of horses (59.4% (61.8% cases, 57.3% controls)). Overall, previous parasitic infection was not significantly different between cases and controls (p = 0.10–0.79).

Table 4. Result of questionnaire data: anthelmintic routines. The p-value for statistical significance was corrected for the number of analyses (n = 4) according to Bonferroni, i.e., $p < 0.01$ (0.05/4).

Information	Cases	Controls	p-Value
Time from last helmintic treatment (months) (%)			
0–3	17.9%	19.7%	0.38
3–6	28.6%	22.7%	0.40
6–12	26.8%	34.8%	0.66
>12	26.8%	22.7%	0.51
For the variable (all time groups) cases/controls:			0.46
Drugs used at last treatment (%)			
FBZ [a]	4.5%	2.3%	0.48
PYR [b]	3.6%	6.9%	0.16
MOX [c]	3.6%	3.1%	0.62
IVM [d]	35.5%	46.9%	0.21
COMB [e]	10%	17.7%	0.13
Unknown	42.7%	23.1%	0.94
For the variable (all drug groups) cases/controls:			0.02
Deworming routines applied on the farm (%)			
(i) after FEC [f]	35.7%	35.8%	0.99
(ii) after FEC and *S. vulgaris* diagnostics	30.2%	33.6%	0.55
(iii) diagnostics for *A. perfoliata* at least once/year	13.5%	18.7%	0.26
(iv) routine deworming once/year	13.5%	11.1%	0.57
(v) routine deworming 2–4 times/year	12.7%	20.1%	0.11
Presence of parasites in fecal sample during last 24 months			
Cyathostomins	29.4%	36.8%	0.13
S. vulgaris	6.9%	2.6%	0.10
A. perfoliata	1.0%	1.7%	0.79
Parascaris spp.	1.0%	1.7%	0.79
Do not know	61.8%	57.3%	0.31

[a] fenbendazole; [b] pyrantel; [c] ivermectin; [d] moxidectin; [e] ivermectin/moxidectin and praziquantel; [f] fecal egg count.

3.7. Pasture Management

Regarding pasture management, it was rare for owners to report the use of mixed or alternate grazing with other species (9.8% (cases 10.6%, controls 9.0%)). However, the majority of owners employed the use of separate winter and summer paddocks/grazing areas (64.0% (67.5% cases, 60.9% controls)) (Table 5). Fecal removal was utilised overall by 46.2% of horse owners, with no statistical difference between cases and controls (54.6% and 45.9%, respectively) (p = 0.17). However, only a small proportion of owners declared that they removed feces at least twice weekly (7.1%), with almost two thirds of owners (65.3%) reporting that they removed feces less often than once weekly.

All horses had access to pasture during the summer months, with the majority (54.7%) kept at pasture 24 h per day during the summer, with no significant difference in time at pasture during the summer found between cases and controls ($p = 0.96$). Similarly, there were no differences found between the two groups in the number of horses kept solely outdoors in the winter ($p = 0.88$). Cases were more likely to be stabled > 12 h per day in the winter, compared to controls (65% compared to 55.6%), although no statistical difference was found ($p = 0.13$).

Stocking density (hectare/horse) calculated from data regarding paddock size and number of horses/paddocks, was similar between cases and controls in both winter and summer ($p = 0.22$–0.68) (Table 5). The majority of horses, both cases and controls, were kept at stocking densities providing < 0.4 hectare/horse (72.8% winter, 64.7% summer). Overall, season (winter/summer) showed a significant difference in both time at pasture and stocking density, with more horses stabled > 12 h/d and kept at stocking densities providing less than 0.4 hectare/horse during the winter as compared to the summer ($p < 0.01$).

Table 5. Results of questionnaire data: pasture management. The *p*-value for statistical significance was corrected for the number of analyses (n = 10) according to Bonferroni, i.e., $p < 0.005$ (0.05/10).

Information>	Cases	Controls	*p*-Value
Access to separate summer/winter paddocks	67.5%	60.9%	0.27
Grazing with other animal species	10.6%	9.0%	0.67
Fecal removal			
Fecal removal from pasture, overall	54.6%	45.9%	0.17
Fecal removal from pasture, ≥2x/week	5.9%	8.1%	0.33
Time stabled/outdoors			
Outdoors 24 h/d, winter	14.2%	14.8%	0.88
Stabled > 12 h/d, winter	65%	55.6%	0.13
Outdoors 24h/d, summer	54.2	54.8%	0.96
Stabled > 12 h/d, summer	26.7%	26.7%	0.93
Stocking density			
Stocking density, summer (hectare/horse)	0.6	0.8	0.68
Stocking density, winter (hectare/horse)	0.4	0.4	0.22

3.8. Multivariate Model for Cases and Controls

The multivariate logistic regression model investigating factors associated with being a case resulted in a final model that included three variables: age, treated for colic during last 24 months and stocking density. There was a significantly higher odds of being a case if treated for colic within the last 24 months (OR 7.87 (95% C.I. 3.04–20.34), $p < 0.001$), when adjusting for stocking density and age.

3.9. Strongyle Fecal Egg Counts

The overall prevalence of strongyle eggs was 44.9% (123/274) and the prevalence was similar between cases and controls, 47.4% and 42.3%, respectively. The majority, 67.5% (185/274), of the horses were classified as low egg shedders (EPG ≤ 200), with 16.4% denoted as high shedders (>500 EPG). The mean strongyle EPG was 238 ± SD 397 in the case group of which 55.5% (76/137) of the horses were shedding ≤ 50 EPG. The mean strongyle EPG was 320 ± SD 688 in the control group of which 52.6% (72/137) of the horses were shedding ≤ 50. Both the case group and the control group revealed similar number of high egg shedding horses (>500 EPG), 18.2% (25/137) and 14.6% (20/137), respectively. No statistical difference in the number of low, medium and high shedding horses between cases and controls was found ($p = 0.92$). In the study population as a whole, age was associated with egg shedding, with increasing age resulting in an increase in the number of low shedding horses and a decrease in the number of high shedders ($p < 0.01$). No statistically significant association with season and EPG level was found ($p = 0.38$).

Horses with a previous history of weight loss or diarrhea had higher average EPG levels (426 ± SD 778 and 400 ± SD 530, respectively) compared to horses without such complaints (268 ± SD

732 and 262 ± SD 544, respectively); however, these differences were not significant ($p = 0.39$–0.49). EPG levels were not associated with gastrointestinal disease in general ($p = 0.92$) (Table 6), nor with a particular colic diagnosis ($p = 0.61$).

EPG levels were lowest in horses most recently dewormed (0–3 months), with time since deworming overall approaching a significant association with EPG levels ($p = 0.01$). However, previously used anthelmintic drugs were not associated with egg shedding ($p = 0.22$). Although horses whose owners employed the use of FECs, as opposed to regular deworming at least once/year, had higher mean EPG levels, anthelmintic routine had no significant association with EPG level ($p = 0.06$). The result of the previous fecal sample had no statistical relationship with EPG level ($p = 0.50$).

Horses kept in smaller enclosures during the winter, often on their own, were more likely to be low shedders, compared with horses kept in larger paddocks but with more companions, although no statistical significance was shown ($p = 0.01$). No such trend was observed during the summer ($p = 0.22$). Time at pasture, either in the winter or summer, did not influence EPG levels ($p = 0.07$–0.49).

Table 6. Fecal and serology results in cases and controls. The p-value for statistical significance was corrected for the number of analyses (n = 4) according to Bonferroni, i.e., $p < 0.01$ (0.05/4).

Investigated Parameter	Cases n (%)	Controls n (%)	p-Value
EPG [a] level LOW	93 (67.9%)	92 (67.1%)	
EPG [a] level MEDIUM	19 (13.9%)	25 (18.2%)	0.92
EPG [a] level HIGH	25 (18.3%)	20 (14.6%)	
S. vulgaris PCR	7 (5.2%)	8 (5.8%)	0.80
S. vulgaris ELISA	75 (61.5%)	86 (62.8%)	0.83
A. perfoliata	17 (12.4%)	26 (19.0%)	0.14

[a] EPG, eggs per gram of feces.

3.10. Strongylus Vulgaris PCR

Seven horses in the case group and eight horses in the control group were PCR positive for *S. vulgaris*. The overall occurrence of *S. vulgaris* in this study was 5.5% (15/274). Seven of these 15 horses (46.7%) were defined as low egg shedders with EPG ≤ 200. Age showed no significant association with the number of positive horses ($p = 0.46$). Fewer positive horses were demonstrated during the winter months, compared to the other seasons, although no statistical association with season was shown ($p = 0.05$).

The majority, 80% (12/15), of the *S. vulgaris*-positive horses were dewormed either between 6–12 months or ≥ 12 months prior to fecal sampling, with almost half (7/15, 46.7%) of the positive horses having been dewormed with an unknown anthelmintic product > 12 months ago. Two horses were dewormed with a macrocyclic lactone (ML) within 3–6 months but were still positive for *S. vulgaris*. In addition, four horses were dewormed either once (3/15) or 2–4 times (1/15) a year but were also positive for *S. vulgaris*. None of the horses that were positive for *S. vulgaris* had reported a positive previous fecal sample for their last sample.

In the *S. vulgaris* PCR positive group, only 20.0% (3/15) of owners (compared to 31.9% overall in the study sample) utilised larval culture at least once yearly, with more horses (46.0%) regularly diagnosed by only performing FECs as compared to the study population as a whole (35.8%). Likewise, horses regularly dewormed once per year were more likely to be positive for *S. vulgaris* compared to horses that received anthelmintic treatment 2–4 times per year and to horses, where specific diagnostics for *S. vulgaris* were used. The majority of owners that performed diagnostics for *A. perfoliata*, also screened for *S. vulgaris* (78.6%). However, the number of horses positive for *S. vulgaris* in each group was low, making true comparisons difficult and no statistically significant association with anthelmintic routine and the number of positive horses was found ($p = 0.32$). There was no association between a positive PCR and gastrointestinal disease ($p = 0.80$) (Table 6) or with a specific diagnosis ($p = 0.76$).

Neither stocking density nor hours at pasture had a significant association with the number of *S. vulgaris*-positive horses ($p = 0.42$–0.80).

3.11. Strongylus vulgaris ELISA

The total *S. vulgaris* seroprevalence in this study was 62.2% (161/259). Serum was unavailable in 15 cases. No statistical differences between cases and controls in a number of horses positive in the ELISA test was shown ($p = 0.83$) (Table 6). However, when the separate gastrointestinal diagnoses were compared to controls matched by time of presentation, horses diagnosed with peritonitis showed significantly higher ELISA scores, expressed as a percentage of positive control, than their matched controls ($p < 0.02$) (Figure 1).

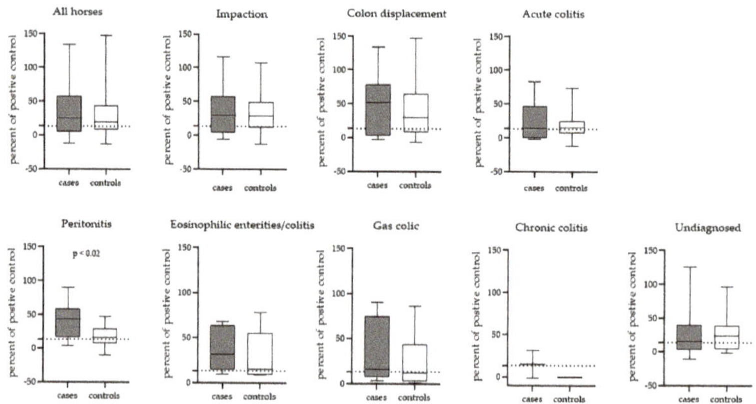

Figure 1. A boxplot illustrating the quantitative enzyme-linked immunosorbent assay (ELISA) values for each colic diagnosis compared with their matched controls. The bold line indicates median, the dotted line indicates the cut-off (13.47) value for positive ELISA.

Overall, there was a trend towards a greater prevalence of positive ELISA tests in older horses (>20) compared to young individuals (≤2 years) (81.3% and 44.4%, respectively), but this difference did not reach significance ($p = 0.04$). Overall, age had no significant association with ELISA score ($p = 0.37$). A lower prevalence was observed during the spring (60%) as compared to the rest of the year (71.4%–81.7%), but overall there was no significant association with season on ELISA score ($p = 0.17$).

Of the ELISA positive horses, 39.8% (64/161) were regularly diagnosed only performing FEC, 29.8% (48/161) of the horses performed FEC combined with *S. vulgaris* diagnostics, 16.1% (26/161) of the horses were dewormed 2–4 times/year and 8.7% (14/161) of the horses were dewormed on a yearly basis. Horses dewormed based on FECs without further diagnostics were more likely to have a positive ELISA score, compared to other anthelmintic routines, although using the conservative Bonferroni correction, statistical significance was not shown ($p = 0.02$). Neither the time from last anthelmintic treatment nor the drug used was associated with a positive ELISA result ($p = 0.37$–0.81).

Stocking density in either season had no significant association with number of positive horses ($p = 0.77$–0.90). Time at pasture during the summer was shown to have a significant association with the outcome of the ELISA test ($p < 0.01$), but no such association was demonstrated during the winter ($p = 0.51$).

3.12. Anoplocephala perfoliata

A total of 15.7% (43/274) of the horses were positive for *A. perfoliata*. Seventeen horses (12.4%) were positive for *A. perfoliata* in the case group and 26 horses (19.0%) in the control group, with no significant difference between the two groups ($p = 0.14$). Age had no significant association with *A. perfoliata* prevalence ($p = 0.12$). Although the highest number of positive horses was found in the spring (44.2%), no statistically significant seasonal differences were observed ($p = 0.05$).

No association between gastrointestinal disease and a positive *A. perfoliata* sample was demonstrated ($p = 0.14$) (Table 6). In addition, no specific gastrointestinal diagnosis was associated with a positive *A. perfoliata* sample ($p = 0.96$). Of the only two cases diagnosed with ileocecal invagination, fecal samples were negative for *A. perfoliata* and no tapeworms were found during abdominal surgery.

The time since last deworming was not associated with a positive test for *A. perfoliata* ($p = 0.92$). Three horses were positive for *A. perfoliata* despite treatment with ML combined with praziquantel within the last 3–6 months and for one horse, within 0–3 months. Another horse had been treated with this combination within 6–12 months of fecal sampling. Previous anthelmintic drug use had no overall significant association with the occurrence of *A. perfoliata* ($p = 0.06$). Eight of the 43 positive horses (18.6%) were normally analysed for *A. perfoliata* on a yearly basis. The group with the highest proportion of individuals positive for *A. perfoliata*, were horses whose owners reported the use of specific diagnostics for tapeworm. However, no horse that had a previous sample positive for *A. perfoliata*, had a positive fecal sample in the study. Overall, there was no significant association with anthelmintic routine used and a positive test for *A. perfoliata* ($p = 0.20$).

Although horses with at least 12 h of pasture/paddock access, regardless of time of year, had a higher prevalence of *A. perfoliata*, no statistically significant association with hours at pasture was shown ($p = 0.04–0.06$). The majority of positive horses were kept in paddocks allowing less than 0.4 hectares/horse, but no significant association with stocking density on *A. perfoliata* prevalence was demonstrated ($p = 0.10–0.67$).

3.13. Multivariate Model for Parasite Outcome

The multivariate ordered logistic regression model investigating factors associated with the fecal egg count in categories of low, medium and high egg shedders, resulted in a final model that included four variables: season, stocking density, any previous history of colic and time since deworming ($p < 0.05$ for all). There was an increased odds of a higher fecal egg category when time since deworming increased, with a 7.3 (95% C.I. 2.2 –24.1; $p = 0.001$) times increase in risk found when a horse had received anthelmintic treatment 6–12 months prior to presentation, compared to being treated within 0–3 months, when adjusting for season, stocking density in the winter and known history of previous colic.

When investigated factors associated with a positive PCR result for *S. vulgaris*, the multivariate ordered logistic regression model resulted in one variable: time since deworming. Deworming more than six months prior to sampling, as compared to treatment less than six months before presentation, resulted in an increased likelihood of a positive *S. vulgaris* PCR of 4.4 (95% CI 1.20–15.80; $p = 0.025$).

A multivariate ordered logistic regression for a positive *S. vulgaris* ELISA outcome resulted in a final model that included three variables: time in ownership, stabling > 12 h/d in the summer and deworming only after FECs. Deworming using only FECs was associated with an increased risk of a positive test, when adjusting for time of ownership and stabling > 12 h/d in the summer (OR 2.3 (95% C.I. 1.3–4.1; $p = 0.025$)).

Factors investigated to be associated with presence of *A. perfoliata*, resulted in a final model that included two variables: season and stabled > 12 h/d in the winter. The odds for the presence of *A. perfoliata* in a fecal sample was 2.8 times higher (95% C.I. 1.2–6.4; $p = 0.035$) in March–May when adjusting for time stabled in the winter.

4. Discussion

The present study demonstrated a high percentage of horses (cases and controls) were positive in the *S. vulgaris* ELISA test, amounting to almost two thirds of the horses included, indicating a high frequency of exposure to large strongyles in the population studied. The ELISA test assesses IgG (T) antibodies in *S. vulgaris* excretory proteins with a relatively high specificity (81%), and appears to reflect current infection or recent exposure, within the previous five months [31,34]. The number of positive horses in this study may be a surprisingly high figure, given the low individual *S. vulgaris* prevalence found using fecal PCR analyses (5.5%), but it is equivalent to the recently shown farm prevalence of *S. vulgaris* (61%) in Sweden [22].

Given the known pathogenicity of *S. vulgaris*, horses diagnosed with peritonitis were considered of particular interest to a possible *S. vulgaris* etiology. One case in this group was excluded due to the reason that peritonitis was known (esophageal perforation). However, two cases of peritonitis were confirmed on necropsy to have been caused by non-strangulating infarction secondary to *S. vulgaris* thrombosis. The remaining eight cases of peritonitis were classified as idiopathic and responded successfully to antimicrobial treatment, but none of these cases underwent exploratory laparotomy or necropsy. In addition to causing non-strangulating intestinal lesions, *S. vulgaris* has also been suggested as a possible cause of *Actinobacillus equuli* peritonitis, the most common bacterium isolated from horses presenting with idiopathic peritonitis [35]. The causal relationship, however, cannot be confirmed using fecal analyses, given that peritonitis associated with *S. vulgaris* would be due to the migrating larval stages and not the adult form of the parasite, and therefore before egg excretion is expected. None of the horses presenting with peritonitis in the present study were PCR positive for *S. vulgaris*. However, in the peritonitis group, the percentage of horses positive for the ELISA test was significantly higher (77.8%) than overall, with only two cases being negative (serum was missing in one case). In addition, horses diagnosed with peritonitis had significantly higher ELISA scores as compared to matched controls, which could be suggestive of an *S. vulgaris* etiology ($p < 0.02$). *S. vulgaris*-specific antibodies have been shown to be strongly associated with non-strangulating intestinal infarction, resulting in septic peritonitis; however, not with other types of colic [6,36]. The cause of eosinophilic intestinal lesions is still under debate, and previous studies do not convincingly demonstrate a parasitic etiology [37,38]. Nonetheless, it is interesting to observe that, although few cases were included, all but one case with an eosinophilic intestinal lesion were also positive in the ELISA test (80%, 4/5). In one case of eosinophilic enteritis, although no *S. vulgaris* larvae were found, thrombosed vessels were observed adjacent to the lesion.

In agreement with other case–control studies, no significant differences in parasitological status could be demonstrated using coprological assays between horses presenting with gastrointestinal disease in general and controls [16,36,39]. Overall, the fecal prevalence of *S. vulgaris* as assessed by PCR analyses and the presence of cyathostomins and *A. perfoliata* eggs were either lower or comparable to previously published prevalences of these equine intestinal parasites, as outlined below.

Although a possible role in colic incidence has been suggested [40], cyathostomins appear to be well tolerated without causing disease in most cases, apart from the well-known syndrome of larval cyathostominosis in young animals [12,13]. As in the present study, a recent publication showed no differences in strongyle EPG between non-surgical colic cases and controls [16]. Previous studies have, however, shown an association of cyathostomins with weight loss and diarrhea [12,15,41,42]. In the present study, although horses with a previous history of diarrhea or weight loss within the last two months had higher average EPG levels compared to the rest of the study population, there was considerable variation in EPG levels and this difference was not significant. In the case group, only two cases (2/14) of acute colitis and one case (1/3) of chronic colitis were high shedders (EPG > 500). There are, however, several causes of colitis in horses [43], and even in diarrhea caused by cyathostomins, fecal egg counts are often negative, with the L4 stage present in the feces instead [12,41]. In addition, it has been shown that neither egg nor larval counts correlate well with the horse´s actual worm burden [44].

The prevalence of *A. perfoliata* reported here is in accordance with a previous Swedish case–control study that showed positive samples in 13% (18/134) of horses [11]. Previous studies have shown conflicting results regarding the association of tapeworm infection and general colic [7]. However, *A. perfoliata* has consistently shown to be associated with ileocecal lesions [8,10,45–49]. The present study included few cases with confirmed ileocecal lesions (2/137) and was thus unable to demonstrate such a relationship.

No differences in parasite control strategies between cases and controls were found in the present study. ML was the most commonly used anthelmintic group overall, in accordance with several previous publications from various countries [50–55]. However, within the study population as a whole, differences in anthelmintic routines were present, with 28.8% of the owners deworming their horses on a routine basis without prior parasitic diagnostics. Likewise, in a recently published study 21% of Swedish horse-owners employed routine anthelmintic treatment 1–4 times yearly, without using coproscopic analyses [22]. However, before the prescription-only legislation was implemented, only 1% of owners were reported to use fecal analyses [25]. National differences in anthelmintic routines were recently described, with Denmark showing a much greater use of FECs than Austria, Netherlands, Germany and the United States [19]. Interestingly, the frequencies of use of FECs were very low (1.8% and 3.1%, respectively) in Germany and Austria, although these countries have allowed only the use of prescription anthelmintic drugs since 1975. In Denmark, where the concept was introduced in 1999, the figure was 50% [19]. One reason for this discrepancy was suggested to be differences in administration of the law and perhaps how rigidly the legislation was adhered to [19]. Thus, although almost a third of the owners in the present study did not use fecal analyses, compared to the above countries, the current use of FECs in Sweden appears comparatively high and shows a clear increase since selective treatment was enforced. However, it has to be noted that only approximately one third of owners declared the use of specific diagnostics for *S. vulgaris*, which is concerning, given that, in the present study, the use of FECs alone to determine the need for anthelmintic treatment, without further diagnostics, was associated with an increased odds of horses being positive in the *S. vulgaris* ELISA test.

Regular deworming, once or twice yearly, has been suggested to provide adequate control of *S. vulgaris* [21,56]. However, in the present study, although *S. vulgaris* was more common in horses dewormed six months or more prior to sampling, with time since deworming increasing the odds of a positive *S. vulgaris* sample, a total of six positive horses (40%) had either been dewormed within six months of sampling or were regularly dewormed at least once yearly. A post-mortem study in Sardinian horses demonstrated parasite-induced lesions in the cranial mesenteric arteries in all 46 horses in the study, and *S. vulgaris* larvae in 39%, despite treatment with broad-spectrum anthelmintic drugs at least three times per year [24]. Also, a recent study in Kentucky showed a high level of exposure to *S. vulgaris*, with 95/128 of horses being ELISA positive, despite treatment with anthelmintics on average twice yearly [57]. Considering clinical signs of disease may occur in as early as 2–4 months after infection [58], re-infection resulting in disease is possible despite regular anthelmintic treatment, and indicates that targeting the infected pasture is critical in management, and that eradication of *S. vulgaris* is unlikely using annual or bi-annual anthelmintic treatment, if horses are still allowed to graze in contaminated pastures. In addition, an experimental study in foals indicated that ivermectin may have little or no effect against migrating L5 larvae [59], suggesting that treatment may be less effective than often assumed.

Similar to a previous study on parasite control strategies in Sweden [25], only a minority of owners used alternating grazing species (9.8%), one method to reduce pasture contamination. This is low compared to some countries, such as Ireland, where co-grazing was used by the majority of owners [55]. However, most owners in the present study employed the use of separate winter and summer grazing areas. The regular use of fecal removal from the pasture can be an effective method to reduce parasite burden [60,61] and may be as effective as regular anthelmintic treatment, if performed twice weekly [61]. In the present study, although almost half of the owners reported removing feces from the pasture or paddock, only a minority did so at least twice weekly. The effect of fecal removal done more seldom than once weekly, which was the case in almost two thirds (65.4%) of owners that removed feces, is most likely poor, given that infective larvae will develop from excreted eggs within 2–3 weeks in spring and fall, and in as short as three days under ideal conditions [62]. In addition, most horses, both cases and controls, were kept at high stocking densities, provided with < 0.4 hectare/horse both in summer and winter. A high stocking density, with a large number of horses in paddocks of limited space, increases the risk of parasitic infections [63], and, in general, 1–2 acres/horse (corresponding to 0.4–0.8 hectare) is recommended [64].

Given the prospective nature, the study population in the present study was dependent on whichever gastrointestinal cases presented to the hospital within the set time-frame and, gastrointestinal diagnoses of particular interest with regards to a parasitic etiology, such as non-strangulating intestinal infarctions and ileo-cecal lesions, were, unfortunately, few. In addition, the study included a relatively small number of severe and/or surgical cases, probably because such cases required more urgent attention, resulting in lower consideration for study participation.

5. Conclusions

In conclusion, the present study demonstrated a high level of exposure of Swedish horses to *S. vulgaris*, as shown by the majority of horses, both cases and controls, being positive in the ELISA test. In addition, horses with peritonitis had significantly higher ELISA values as compared to controls and other gastrointestinal diagnoses, suggestive of an *S. vulgaris* etiology in this case group. Future studies addressing the association of *S. vulgaris* and peritonitis in the Swedish horse population is highly desirable, in order to further elucidate the clinical implication of the increasing prevalence of *S. vulgaris* in this country.

Furthermore, the present study revealed important information regarding the current use of anthelmintic treatment regimes in Sweden. For example, although the study showed a clear increase in the use of fecal analyses compared with before the prescription-only law was implemented, such legislation does not, per se, result in all owners using fecal diagnostics prior to treatment. The use of specific diagnostics is still low, indicating a need for education, of both owners and the veterinary profession, on how to best apply the diagnostic tools available. Treating horses annually or bi-annually, did not ensure a negative test result for *S. vulgaris*. Moreover, the study revealed that stocking intensity is often high and practices such as frequent fecal removal or co-grazing with other species were rarely performed. Thus, in addition, educating owners regarding optimal pasture management appears vital. In the advent of the increasing resistance to the anthelmintic drugs available together with the increase in prevalence of *S. vulgaris*, strategies for sustainable control of equine internal parasites are urgently needed, including pasture management.

Author Contributions: This study was performed within a larger project, where E.T. and E.O.-L. are PI. The individual contributions in the present study were as follows: conceptualization Y.H.-A., E.T., J.P., M.R., E.O.-L. and methodology E.T., Y.H.-A., J.P., E.O.-L., M.K.N., M.R.; software Y.H.-A., J.P., E.T.; validation Y.H.-A., J.P., E.T.; formal analysis Y.H.-A., J.P., E.T.; investigation Y.H.-A., E.T., E.O.-L., M.R.; resources E.T. E.O.-L.; data curation Y.H.-A., J.P., E.T.; writing—original draft preparation Y.H.-A., E.T., J.P.; writing—Y.H.-A., E.T., J.P.; M.K.N., E.O.-L. visualization Y.H.-A., E.T., and J.P.; project administration E.T., M.R., Y.H.-A.; funding acquisition E.T., E.O.-L. All authors have read and agreed to the published version of the manuscript.

Funding: This study was supported by the Foundation for Swedish and Norwegian Equine Research H-15-47-097.

Acknowledgments: We are greteful for the collaboration with the personel at the Equine Veterinary Clinic at SLU and all horseowners participating in this study.

Conflicts of Interest: The authors declare no conflict of interest. The funders played no role in the design of the study; in the collection, analyses, or interpretation of data; in writing the manuscript; or in the decision to publish the results.

References

1. Tinker, M.K.; White, N.A.; Lessard, P.; Thatcher, C.D.; Pelzer, K.D.; Davis, B.; Carmel, D.K. Prospective study of equine colic risk factors. *Equine Vet. J.* **1997**, *29*, 454–458. [CrossRef]
2. Reinemeyer, C.R.; Nielsen, M.K. Parasitism and colic. *Vet. Clin. N. Am. Equine Pract.* **2009**, *25*, 233–245. [CrossRef]
3. Curtis, L.; Burford, J.H.; England, G.C.W.; Freeman, S.L. Risk factors for acute abdominal pain (colic) in the adult horse: A scoping review of risk factors, and a systematic review of the effect of management-related changes. *PLoS ONE* **2019**, *14*, e0219307. [CrossRef]

4. Drudge, J.H.; Lyons, E.T. Control of internal parasites of the horse. *J. Am. Vet. Med. Assoc.* **1966**, *148*, 378–383.

5. Duncan, J.L.; Pirie, H.M. The pathogenesis of single experimental infections with *Strongylus vulgaris* in foals. *Res. Vet. Sci.* **1975**, *18*, 82–93. [CrossRef]

6. Pihl, T.H.; Nielsen, M.-K.; Olsen, S.N.; Leifsson, P.S.; Jacobsen, S. Nonstrangulating intestinal infarctions associated with *Strongylus vulgaris*: Clinical presentation and treatment outcome of 30 horses (2008–2016). *Equine Vet. J.* **2018**, *50*, 474–480. [CrossRef]

7. Nielsen, M.K. Equine tapeworm infections—Disease, diagnosis, and control. *Equine Vet. Educ.* **2016**, *28*, 388–395. [CrossRef]

8. Pavone, S.; Veronesi, F.; Piergili Fioretti, D.; Mandara, M.T. Pathological changes caused by *Anoplocephala perfoliata* in the equine ileocecal junction. *Vet. Res. Commun.* **2010**, *34*, 53–56. [CrossRef]

9. Pavone, S.; Veronesi, F.; Genchi, C.; Fioretti, D.P.; Brianti, E.; Mandara, M.T. Pathological changes caused by *Anoplocephala perfoliata* in the mucosa/submucosa and in the enteric nervous system of equine ileocecal junction. *Vet. Parasitol.* **2011**, *176*, 43–52. [CrossRef]

10. Proudman, C.J.; French, N.P.; Trees, A.J. Tapeworm infection is a significant risk factor for spasmodic colic and ileal impaction colic in the horse. *Equine Vet. J.* **1998**, *30*, 194–199. [CrossRef]

11. Back, H.; Nyman, A.; Osterman Lind, E. The association between *Anoplocephala perfoliata* and colic in Swedish horses–a case control study. *Vet. Parasitol.* **2013**, *197*, 580–585. [CrossRef] [PubMed]

12. Lyons, E.T.; Drudge, J.H.; Tolliver, S.C. Larval cyathostomiasis. *Vet. Clin. N. Am. Equine Pract.* **2000**, *16*, 501–513. [CrossRef]

13. Mair, T.S. Outbreak of larval cyathostomiasis among a group of yearling and two-year-old horses. *Vet. Rec.* **1994**, *135*, 598–600.

14. Murphy, D.; Love, S. The pathogenic effects of experimental cyathostome infections in ponies. *Vet. Parasitol.* **1997**, *70*, 99–110. [CrossRef]

15. Love, S.; Murphy, D.; Mellor, D. Pathogenicity of cyathostome infection. *Vet. Parasitol.* **1999**, *85*, 113–122. [CrossRef]

16. Stancampiano, L.; Usai, F.; Marigo, A.; Rinnovati, R. Are small strongyles (Cyathostominae) involved in horse colic occurrence? *Vet. Parasitol.* **2017**, *247*, 33–36. [CrossRef]

17. Kaplan, R.M. Anthelmintic resistance in nematodes of horses. *Vet. Res.* **2002**, *33*, 491–507. [CrossRef]

18. Von Samson-Himmelstjerna, G. Anthelmintic resistance in equine parasites—Detection, potential clinical relevance and implications for control. *Vet. Parasitol.* **2012**, *185*, 2–8. [CrossRef]

19. Becher, A.M.; van Doorn, D.C.; Pfister, K.; Kaplan, R.M.; Reist, M.; Nielsen, M.K. Equine parasite control and the role of national legislation—A multinational questionnaire survey. *Vet. Parasitol.* **2018**, *15*, 259–612. [CrossRef]

20. SJVFS 2010:17 24 § Legislations (In Swedish) Swedish Board of Agriculture. Available online: http://www.jordbruksverket.se/forfattningar/forfattningssamling (accessed on 26 March 2019).

21. Nielsen, M.K.; Vidyashankar, A.N.; Olsen, S.N.; Monrad, J.; Thamsborg, S.M. *Strongylus vulgaris* associated with usage of selective therapy on Danish horse farms-is it reemerging? *Vet. Parasitol.* **2012**, *189*, 260–266. [CrossRef]

22. Tydén, E.; Larsen Enemark, H.; Andersson Franko, M.; Höglund, J.; Osterman-Lind, E. Prevalence of *Strongylus vulgaris* in horses after ten years of prescription usage of anthelmintics in Sweden. *Vet. Parasitol.* **2019**, *2*, 100013. [CrossRef]

23. Stancampiano, L.; Mughini Gras, L.; Poglayen, G. Spatial niche competition among helminth parasites in horse's large intestine. *Vet. Parasitol.* **2010**, *170*, 88–95. [CrossRef] [PubMed]

24. Pilo, C.; Altea, A.; Pirino, S.; Nicolussi, P.; Varcasia, A.; Genchi, M.; Scala, A. *Strongylus vulgaris* (Looss, 1900) in horses in Italy: Is it still a problem? *Vet. Parasitol.* **2012**, *184*, 161–167. [CrossRef] [PubMed]

25. Osterman Lind, E.; Rautalinko, E.; Uggla, A.; Waller, P.J.; Morrison, D.A.; Höglund, J. Parasite control practices on Swedish horse farms. *Acta Vet. Scand.* **2007**, *49*, 25. [CrossRef]

26. Swedish National Veterinary Institute. 2018. Avmaskning av häst. (In Swedish). Available online: http://www.sva.se/djurhalsa/hast/parasiter-hos-hast/avmaskning-av-hast (accessed on 3 March 2020).

27. Coles, G.C.; Bauer, C.; Borgsteede, F.H.; Geerts, S.; Klei, T.R.; Taylor, M.A.; Waller, P.J. World Association for the Advancement of Veterinary Parasitology (W.A.A.V.P.) methods for the detection of anthelmintic resistance in nematodes of veterinary importance. *Vet. Parasitol.* **1992**, *44*, 35–44. [CrossRef]

28. Bellaw, J.L.; Nielsen, M.K. Evaluation of Baermann apparatus sedimentation time on recovery of *Strongylus vulgaris* and *Strongylus edentatus* third stage larvae from equine coprocultures. *Vet. Parasitol.* **2015**, *211*, 99–101. [CrossRef]

29. Van Wyk, J.A.; Mayhew, E. Morphological identification of parasitic nematode infective larvae of small ruminants and cattle: A practical lab guide. *Onderstepoort J. Vet. Res.* **2013**, *80*, 539. [CrossRef]

30. Nielsen, M.K.; Peterson, D.S.; Monrad, J.; Thamsborg, S.M.; Olsen, S.N.; Kaplan, R.M. Detection and semi-quantification of *Strongylus vulgaris* DNA in equine faeces by real-time quantitative PCR. *Int. J. Parasitol.* **2008**, *38*, 443–453. [CrossRef]

31. Beroza, G.A.; Marcus, L.C.; Williams, R.; Bauer, S.M. Laboratory diagnosis of *Anoplocephala perfoliata* infection in horses. *Proc. Am. Assoc. Equine Pract.* **1987**, *32*, 435–439.

32. Andersen, U.V.; Howe, D.K.; Dangoudoubiyam, S.; Toft, N.; Reinemeyer, C.R.; Lyons, E.T.; Olsen, S.N.; Monrad, J.; Nejsum, P.; Nielsen, M.K. SvSXP: A *Strongylus vulgaris* antigen with potential for prepatent diagnosis. *Parasites Vectors* **2013**, *6*, 84. [CrossRef]

33. Nielsen, M.; Mittel, L.; Grice, A.; Erskine, M.; Graves, E.; Vaala, W.; Tully, R.; French, D.; Bowman, R.; Kaplan, R. *AAEP Parasite Control Guidelines*; American Association of Equine Practitioners: Lexington, KY, USA, 2016.

34. Nielsen, M.K.; Vidyashankar, A.N.; Bellaw, J.; Gravatte, H.S.; Cao, X.; Rubinson, E.F.; Reinemeyer, C.R. Serum *Strongylus vulgaris*-specific antibody responses to anthelmintic treatment in naturally infected horses. *Parasitol Res.* **2015**, *114*, 445–451. [CrossRef]

35. Matthews, S.; Dart, A.J.; Dowling, B.A.; Hodgson, J.L.; Hodgson, D.R. Peritonitis associated with *Actinobacillus equuli* in horses: 51 cases. *Aust. Vet. J.* **2001**, *79*, 536–539. [CrossRef]

36. Nielsen, M.K.; Jacobsen, S.; Olsen, S.N.; Bousquet, E.; Pihl, T. Nonstrangulating intestinal infarction associated with *Strongylus vulgaris* in referred Danish equine cases. *Equine Vet. J.* **2016**, *48*, 376–379. [CrossRef]

37. Rötting, A.K.; Freeman, D.E.; Constable, P.D.; Moore, R.M.; Eurell, J.C.; Wallig, M.A.; Hubert, J.D. The effects of *Strongylus vulgaris* parasitism on eosinophil distribution and accumulation in equine large intestinal mucosa. *Equine Vet. J.* **2008**, *40*, 379–384. [CrossRef]

38. Olofsson, K.M. Immunopathological Aspects of Equine Inflammatory Bowel Disease. Ph.D. Thesis, Swedish University of Agricultural Sciences, Uppsala, Sweden, 2016.

39. Trotz-Williams, L.; Physick-Sheard, P.; McFarlane, H.; Pearl, D.L.; Martin, S.W.; Peregrine, A.S. Occurrence of *Anoplocephala perfoliata* infection in horses in Ontario, Canada and associations with colic and management practices. *Vet. Parasitol.* **2008**, *153*, 73–84. [CrossRef]

40. Uhlinger, C. Effects of three anthelmintic schedules on the incidence of colic in horses. *Equine Vet. J.* **1990**, *22*, 251–254. [CrossRef]

41. Smets, K.; Shaw, D.J.; Deprez, P.; Vercruysse, J. Diagnosis of larval cyathostominosis in horses in Belgium. *Vet. Rec.* **1999**, *144*, 665–668. [CrossRef]

42. Steinbach, T.; Bauer, C.; Sasse, H.; Baumgärtner, W.; Rey-Moreno, C.; Hermosilla, C.; Damriyasa, I.M.; Zahner, H. Small strongyle infection: Consequences of larvicidal treatment of horses with fenbendazole and moxidectin. *Vet. Parasitol.* **2006**, *139*, 115–131. [CrossRef]

43. Larsen, J. Acute colitis in adult horses. A review with emphasis on aetiology and pathogenesis. *Vet. Q.* **1997**, *19*, 72–80. [CrossRef] [PubMed]

44. Nielsen, M.K.; Baptiste, K.E.; Tolliver, S.C.; Collins, S.S.; Lyons, E.T. Analysis of multiyear studies in horses in Kentucky to ascertain whether counts of eggs and larvae per gram of feces are reliable indicators of numbers of strongyles and ascarids present. *Vet. Parasitol.* **2010**, *74*, 77–84. [CrossRef]

45. Nilsson, O.; Ljungström, B.L.; Höglund, J.; Lundquist, H.; Uggla, A. *Anoplocephala perfoliata* in horses in Sweden: Prevalence, infection levels and intestinal lesions. *Acta Vet. Scand.* **1995**, *36*, 319–328. [PubMed]

46. Proudman, C.J.; Edwards, G.B. Validation of a centrifugation/flotation technique for the diagnosis of equine cestodiasis. *Vet. Rec.* **1992**, *131*, 71–72. [CrossRef]

47. Proudman, C.J.; Edwards, G.B. Are tapeworms associated with equine colic? A case control study. *Equine Vet. J.* **1993**, *25*, 224–226. [CrossRef] [PubMed]

48. 48 Ryu, S.H.; Bak, U.B.; Kim, J.G.; Yoon, H.J.; Seo, H.S.; Kim, J.T.; Park, J.Y.; Lee, C.W. Cecal rupture by *Anoplocephala perfoliata* infection in a thoroughbred horse in Seoul Race Park, South Korea. *J. Vet. Sci.* **2001**, *3*, 189–193. [CrossRef]

49. Little, D.; Blikslager, A.T. Factors associated with development of ileal impaction in horses with surgical colic: 78 cases (1986–2000). *Equine Vet. J.* **2002**, *34*, 464–468. [CrossRef] [PubMed]

50. Hinney, B.; Wirtherle, N.C.; Kyule, M.; Miethe, N.; Zessin, K.H.; Clausen, P.H. A questionnaire survey on helminth control on horse farms in Brandenburg, Germany and the assessment of risks caused by different kinds of management. *Parasitol. Res.* **2011**, *109*, 1625–1635. [CrossRef]

51. Robert, M.; Hu, W.; Nielsen, M.K.; Stowe, C.J. Attitudes towards implementation of surveillance-based parasite control on Kentucky Thoroughbred farms—Current strategies, awareness and willingness-to-pay. *Equine Vet. J.* **2015**, *476*, 694–700. [CrossRef]

52. Stratford, C.H.; Lester, H.E.; Morgan, E.R.; Pickles, K.J.; Relf, V.; McGorum, B.C.; Matthews, J.B. A questionnaire study of equine gastrointestinal parasite control in Scotland. *Equine Vet. J.* **2014**, *46*, 25–31. [CrossRef]

53. Sallé, G.; Cabaret, J. A survey on parasite management by equine veterinarians highlights the need for a regulation change. *Vet. Rec. Open.* **2015**, *2*, e000104. [CrossRef]

54. Nielsen, M.K.; Branan, M.A.; Wiedenheft, A.M.; Digianantonio, R.; Scare, J.A.; Bellaw, J.L.; Garber, L.P.; Kopral, C.A.; Phillippi-Taylor, A.M.; Traub-Dargatz, J.L. Anthelmintic efficacy against equine strongyles in the United States. *Vet. Parasitol.* **2018**, *259*, 53–60. [CrossRef]

55. Elghryani, N.; Duggan, V.; Relf, V.; de Waal, T. Questionnaire survey on helminth control practices in horse farms in Ireland. *Parasitology* **2019**, *146*, 873–882. [CrossRef] [PubMed]

56. *A Guide to the Treatment and Control of Equine Gastrointestinal Parasite Infections ESCCAP Guidelines*, 1st ed.; ESCCAP: Malvern, UK, 2018; ISBN 978-1-907259-70-8.

57. Scare, J.A.; Steuer, A.E.; Gravatte, H.; Kálmán, C.S.; Ramires, L.; Dias de Castro, L.L.; Norris, J.; Miller, F.; Camargo, F.; Lawyer, A.; et al. Management practices associated with strongylid parasite prevalence on horse farms in rural counties of Kentucky. *Vet. Parasitol.* **2018**, *14*, 25–31. [CrossRef] [PubMed]

58. Borji, H.; Moosavi, Z.; Ahmadi, F. Cranial mesenteric arterial obstruction due to *Strongylus vulgaris* larvae in a Donkey (*Equus asinus*). *Iran. J. Parasitol.* **2014**, *9*, 441–444. [PubMed]

59. Nielsen, M.K.; Scare, J.; Gravatte, H.S.; Bellaw, J.L.; Prado, J.C.; Reinemeyer, C.R. Changes in Serum *Strongylus vulgaris*-Specific Antibody Concentrations in Response to Anthelmintic Treatment of Experimentally Infected Foals. *Front. Vet. Sci.* **2015**, *2*, 17. [CrossRef]

60. Herd, R.P. Epidemiology and control of equine strongylosis at Newmarket. *Equine Vet. J.* **1986**, *18*, 447–452. [CrossRef] [PubMed]

61. Corbett, C.J.; Love, S.; Moore, A.; Burden, F.A.; Matthews, J.B.; Denwood, M.J. The effectiveness of faecal removal methods of pasture management to control the cyathostomin burden of donkeys. *Parasit Vectors* **2014**, *7*, 48. [CrossRef]

62. Ogbourne, C.P. Observations on the free-living stages of strongylid nematodes of the horse. *Parasitology* **1972**, *64*, 461–477. [CrossRef]

63. Proudman, C.; Matthews, J. Control of intestinal parasites in horses. *Pract. BMJ* **2000**, *22*, 90–97. [CrossRef]

64. Singer, J.W.; Bamka, W.J.; Kluchinski, D.; Govindasamy, R. Using the recommended stocking density to predict equine pasture management. *J. Equine Vet. Sci.* **2002**, *22*, 73–76. [CrossRef]

Article

Anthelmintic Activity of Wormwood (*Artemisia absinthium* L.) and Mallow (*Malva sylvestris* L.) against *Haemonchus contortus* in Sheep

Dominika Mravčáková [1], Michaela Komáromyová [2], Michal Babják [2], Michaela Urda Dolinská [2], Alžbeta Königová [2], Daniel Petrič [1], Klaudia Čobanová [1], Sylwester Ślusarczyk [3], Adam Cieslak [4], Marián Várady [2,*] and Zora Váradyová [1,*]

1 Institute of Animal Physiology, Centre of Biosciences of Slovak Academy of Sciences, 040 01 Košice, Slovakia; mravcakova@saske.sk (D.M.); petric@saske.sk (D.P.); boldik@saske.sk (K.Č.)
2 Institute of Parasitology of Slovak Academy of Sciences, 040 01 Košice, Slovakia; komaromyova@saske.sk (M.K.); babjak@saske.sk (M.B.); dolinska@saske.sk (M.U.D.); konig@saske.sk (A.K.)
3 Department of Pharmaceutical Biology with Botanical Garden of Medicinal Plants, Medical University of Wroclaw, 50-556 Wroclaw, Poland; sylwester.slusarczyk@umed.wroc.pl
4 Department of Animal Nutrition, Poznan University of Life Sciences, 60-637 Poznan, Poland; adam.cieslak@up.poznan.pl
* Correspondence: varady@saske.sk (M.V.); varadyz@saske.sk (Z.V.); Tel.: +421-55-633-1411-13 (M.V.); +421-55-792-2972 (Z.V.)

Received: 8 December 2019; Accepted: 26 January 2020; Published: 29 January 2020

Simple Summary: The gastrointestinal parasitic nematode *Haemonchus contortus* of small ruminants is an important target for chemoprophylaxis. Repeated use of anthelmintics in the form of synthetic drugs increases the risk of residues in food products and the development of anthelmintic resistance. However, the use of combinations of dry traditional medicinal plants as nutraceuticals is an alternative to chemotherapeutics for controlling haemonchosis in ruminants. Therefore, the aim of this study is to determine the effect of dietary supplementation with wormwood, mallow and their mix on parasitological status and inflammatory response in lambs experimentally infected with *H. contortus*. Simultaneously, the present study evaluated by the egg hatch test the in vitro anthelminthic effects of different concentrations (50–1.563 mg/mL) of the aqueous extracts of these plants. Our results revealed that the strong anthelmintic effect of both medicinal plants observed in vitro was not fully confirmed in vivo. This knowledge builds on our previously published findings and highlights that the effect of dry medicinal plants depends on the variety and synergy of plant polyphenols and the combination of bioactive compounds that together have an effect and contribute to a certain pharmacological efficacy.

Abstract: The objective of this study is to evaluate the effect of dry wormwood and mallow on the gastrointestinal parasite of small ruminants *Haemonchus contortus*. Twenty-four experimentally infected lambs were randomly divided into four groups of six animals each: unsupplemented lambs, lambs supplemented with wormwood, lambs supplemented with mallow and animals supplemented with a mix of both plants. Faecal samples from the lambs were collected on day 23, 29, 36, 43, 50, 57, 64 and 75 post-infection for quantification of the number of eggs per gram (EPG). The mix of both plants contained phenolic acids (10.7 g/kg DM) and flavonoids (5.51 g/kg DM). The nematode eggs were collected and in vitro egg hatch test was performed. The aqueous extracts of both plants exhibited strong ovicidal effect on *H. contortus*, with ED50 and ED99 values of 1.40 and 3.76 mg/mL and 2.17 and 5.89 mg/mL, respectively, in the in vitro tests. Despite the great individual differences between the treated lambs in eggs reduction, the mean EPG of the untreated and treated groups did not differ ($p > 0.05$). Our results indicate that using wormwood and mallow as dietary supplements do not have a sufficient effect on lambs infected with *H. contortus*.

Keywords: dietary treatments; plant bioactive compounds; egg counts; UHRMS; *Haemonchus contortus*

1. Introduction

The gastrointestinal nematode (GIN) infection haemonchosis is a prevalent parasitic disease associated with economic losses, lowered productivity, morbidity and mortality. *Haemonchus contortus* is highly prevalent in sheep and goats worldwide and is mainly controlled by chemoprophylaxis through the repeated application of chemotherapeutics with the risk of development of anthelmintic resistance [1].

The screening of traditional medicinal plants containing promising contents of bioactive compounds with anthelmintic activity has great potential as an alternative source of natural anthelmintics and antioxidants that may be sustainable and environmentally acceptable. Various bioactive compounds (i.e., polyphenols, flavonoids, condensed tannins) that possess an anthelmintic effect [2,3] and antibacterial and antioxidant activities have been isolated from wormwood (*Artemisia absinthium* L.) [4,5]. Many authors have reported the antioxidant and antimicrobial properties of wormwood essential oils [6,7] and the anthelmintic activity of the flavonoids quercetin and apigenin [3]. Diets containing dried wormwood as a 5%–10% replacement for rice straw also provide better quality roughage with a considerable content of crude protein [8–10]. The high pharmacological activity of the medicinal plant mallow (*Malva sylvestris* L.), due to the presence of amino acids, flavonoids, mucilages, terpenoids, phenol derivatives, enzymes, coumarins and sterols, is known [11,12]. Mallow has antimicrobial, antifungal and anti-inflammatory properties [13,14].

In traditional medicine, whole plants or mixtures of plants are used rather than isolated compounds, and therefore more research is needed on all types of interaction between plant constituents. The ultra-high-resolution mass spectrometry (UHRMS) analyses of dry medicinal plants or plant mixtures in our recent studies with *H. contortus* [15–17] identified a wide range of bioactive compounds with important pharmacological activities, mainly flavonoids, phenolic acids, diterpenes, and alkaloids. These experiments, which combined chromatographic analyses with the determination of antioxidant capacity, are helpful in identifying plants with consistent concentrations of anthelmintic and antioxidant compounds for in vitro and in vivo studies. Our previous studies showed that medicinal plant mixtures are multicomponent mixes that possess effects via a multitarget additive and synergistic mode [16–18].

Therefore, in the present study, we hypothesize that some medicinal plants from these mixtures are by themselves multicomponent mixes and can elicit effects via pharmacological activity on *H. contortus* infected lambs. The medicinal plants, wormwood and mallow, were chosen based on their previously described best phytotherapeutic properties and anthelmintic activity in vitro [15]. The goal is to determine the effect of dietary supplementation with wormwood and mallow on parasitological status and inflammatory parameters of lambs experimentally infected with *H. contortus*.

2. Material and Methods

2.1. Ethics Statement

All procedures and animals were cared for under European Community guidelines (EU Directive 2010/63/EU). The experimental protocol was approved by the Ethical Committee of the Institute of Parasitology of the Slovak Academy of Sciences, in accordance with national legislation in Slovakia.

2.2. Analysis of Bioactive Compounds

Wormwood and mallow were ground to a fine powder, and 100 mg were extracted three times in 80% MeOH for 30 min at 40 °C. The extracts were evaporated to dryness, dissolved in 2 mL of Milli-Q water (acidified with 0.2% formic acid) and purified by Solid Phase Extraction (SPE) using Oasis HLB 3cc Vac Cartrige (60 mg, Waters Corp., Milford, MA, USA). The cartridges were washed with 0.5% methanol to remove carbohydrates and then washed with 80% methanol to elute phenolics. The phenolic fraction was re-evaporated and dissolved in 1 mL of 80% methanol (acidified with 0.1% formic acid). The sample was then centrifuged (23,000× g, 5 min) before spectrometric analysis. All analyses were performed in triplicate for three independent samples and stored in a freezer at −20 °C before analysis. The phenolic acids and flavonoids of the plant materials were analysed by a ultra-high-resolution mass spectrometry (Dionex UltiMate 3000RS, Thermo Scientific, Darmstadt, Germany) system with a charged aerosol detector interfaced with a high-resolution quadrupole time-of-flight mass spectrometer (HR/Q-TOF/MS, Compact, Bruker Daltonik GmbH, Bremen, Germany). The metabolomes of the samples were chromatographically separated, as was described [19]. The flow rate, spectra, operating parameters, collision energy, data calibration and spectra processing were previously described [20]. The amount of the particular phenolic acids in the samples was calculated as the chlorogenic acid (CAS 327-97-9, 3-Caffeoylquinic acid) equivalent, and hyperoside (CAS 482-36-0, quercetin 3-galactoside) was used for calculating the amount of identified flavonoids. Stock solutions of hyperoside and chlorogenic acids were prepared in MeOH, as was described previously [16].

2.3. In Vitro Test

The in vitro egg hatch test (EHT) was performed in order to assess the ovicidal effect of aqueous extracts of wormwood and mallow and compared with the chemotherapeutic effect of thiabendazole anthelmintic drug. The nematode eggs for in vitro EHT were obtained from the untreated UNS group. The concentrations of aqueous extracts used and EHT has been previously described [15].

Chemical tests for the screening of main constituents in the medicinal plants under study were carried out in the aqueous extracts using standard procedures [21,22]. Qualitative phytochemical screening revealed the active compounds mainly tannins, flavonoids, glycosides, saponins, alkaloids, and terpenoids (Table 1).

Table 1. Chemical composition of the aqueous plant extracts.

Plant Species	Tannins	Flavonoids	Glycosides	Saponins	Alkaloids	Terpenoids
A. absinthium	+	-	-	+	-	+
M. sylvestris	+	+	+	-	+	+

(+): the presence of phytochemicals; (-): the absence of phytochemicals.

2.4. Experiment In Vivo

The experiment was conducted on 24 3–4-month-old female lambs (Improved Valachian) with initial body weights of 18.67 ± 0.55 kg. The lambs were housed in common stalls on a sheep farm (Hodkovce, Slovak Republic) with free access to water. After a period of adaptation, all parasite-free lambs were infected by L3 larvae of *H. contortus* MHCo1 strain [16]. The diet of each animal consisted of oats (500 g DM/d) and meadow hay (ad libitum). Four groups of six animals based on their live-weight were established: unsupplemented lambs (UNS), lambs supplemented with stem of *A. absinthium* (ART, 1 g DM/d/lamb), lambs supplemented with flower of *M. sylvestris* (MAL, 15 g DM/d/lamb) and animals supplemented with mix of *A. absinthium* and *M. sylvestris* (ARTMAL, 16 g DM/d/lamb). The doses of plant supplements were based on the plant proportions used in our previous study [18]. The dry plants from commercial sources (AGROKARPATY, Plavnica, Slovak Republic) were mixed daily with the oats during the experimental period (75 days, D). The lambs were weighed on D0, D15, D30, D45 and D70. Faecal samples from the rectum of lambs were collected on D23, D29, D36, D43, D50, D57,

D64 and D75 post-infection for quantification of the eggs per gram (EPG). The detection of strongylid eggs was performed by McMaster technique, as was previously described [23]. The blood sera samples of each animal were obtained from D15, D30, D45 and D70 [18]. Helminthological autopsy were done after 75 days of infection [16].

2.5. Blood Sera Analysis

Sheep immunoglobulin G (IgG), sheep immunoglobulin A (IgA) and sheep eosinophil peroxidase (EPX) were measured by ELISA kits (MyBioSource Ltd., San Diego, CA, USA). The sensitivity of the IgG, IgA and EPX kits were 0.938 ng/mL, 1.875 ng/mL and 1.0 ng/mL, respectively.

2.6. Statistical Analysis

Statistical analysis was performed using analysis variance (GraphPad Prism 8, GraphPad Software, Inc., San Diego, CA, USA) as repeated-measures mixed models representing the four animal groups (UNS, ART, MAL, ARTMAL) and sampling days. Differences between the animal groups were analysed by a two-way ANOVA with a Bonferroni post hoc test. Differences between the arithmetic EPG means between groups and between worm counts at dissection were analysed by Student's t-tests. A logistic regression model was used to determine the ED_{50} and ED_{99} [24].

3. Results

3.1. Bioactive Compounds

The *A. absinthium* contained 6.48 g/kg DM of phenolic acids and 0.35 g/kg DM of flavonoids with greater concentrations of chlorogenic acid (3.42 g/kg DM) and 1,5-dicaffeoylquinic acid (2.12 g/kg DM) (Table 2). The *M. sylvestris* contained 0.65 g/kg DM of phenolic acids and 6.48 g/kg DM of flavonoids with higher concentrations of delphinidin-5-glucoside 3-lathyroside (1.64 g/kg DM), kaempferol-3-O-rutinoside (0.82 g/kg DM), apigenin-o-hex (1.56 g/kg DM) and coumarinic acid (0.47 g/kg DM). The mix of both plants contained 10.7 g/kg DM phenolic acids and 5.51 g/kg DM of flavonoids with greater concentrations of methyl-4-O-beta-d-glucopyranosylcaffeate (2.23 g/kg DM), 1,5-dicaffeoylquinic acid (1.64 g/kg DM), kaempferol-O-Hex (1.40 g/kg DM), apigenin-O-Hex (1.29 g/kg DM) and luteolin-O-Hex (0.70 g/kg DM).

Table 2. Contents of the flavonoids and phenolic acids (g/kg DM) identified in the plants and mix.

No.	RT (min)	λ_{max} (nm)	m/z [M-H]⁻	MS²	MS² Fragments	Formula	Compound	Flavonoids	Phenolic Acids
						Artemisia absinthium			
1	2.80		189.0759	127/0759	171/145/115	$C_8H_{14}O_5$	L-(-)Malic acid diethyl ester		0.22
2	4.10	215.325	353.0877	191/0567	179/161/135	$C_{16}H_{18}O_9$	Chlorogenic acid		3.42
3	7.80		281.1023			$C_{14}H_{18}O_6$	ND	0.01	
4	8.00		367.1031	191/0546	173	$C_{17}H_{20}O_9$	3-O-Feruloylquinic acid		0.08
5	8.90		279.1223	234/1009	261/217/177/199	$C_{15}H_{20}O_5$	Artabsinolide		
6	9.10		325.1283	163	279/235	$C_{16}H_{22}O_7$	ND		0.03
7	9.20		327.1440	279	235	$C_{16}H_{24}O_7$	ND		
8	10.00	289.000	263.1282	201/1271	245/219/149/161/177	$C_{15}H_{20}O_4$	Tanacetin		
9	10.20		281.1386	219/373	263/237/201	$C_{16}H_{24}O_7$	Artabsinolide D		
10	11.00		515.1193	353/0867	191/179/135	$C_{25}H_{24}O_{12}$	1,5-Dicaffeoylquinic acid		2.12
11	11.20		653.1719	345/0595	330/302	$C_{29}H_{34}O_{17}$	Spinacetin 3-rutinoside	0.24	
12	11.40		477.1032	314/0415	357	$C_{22}H_{22}O_{12}$	Isorhamnetin 7-glucoside	0.10	
13	11.70		515.1192	353/0869	173/179/191/155	$C_{25}H_{24}O_{12}$	4,5-Dicaffeoylquinic acid		0.61
14	14.90		507.1502	413/1246	101/324/259	$C_{24}H_{28}O_{12}$	Hedycoryside B		
15	15.00		511.2698	467/2775	405	$C_{30}H_{40}O_7$	Anabsin		
16	15.50		345.1344	301/1433	257/213/187	$C_{19}H_{22}O_6$	Diosbulbin E		
17	15.60		511.2698	245/1175	263/201	$C_{30}H_{40}O_7$	Anabsin		
18	16.50		329.2323	211/1324	229/171/183/139	$C_{18}H_{34}O_5$	Pinellic acid		
						Total flavonoids and phenolic acids		0.35	6.48
						Malva sylvestris			
1	1.60	250.320	517.1195	355/0667	193	$C_{21}H_{26}O_{15}$	Ferullo-O-Hex-O-Hex		0.02
2	1.80	250.301	206.0443	144/0437		$C_{10}H_9NO_4$	ND		0.17
3	7.00	523	757.1846	347/0761	329/261/509	$C_{32}H_{39}O_{21}$	Delphinidin 5-glucoside 3-lathyroside	1.64	
4	7.90	308	163.0381	119/0502		$C_9H_8O_3$	Coumaric acid		0.47
5	8.00	288	465.1046	303/0505	285/275/177	$C_{21}H_{22}O_{12}$	Xeractinol	0.17	
6	8.20	520	449.1094	287/0555	259/243	$C_{21}H_{22}O_{11}$	Cyanidin-O-Hex	0.28	
7	8.50	518	593.1645	431/0982	269/0460	$C_{27}H_{31}O_{15}$	Pelargonidin-O-Hex-O-Hex	0.14	
8	8.70	283	687.1784	507/1142	345/0629/165		ND	0.18	
9	9.00	283	525.1246	345/0815	165/197/139	$C_{23}H_{25}O_{14}$	ND	0.07	
10	9.20	287	303.0498	153/0169	125/217	$C_{15}H_{12}O_7$	ND	0.04	
11	9.50	283	773.1781	507/1124	345/165	$C_{32}H_{38}O_{22}$	ND	0.22	
12	10.00		609.1458	301/0330		$C_{27}H_{31}O_{16}$	Quercetin-3-O-rutinoside	0.40	

Table 2. *Cont.*

No.	RT (min)	λ$_{max}$ (nm)	m/z [M-H]$^-$	MS^2	MS^2 Fragments	Formula	Compound	Flavonoids	Phenolic Acids
13	10.20	268.343	447.0928	285/0386		$C_{21}H_{20}O_{11}$	Kaempferol-O-Hex	0.49	
14	10.50	346	505.0981	343/0442		$C_{23}H_{22}O_{13}$	Quercetin 3′-glucoside-7-acetate	0.03	
15	10.90	266.343	593.1504	285/0395		$C_{27}H_{30}O_{15}$	Kaempferol-3-O-rutinoside	0.82	
16	11.10	291.346	433.1124	271/0599	151	$C_{21}H_{22}O_{10}$	Naringenin-O-Hex	0.13	
17	11.25	291	287.0550	259/0596	152/201/243	$C_{15}H_{12}O_6$	Tetrahydroxyflavone	0.30	
18	11.40	268.336	431.0978	269/0435		$C_{21}H_{20}O_{10}$	Apigenin-O-Hex	1.56	
19	15.00	285.340	271.0595	151/0012	177/119	$C_{15}H_{12}O_5$	Naringenin	0.01	
20	15.40		327.2169			$C_{18}H_{32}O_5$	(E)-10-(8-Hydroxyoctanoyloxy)-enoic acid		
21	15.70	215.334	269.0443	151/0016	225	$C_{15}H_{10}O_5$	Trihydroxyflavone	0.02	
							Total flavonoids and phenolic acids	6.50	0.66
							Mix of *A. absinthium* and *M. sylvestris*		
1	4.00	215.325	353.0883	191/0561	173/179	$C_{16}H_{18}O_9$	4-O-Caffeoylquinic acid		0.61
2	4.10	215.325	353.0883	191/0561	179/173	$C_{16}H_{18}O_9$	3-O-Caffeoylquinic acid		0.74
3	5.90	215.287	355.1035	193/0498	149/134	$C_{16}H_{20}O_9$	1-O-2′-Hydroxy-4′-methoxycinnamoyl-b-D-glucose		0.38
4	6.10	215.302	355.1038	149/0598	193/134	$C_{16}H_{20}O_9$	1-O-Feruloylglucose		0.70
5	7.90		161.0225	133/0282		$C_9H_6O_3$	Umbeliferone		0.40
6	8.00		323.0760	161/0221		$C_{15}H_{16}O_8$	Mahaleboside		0.02
7	8.10	225.287	465.1033	303/177	285/0399	$C_{21}H_{22}O_{12}$	Xeractinol	0.04	
8	8.30	520.000	449.1094	287/0555	259/243	$C_{21}H_{22}O_{11}$	Cyanidin-O-Hex	0.03	
9	8.50		367.1025	173/0433	193/155/134	$C_{17}H_{20}O_9$	Feruloylquinic acid		0.25
10	9.00	233.294.318	355.1034	193/0507	149/134	$C_{16}H_{20}O_9$	Methyl-4-O-beta-D-glucopyranosylcaffeate		2.23
11	9.80	255.354	463.0882	301/0337	343	$C_{21}H_{20}O_{12}$	Quercetin O-Hex	0.44	
12	9.90	252.351	609.1472	301/0331	285/0415	$C_{27}H_{30}O_{16}$	Isoquercitrin O-Dhex	0.42	
13	10.30	257,4	447.0920	285/0386		$C_{21}H_{20}O_{11}$	Kaempferol-O-Hex	1.40	
14	10.70	217.291.325	515.1189	353/0877	179/191	$C_{25}H_{24}O_{12}$	3,5-Dicaffeoylquinic acid		0.80
15	10.80		187.0958	125/0968	169	$C_9H_{16}O_4$	ND		
16	10.90	221.329	593.1520	285/0397		$C_{27}H_{30}O_{15}$	Kaempferol-3-O-rutinoside	0.37	
17	11.10	217.291.325	515.1197	353/0869	191/179	$C_{25}H_{24}O_{12}$	1,5-Dicaffeoylquinic acid		1.64
18	11.15	291.346	433.1124	271/0599	151	$C_{21}H_{22}O_{10}$	Naringenin-O-Hex	0.19	

Table 2. *Cont.*

No.	RT (min)	λ_{max} (nm)	m/z [M-H]⁻	MS²	MS² Fragments	Formula	Compound	Flavonoids	Phenolic Acids
19	11.40	266.3	431.0976	269/0434		$C_{21}H_{20}O_{10}$	Apigenin O-Hex	0.56	
20	11.50	268.343	447.0928	285/0386		$C_{21}H_{20}O_{11}$	Luteolin O-Hex	0.70	
21	11.60	266.3	431.0976	269/0434		$C_{21}H_{20}O_{10}$	Apigenin O-Hex	0.73	
22	11.70	215.290.325	515.119	353/0868	173/179/191	$C_{25}H_{24}O_{12}$	4,5-Dicaffeoylquinic acid		0.68
23	12.40	325.0	517.1342	355/1022	353/193/149/161	$C_{25}H_{26}O_{12}$	3-caffeoyl-4-dihydrocaffeoyl quinic acid		0.35
24	12.90	268.320	639.3176	519/2604	476/373/145	$C_{37}H_{44}N_4O_6$	Tris-trans-p-coumaroylspermine		0.50
25	13.10	218.268.339	473.1083	269/0426	413	$C_{23}H_{22}O_{11}$	Apigenin -O-(Hex-Ac)	0.12	
26	13.70	325.0	517.1330	323/0759	353/193/149/161	$C_{25}H_{26}O_{12}$	4-caffeoyl-3-dihydrocaffeoyl quinic acid		0.07
27	14.20	218.268.339	473.1083	269/0426	413	$C_{23}H_{22}O_{11}$	Apigenin -O-(Hex-Ac)	0.22	
28	14.40	266.336	515.1187	269/0444		$C_{25}H_{24}O_{12}$	Formononetin 7-O-glucoside-6″-malonate		0.22
29	15.00	285.340	271.0595	151/0012	177/119	$C_{15}H_{12}O_5$	Naringenin	0.07	
30	15.40		327.2169			$C_{18}H_{32}O_5$	(E)-10-(8-Hydroxyoctanoyloxy)dec-2-enoic acid	0.03	
31	15.70	215.334	269.0443	151/0016	225	$C_{15}H_{10}O_5$	Trihydroxyflavone	0.03	
32	17.10	222.309	785.3554	545/2397	665/502/399/145	$C_{46}H_{50}N_4O_8$	Tetra-trans-p-coumaroylspermine		0.47
33	19.00		373.0914	358/0681	343/329/315	$C_{19}H_{18}O_8$	Dihydroxy—tetramethoxyflavone	0.03	
34	20.30	267.334	559.1069	269/0443	515/1172	$C_{26}H_{24}O_{14}$	Apigenin 7-(2″-acyl-6″maloylglycosyl)	0.13	0.65
							Total flavonoids and phenolic acids	5.51	10.7

No: peak numbers from UV chromatograms; RT: retention time; λ_{max}: wavelengths of maximum absorption in the visible region; MS²: main ion; ND: not determined.

3.2. In Vitro Test (EHT)

The dose-response relationships of aqueous extracts of *A. absinthium* or *M. sylvestris*, respectively, against *H. contortus* in the egg hatch test (EHT) are shown in Figure 1a,b. Both aqueous plant extracts exhibited a strong ovicidal effect on *H. contortus* in in vitro EHT. The ED_{50} and ED_{99} values were 1.40 and 3.76 mg/mL in *A. absinthium* (Figure 1a) and 2.17 and 5.89 mg/mL in *M. sylvestris* (Figure 1b), respectively. Thiabendazole at a concentration of 1.0 µg/mL has a 100% ovicidal effect.

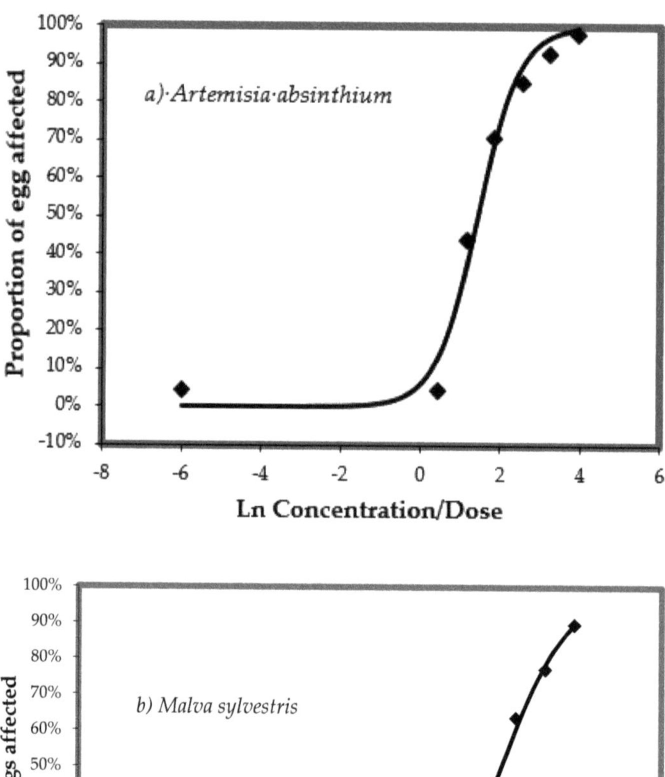

Figure 1. (**a,b**) Dose-response relationship of plant aqueous extracts against *Haemonchus contortus* in the egg hatch test (EHT) after 24 h of incubation at 26 °C.

3.3. Parasitological Status

The patterns of egg shedding for UNS, ART, MAL and ARTMAL are shown in Figure 2. Data from D36 were statistically compared and used to determine the reduction in egg output for ART, MAL and ARTMAL relative to UNS. Mean faecal eggs per gram (EPGs) were influenced by time from infection ($p < 0.05$), and for all groups, EPGs increased until D50 or D57, respectively. The EPGs in the lambs treated with MAL, ART and ARTMAL compared with UNS group did not differ ($p > 0.05$). The necropsy on D75 found a numerical decrease ($p > 0.05$) in the abomasal worm counts for the ARTMAL groups compared to the other groups (Figure 3).

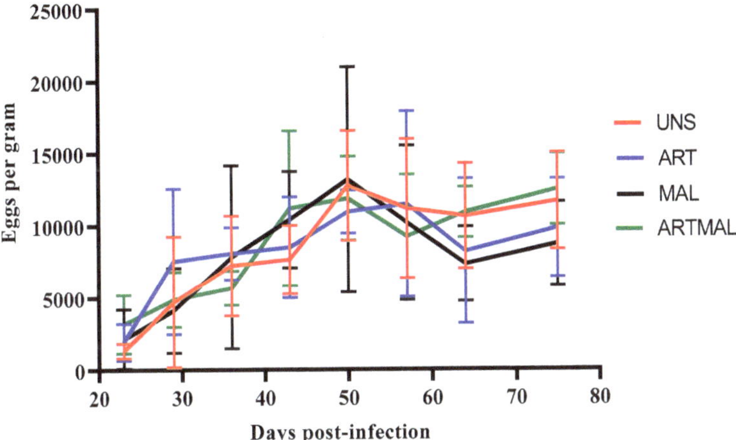

Figure 2. Mean faecal egg counts for the groups of lambs infected with *Haemonchus contortus* (Treatment: $p > 0.05$; time: $p < 0.001$; treatment × time: $p > 0.05$). UNS: unsupplemented; ART: *A. absinthium*; MAL: *M. sylvestris*; ARTMAL: ART plus MAL.

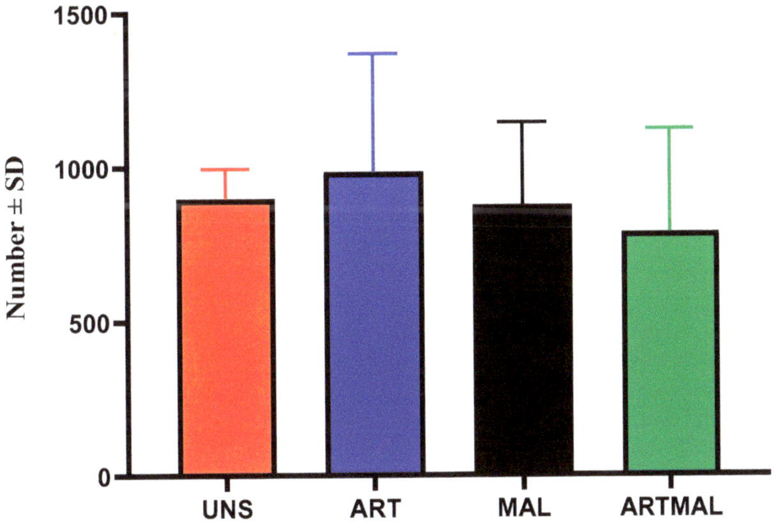

Figure 3. Abomasal worm counts of *Haemonchus contortus* in the lambs of each treatment at the end of the experiment ($p > 0.05$). UNS: unsupplemented; ART: *A. absinthium*; MAL: *M. sylvestris*; ARTMAL: ART plus MAL.

3.4. Inflammatory Response

Table 3 shows the inflammatory IgG, IgA and EPX response. Serum IgG and IgA values were not influenced by treatment, time and treatment × time ($p > 0.05$). Mean serum EPX values ranged in the treated groups from 24.1 to 73.4 ng/mL, and the values were influenced by treatment and time ($p < 0.05$).

Table 3. Inflammatory responses of the experimental groups ($n = 6$).

Item	Day	UNS	ART	MAL	ARTMAL	SD	Significance of Effects		
							Treatment (T)	Time	T × Time
IgG (ng/mL)	15	0.627	2.15	2.31	2.23	2.54			
	30	1.03	0.780	0.986	1.61	1.61	NS	NS	NS
	45	1.33	0.639	0.378	1.04	1.07			
	70	1.16	0.689	3.53	1.39	1.66			
IgA (ng/mL)	15	0.434	1.28	0.811	0.767	0.522			
	30	0.825	0.666	0.519	0.589	0.245	NS	NS	NS
	45	0.529	0.438	0.787	0.485	0.233			
	70	0.813	0.696	0.726	0.626	0.195			
EPX (ng/mL)	15	37.5	66.6	52.2	50.0	17.9			
	30	44.0	36.9	29.4	44.8	13.9	*	*	NS
	45	31.4	38.6	24.1	51.4	17.5			
	70	49.2	33.4	37.8	73.4	21.1			

UNS: unsupplemented; ART: *A. absinthium*; MAL: *M. sylvestris*; ARTMAL: ART plus MAL; EPX: eosinophil peroxidase; NS: not significant; SD: standard deviation. * $p < 0.05$.

4. Discussion

In the present study, phenolic compounds, flavonoids and phenolic acids among them, were detected in wormwood, mallow and a mix of both plants. Phenolic acids, including chlorogenic acid, caffeoylquinic acid derivatives, coumaric acid and methyl 4-O-beta-D-glucopyranosylcaffeate, were identified in a range from 0.65 to 10.7 g/kg DM. Chlorogenic acid and 1,5- and 4,5-dicaffeoylquinic acid possess well-known antibacterial, anthelmintic, anti-inflammatory, and antioxidant biological activities in vitro and in vivo [25,26]. Similarly, coumaric acid in *Senegalia gaumeri* leaf extract compounds has shown potential anthelmintic effects against *H. contortus* larvae [27]. The phenolic acid contents for wormwood and the mix (but not for mallow) were within the range of 3.6 to 57.3 g/kg DM, as was reported for plant mixtures used previously in infected lambs [16,17]. In relation to flavonoids with antioxidant properties, we identified mainly flavones (apigenin and luteolin), flavonols (kaempferol and quercetin) and flavanones (naringenin) [28], which may also have anthelmintic activity [3,29]. However, the total content of flavonoids in wormwood in the present study was lower (0.35 g/kg DM) versus mallow or the mix (6.48 or 5.51 g/kg DM, respectively) or compared to previous studies (9.96 and 29.5 or 41.5 g/kg DM, respectively) [16,17].

It is clear that medicinal plants that have an anthelmintic effect in vitro are often not equally effective in vivo, because there is different bioavailability, pharmacology of host animals, metabolism of bioactive compounds by rumen microflora and experimental conditions [30]. In the present experiment, the aqueous plant extracts of both medicinal plants, *A. absinthium* and *M. sylvestris*, exhibited a strong ovicidal effect on *H. contortus* in vitro, similar to the extracts of the species *Artemisia* against sheep nematodes [31,32]. However, the mean egg outputs of the UNS group compared to ART, MAL and ARTMAL groups showed no significant differences in egg reduction in lambs. Egg production by *H. contortus* females remained high (i.e., thousands EPG) until D50 post-inoculation and then decreased similarly as during the patent period of *H. contortus* in sheep [33]. The rapid reduction in egg excretion after D50 (MAL and ARTMAL) or D57 (ART), respectively, was accompanied by a lower number (not statistically) of adult *H. contortus* worms in the ARTMAL group. No significant differences in egg excretion in the treated groups may have been due to the lower content of plant biologically active compounds, especially flavonoids, compared to our previous studies [16,17]. However, relatively high SD of the means of the treated groups in the present experiment point to a potentially different treatment effect between lambs. This suggests that these plant materials could have an indirect antiparasitic effect and may promote a host's resistance to parasitic infection in the longer term. However, the anthelmintic mechanism of action is unknown. It seems that flavonoids with antioxidant capacity, in particular, contribute to the indirect antiparasitic activity [34,35]. However, not only the content of flavonoids appears to play an important role in the anthelmintic activity of medicinal plants and their mixtures. Our results indicate that wormwood and mallow themselves are not responsible for egg and worm reduction in the lambs but probably acting in synergy with other medicinal plants and their bioactive compounds, as was done in previous mixes [17,18]. The effect of dry medicinal plants on the health of animals depends on the source of the multitarget complex bioactive compounds that work synergistically [36,37] and antagonistically [38]. An increase in the resistance of infected lambs to *H. contortus* infection was shown after the administration of mixtures of dry medicinal plants composed already of 9–13 species [16–18]. However, a generally great contribution to the discovery and development of new drugs is a widely applicable strategy for identifying the combinatory compounds responsible for a certain pharmacological activity of plant medicines followed by in vitro and in vivo validation [39].

Ruminal and intestinal fermentation parameters can be manipulated by supplementing a diet with medicinal plants [40]. No adverse effects of wormwood and mallow on the ruminal fermentation parameters (i.e., pH, ammonia N, methane, gas production, and volatile fatty acids) were found [18]. Mainly, pH and ammonia N can affect the release of phenolic compounds from plant materials and the growth of ruminal microbes for microbial protein synthesis [41,42]. Additionally, phenolic compounds, especially flavonoids, can improve body weight gain, growth and the quality of animal products [43]. In the present experiment, polyphenols as dietary supplements did not significantly affect the body weights or live-weight gains of the infected lambs. Dry medicinal plants in the diets of the infected lambs may [18] or may not have influenced the body weights or live-weight gains, which is consonant with a meta-analysis of gastro-intestinal nematode infection in sheep [44].

Our recent studies [16–18] of lambs infected with the gastrointestinal nematode *H. contortus* showed the potential value of medicinal plant mixtures to decrease egg output and worm numbers in parasitic infections of the digestive tract. However, this effect is probably not a consequence of a direct anthelmintic impact on the viability of nematodes, but an increase in the resistance of lambs to nematode infections. Additionally, a recent study also showed that a complementary vegetable mixture of plants belonging to the *Compositae, Cesalpinacae, Liliacae, Bromeliaceae* and *Labiatae* families used as feed at two different dosages was ineffective against gastrointestinal nematode infection [45]. It seems that mainly a combination of medicinal plants belonging to different botanical families with beneficial bioactive compounds probably contributes to slowing the dynamics of infection. In the present study, because of the low variety and synergy of plant polyphenols and the combination of bioactive compounds of wormwood and mallow, the reduction in parasitic infection intensity in the treated infected lambs was not sufficient during the 75 days of infection compared to previous studies [16–18]. However, it seems that mixtures of dry medicinal plants may affect the host over the longer term. Therefore, more research is needed on combinations of medicinal plants and interactions between compositions of plant mixtures for longer (90-120 days) experimental periods.

5. Conclusions

The data in the present study showed additional new knowledge on the anthelmintic effects of dry medicinal plants as dietary supplements. Our results indicate that using medicinal plants, even those with the best anthelmintic properties in vitro, may not have sufficient effects in vivo on *H. contortus* infected lamb.

Author Contributions: Conceptualization, Z.V., M.V.; Investigation, D.M., Z.V., M.V.; Supervision, M.V.; Formal analysis, D.M., M.K., K.Č., M.B., M.U.D., A.K., D.P., S.Š.; Data Curation, Z.V., M.V., A.C.; Draft Preparation, Z.V.; Writing-Review & Editing, Z.V., M.V. All authors have read and agreed to the published version of the manuscript.

Funding: This study was supported by funds from the Slovak Research and Development Agency (APVV 18-0131).

Acknowledgments: The authors are grateful to Valéria Venglovská, Silvia Spišáková, Peter Jerga and Gabriel Benkovský for laboratory and technical assistance.

Conflicts of Interest: The authors declare no conflict of interest.

References

1. Lamb, J.; Elliott, T.; Chambers, M.; Chick, B. Broad spectrum anthelmintic resistance of *Haemonchus contortus* in Northern NSW of Australia. *Vet. Parasitol.* **2017**, *24*, 48–51. [CrossRef]
2. Tariq, K.A.; Chishti, M.Z.; Ahmad, F.; Shawl, A.S. Anthelmintic activity of extracts of *Artemisia absinthium* against ovine nematodes. *Vet. Parasitol.* **2009**, *160*, 83–88. [CrossRef]
3. Akkari, H.; Rtibi, K.; B'chir, F.; Rekik, M.; Darghouth, M.A.; Gharbi, M. In vitro evidence that the pastoral *Artemisia campestris* species exerts an anthelmintic effect on *Haemonchus contortus* from sheep. *Vet. Res. Commun.* **2014**, *38*, 249–255. [CrossRef]
4. Kordali, S.; Kotan, R.; Mavi, A.; Cakir, A.; Ala, A.; Yildirim, A. Determination of the chemical composition and antioxidant activity of the essential oil of *Artemisia dracunculus* and of the antifungal and antibacterial activities of Turkish *Artemisia absinthium, A. dracunculus, Artemisia santonicum* and *Artemisia spicigera* essential oils. *J. Agric. Food Chem.* **2005**, *53*, 9452–9458.
5. Kharoubi, O.; Slimani, M.; Krouf, D.; Seddik, L.; Aoues, A. Role of wormwood (*Artemisia absinthium*) extract on oxidative stress in ameliorating lead induced haematotoxicity. *Afr. J. Tradit. Complement. Altern. Med.* **2008**, *5*, 263–270. [CrossRef]
6. Msaada, K.; Salem, N.; Bachrouch, O.; Bousselmi, S.; Tammar, S.; Alfaify, A.; Al Sane, K.; Ammar, W.B.; Azeiz, S.; Brahim, A.H.; et al. Chemical Composition and Antioxidant and Antimicrobial Activities of Wormwood (*Artemisia absinthium* L.) Essential Oils and Phenolics. *J. Chem.* **2015**, *804658*. [CrossRef]
7. Nguyen, H.T.; Németh Zámboriné, É. Sources of variability of wormwood (*Artemisia absinthium* L.) essential oil. *J. Appl. Res. Med. Aromat. Plants* **2016**, *3*, 143–150. [CrossRef]
8. Kim, J.H.; Kim, C.H.; Ko, Y.D. Influence of dietary addition of dried wormwood (Artemisia sp.) on the performance and carcass characteristics of Hanwoo steers and the nutrient digestibility of sheep. *Asian–Aust. J. Anim. Sci.* **2002**, *15*, 390–395. [CrossRef]
9. Kim, Y.M.; Kim, J.H.; Kim, S.C.; Ha, H.M.; Ko, Y.D.; Kim, C.H. Influence of dietary addition of dried wormwood (Artemisia sp.) on the performance, carcass characteristics and fatty acid composition of muscle tissues of Hanwoo heifers. *Asian–Aust. J. Anim. Sci.* **2002**, *15*, 549–554. [CrossRef]
10. Ko, Y.D.; Kim, J.H.; Adesogan, A.T.; Ha, H.M.; Kim, S.C. The effect of replacing rice straw with dry wormwood (Artemisia sp.) on intake, digestibility, nitrogen balance and ruminal fermentation characteristics in sheep. *Anim. Feed Sci. Technol.* **2006**, *125*, 99–110. [CrossRef]
11. Cutillo, F.; D'Abrosca, B.; Dellagreca, M.; Fiorentino, A.; Zarrelli, A. Terpenoids and phenol derivatives from *Malva silvestris. Phytochemistry* **2006**, *67*, 481–485. [CrossRef] [PubMed]
12. Gasparetto, J.C.; Martins, C.A.; Hayashi, S.S.; Otuky, M.F.; Pontarolo, R. Ethnobotanical and scientific aspects of *Malva sylvestris* L.: A millennial herbal medicine. *J. Pharm. Pharmacol.* **2012**, *64*, 172–189. [CrossRef] [PubMed]
13. Zohra, S.F.; Meriem, B.; Samira, S. Some Extracts of Mallow Plant and its Role in Health. *APCBEE Procedia* **2013**, *5*, 546–550. [CrossRef]
14. Prudente, A.S.; Loddi, A.M.; Duarte, M.R.; Santos, A.R.; Pochapski, M.T.; Pizzolatti, M.G.; Hayashi, S.S.; Campos, F.R.; Pontarolo, R.; Santos, F.A.; et al. Pre-clinical anti-inflammatory aspects of a cuisine and medicinal millennial herb: *Malva sylvestris* L. *Food Chem. Toxicol.* **2013**, *58*, 324–331. [CrossRef]
15. Váradyová, Z.; Pisarčíková, J.; Babják, M.; Hodges, A.; Mravčáková, D.; Kišidayová, S.; Königová, A.; Vadlejch, J.; Várady, M. Ovicidal and larvicidal activity of extracts from medicinal-plants against *Haemonchus contortus. Exp. Parasitol.* **2018**, *195*, 71–77. [CrossRef]
16. Váradyová, Z.; Mravčáková, D.; Babják, M.; Bryszak, M.; Grešáková, Ľ.; Čobanová, K.; Kišidayová, S.; Plachá, I.; Königová, A.; Cieslak, A.; et al. Effects of herbal nutraceuticals and/or zinc against *Haemonchus contortus* in lambs experimentally infected. *BMC Vet. Res.* **2018**, *14*, 78. [CrossRef]
17. Mravčáková, D.; Váradyová, Z.; Kopčáková, A.; Čobanová, K.; Grešáková, Ľ.; Kišidayová, S.; Babják, M.; Urda Dolinská, M.; Dvorožňáková, E.; Königová, A.; et al. Natural chemotherapeutic alternatives for controlling of haemonchosis in sheep. *BMC Vet. Res.* **2019**, *15*, 302. [CrossRef]

18. Váradyová, Z.; Kišidayová, S.; Čobanová, K.; Grešáková, Ľ.; Babják, M.; Königová, A.; Urda Dolinská, M.; Várady, M. The impact of a mixture of medicinal herbs on ruminal fermentation, parasitological status and hematological parameters of the lambs experimentally infected with *Haemonchus contortus*. *Small Rumin. Res.* **2017**, *151*, 124–132. [CrossRef]

19. Rodrigues, M.J.; Matkowski, A.; Ślusarczyk, S.; Magné, C.; Poleze, T.; Pereira, C.; Custódio, L. Sea knotgrass (*Polygonum maritimum* L.) as a potential source of innovative industrial products for skincare applications. *Ind. Crop. Prod.* **2019**, *128*, 391–398. [CrossRef]

20. Rodrigues, M.J.; Monteiro, I.; Placines, C.; Castañeda-Loaiza, V.; Ślusarczyk, S.; Matkowski, A.; Pereira, C.; Pousão-Ferreira, P.; Custódio, L. The irrigation salinity and harvesting affect the growth, chemical profile and biological activities of *Polygonum maritimum* L. *Ind. Crop. Prod.* **2019**, *139*, 111510. [CrossRef]

21. Yadav, R.N.S.; Agarwala, M. Phytochemical analysis of some medicinal plants. *J. Phytol.* **2011**, *3*, 10–14.

22. Jaradat, N.; Hussen, F.; Al Ali, A. Preliminary Phytochemical Screening, Quantitative Estimation of Total Flavonoids, Total Phenols and Antioxidant Activity of *Ephedra alata* Decne. *J. Mater. Environ. Sci.* **2015**, *6*, 1771–1778.

23. Coles, G.C.; Bauer, C.; Borgsteede, F.H.M.; Geerts, S.; Klei, T.R.; Taylor, M.A.; Waller, P.J. World Association for the Advancement of Veterinary Parasitology (W.A.A.V.P) methods for the detection of anthelmintic resistance in nematodes of veterinary importance. *Vet. Parasitol.* **1992**, *44*, 35–44. [CrossRef]

24. Babják, M.; Königová, A.; Urda Dolinská, M.; Vadlejch, J.; Várady, M. Anthelmintic resistance in goat herds—In vivo versus in vitro detection methods. *Vet. Parasitol.* **2018**, *139*, 10–14. [CrossRef] [PubMed]

25. Sato, Y.; Itagaki, S.; Kurokawa, T.; Ogura, J.; Kobayashi, M.; Hirano, T.; Sugawara, M.; Iseki, K. In vitro and in vivo antioxidant properties of chlorogenic acid and caffeic acid. *Int. J. Pharm.* **2011**, *403*, 136–138. [CrossRef] [PubMed]

26. Tajik, N.; Tajik, M.; Mack, I.; Enck, P. The potential effects of chlorogenic acid, the main phenolic components in coffee, on health: A comprehensive review of the literature. *Eur. J. Nutr.* **2017**, *56*, 2215–2244. [CrossRef] [PubMed]

27. Castañeda-Ramírez, G.S.; Torres-Acosta, J.F.J.; Sandoval-Castro, C.A.; Borges-Argáez, R.; Cáceres-Farfán, M.; Mancilla-Montelongo, G.; Mathieu, C. Bio-guided fractionation to identify *Senegalia gaumeri* leaf extract compounds with anthelmintic activity against *Haemonchus contortus* eggs and larvae. *Vet. Parasitol.* **2019**, *270*, 13–19. [CrossRef]

28. Kumar, S.; Pandey, A.K. Chemistry and biological activities of flavonoids: An overview. *Sci. World J.* **2013**, *162750*. [CrossRef]

29. Kozan, E.; Anul, S.A.; Tatli, I.I. In vitro anthelmintic effect of *Vicia pannonica* var. *purpurascens* on trichostrongylosis in sheep. *Exp. Parasitol.* **2013**, *134*, 299–303. [CrossRef]

30. Ferreira, L.E.; Castro, P.M.; Chagas, A.C.; França, S.C.; Beleboni, R.O. In vitro anthelmintic activity of aqueous leaf extract of *Annona muricata* L. (Annonaceae) against *Haemonchus contortus* from sheep. *Exp. Parasitol.* **2013**, *134*, 327–332. [CrossRef]

31. Iqbal, Z.; Lateef, M.; Ashraf, M.; Jabbar, A. Anthelmintic activity of *Artemisia brevifolia* in sheep. *J. Ethnopharmacol.* **2004**, *93*, 265–268. [CrossRef] [PubMed]

32. Irum, S.; Ahmed, H.; Mirza, B.; Donskow-Łysoniewska, K.; Muhammad, A.; Qayyum, M.; Simsek, S. In vitro and in vivo anthelmintic activity of extracts from *Artemisia parviflora* and *A. sieversiana*. *Helminthologia* **2017**, *54*, 218–224. [CrossRef]

33. Borgsteede, H.M.; Couwenberg, T. Changes in LC50 in an in vitro egg development assay during the patent period of *Haemonchus contortus* in sheep. *Res. Vet. Sci.* **1987**, *42*, 413–414. [CrossRef]

34. Akkari, H.; B'chir, F.; Hajaji, S.; Rekik, M.; Sebai, E.; Hamza, H.; Darghouth, M.A.; Gharbi, M. Potential anthelmintic effect of *Capparis spinose* (Capparidaceae) as related to its polyphenolic content and antioxidant activity. *Vet. Med. Czech.* **2016**, *61*, 308–316. [CrossRef]

35. Spiegler, V.; Liebau, E.; Hensel, A. Medicinal plant extracts and plant-derived polyphenols with anthelmintic activity against intestinal nematodes. *Nat. Prod. Rep.* **2017**, *34*, 627–643. [CrossRef] [PubMed]

36. Brusotti, G.; Cesari, I.; Dentamaro, A.; Caccialanza, G.; Massolini, G. Isolation and characterization of bioactive compounds from plant resources: The role of analysis in the ethnopharmacological approach. *J. Pharm. Biomed. Anal.* **2014**, *87*, 218–228. [CrossRef] [PubMed]

37. Klongsiriwet, C.; Quijada, J.; Williams, A.R.; Mueller-Harvey, I.; Williamson, E.M.; Hoste, H. Synergistic inhibition of *Haemonchus contortus* exsheathment by flavonoid monomers and condensed tannins. *Int. J. Parasitol. Drugs Drug Resist.* **2015**, *5*, 127–134. [CrossRef] [PubMed]

38. Xutian, S.; Zhang, J.; Louise, W. New exploration and understanding of traditional Chinese medicine. *Am. J. Chinese Med.* **2009**, *37*, 411–426.

39. Long, F.; Yang, H.; Xu, Y.; Hao, H.; Li, P. A strategy for the identification of combinatorial bioactive compounds contributing to the holistic effect of herbal medicines. *Sci. Rep.* **2015**, *5*, 12361. [CrossRef]

40. Váradyová, Z.; Mravčáková, D.; Holodová, M.; Grešáková, Ľ.; Pisarčíková, J.; Barszcz, M.; Taciak, M.; Tuśnio, A.; Kišidayová, S.; Čobanová, K. Modulation of ruminal and intestinal fermentation by medicinal plants and zinc from different sources. *J. Anim. Physiol. Anim. Nutr. (Berl)* **2018**, *102*, 1131–1145.

41. Gaillard, B.D.; Richards, G.N. Presence of soluble lignin-carbohydrate complexes in the bovine rumen. *Carbohydr. Res.* **1975**, *42*, 135–145. [CrossRef]

42. Pisulewski, P.M.; Okorie, A.U.; Buttery, P.J.; Haresign, W.; Lewis, D. Ammonia concentration and protein synthesis in the rumen. *J. Sci. Food Agric.* **1981**, *32*, 759–766. [CrossRef] [PubMed]

43. Tufarelli, V.; Casalino, E.; D'Alessandro, A.G.; Laudadio, V. Dietary phenolic compounds: Biochemistry, metabolism and significance in animal and human health. *Curr. Drug Metab.* **2017**, *18*, 905–913. [CrossRef] [PubMed]

44. Mavrot, F.; Hertzberg, H.; Torgerson, P. Effect of gastro-intestinal nematode infection on sheep performance: A systematic review and meta-analysis. *Parasit. Vectors* **2015**, *8*, 557. [CrossRef]

45. Castagna, F.; Palma, E.; Cringoli, G.; Bosco, A.; Nistico, N.; Caligiuri, G.; Britti, D.; Musella, V. Use of complementary natural feed for gastrointestinal nematodes control in sheep: Effectiveness and benefits for animals. *Animals* **2019**, *9*, 1037. [CrossRef]

Article

Antibacterial and Hemolytic Activity of *Crotalus triseriatus* and *Crotalus ravus* Venom

Adrian Zaragoza-Bastida [1], Saudy Consepcion Flores-Aguilar [2], Liliana Mireya Aguilar-Castro [2], Ana Lizet Morales-Ubaldo [1], Benjamín Valladares-Carranza [3], Lenin Rangel-López [1], Agustín Olmedo-Juárez [4], Carla E. Rosenfeld-Miranda [5] and Nallely Rivero-Pérez [1,*]

[1] Universidad Autónoma del Estado de Hidalgo, Área Académica de Medicina Veterinaria y Zootecnia, Instituto de Ciencias Agropecuarias, Rancho Universitario Av. Universidad km 1, EX-Hda de Aquetzalpa, Tulancingo, Hidalgo 43600, Mexico; adrian_zaragoza@uaeh.edu.mx (A.Z.-B.); ubaldolizet8@gmail.com (A.L.M.-U.); ralolenin@gmail.com (L.R.-L.)

[2] Universidad Autónoma del Estado de Hidalgo, Área Académica de Biología, Instituto de Ciencias Básicas e Ingeniería. Carretera Pachuca-Tulancingo S/N Int. 22 Colonia Carboneras, Mineral de la Reforma, Hidalgo 42180, Mexico; saudy.fa@gmail.com (S.C.F.-A.); profe_3192@uaeh.edu.mx (L.M.A.-C.)

[3] Facultad de Medicina Veterinaria y Zootecnia Universidad Autónoma del Estado de México, El Cerrillo Piedras Blancas, Toluca 50295, Mexico; benvac2004@yahoo.com.mx

[4] Centro Nacional de Investigación Disciplinaria en Salud Animal e Inocuidad (CENID SAI-INIFAP), Carretera Federal Cuernavaca-Cuautla No. 8534 / Col. Progreso, Jiutepec 62550, Morelos, Mexico; aolmedoj@gmail.com

[5] Facultad de Ciencias Veterinarias, Universidad Austral de Chile, Isla Teja s/n, Casilla 567, Valdivia, Chile; crosenfe@uach.cl

*   Correspondence: nallely_rivero@uaeh.edu.mx; Tel.: +52-771-717-2000

Received: 14 January 2020; Accepted: 7 February 2020; Published: 11 February 2020

Simple Summary: Rattlesnakes (*Crotalus ravus* and *Crotalus triseriatus*) have some compounds that resemble polypeptides and proteins in their venoms which can be used in therapeutic treatment as antibacterial compounds. The aim of the present study is to evaluate the antibacterial and hemolytic activity of two rattlesnake venoms. The results of the present study indicate that the evaluated venoms have bactericidal activity against *Pseudomonas aeruginosa*, an important bacterium that affects animals and humans, thereby providing a new and efficient treatment alternative against this pathogenic bacterium.

Abstract: Rattlesnakes have venoms with a complex toxin mixture comprised of polypeptides and proteins. Previous studies have shown that some of these polypeptides are of high value for the development of new medical treatments. The aim of the present study is to evaluate, in vitro, the antibacterial and hemolytic activity of *Crotalus triseriatus* and *Crotalus ravus* venoms. A direct field search was conducted to obtain *Crotalus triseriatus* and *Crotalus ravus* venom samples. These were evaluated to determine their antibacterial activity against *Escherichia coli*, *Staphylococcus aureus* and *Pseudomonas aeruginosa* through the techniques of Minimum Inhibitory Concentration (MIC) and Minimum Bactericidal Concentration (MBC). Hemolytic activity was also determined. Antibacterial activity was determined for treatments (*Crotalus triseriatus* 2) CT2 and (*Crotalus ravus* 3) CR3, obtaining a Minimum Inhibitory Concentration of 50 µg/mL and a Minimum Bactericidal Concentration of 100 µg/mL against *Pseudomonas aeruginosa*. CT1 (*Crotalus triseriatus* 1), CT2, and CR3 presented hemolytic activity; on the other hand, *Crotalus ravus* 4 (CR4) did not show hemolytic activity. The results of the present study indicate for the first time that *Crotalus triseriatus* and *Crotalus ravus* venoms contain some bioactive compounds with bactericidal activity against *Pseudomonas aeruginosa* which could be used as alternative treatment in diseases caused by this pathogenic bacterium.

Keywords: *Crotalus ravus*; Crotalus triseriatus; venom; antibacterial activity; *Pseudomonas aeruginosa*; hemolytic activity

---

1. Introduction

Rattlesnakes are a species widely distributed through Mexico, occupying practically the whole territory. There exists a great variety of these species, among them, are *Crotalus triseriatus*, distributed in the States of Veracruz, Puebla, Tlaxcala, México, Morelos, and Michoacán and *Crotalus ravus*, which occupies the States of Morelos, México, Puebla, Tlaxcala, Guerrero, Oaxaca, and Hidalgo. These species are primarily recognized for their characteristic hemotoxic venoms [1–3].

Crotalid venoms are comprised mainly of enzymes that cause severe local inflammation, necrosis, hemorrhagic syndromes, and neurological manifestations. These responses would typically help rapid prey subjugation or capture, as well as serve as a defense mechanism [4].

Animal venoms, including that of snakes, are complex mixtures of bioactive compounds that contain large amounts of proteins, peptides, and small molecules that can be considered for use in a wide range of medical applications [5,6].

There are several examples in the development of treatments derived from snake venom compounds. One of the most widely known is Capoten®, a hypotensive agent, used for the treatment of congestive heart failure, diabetic nephropathy, and heart attacks. Another known example is Viprinex®, developed to treat acute strokes [7,8].

Aside from their qualities as potential therapeutic agents, venoms are currently considered as possible sources of molecules with antibacterial activity [9]. This, in fact, has a great impact on public health especially due to the increase of antibacterial resistant bacteria.

In 2017 the World Health Organization (WHO) compiled a list of antibiotic-resistant priority pathogens, among which, were the Gram-negative bacteria *Escherichia coli* and *Pseudomonas aeruginosa* bacteria resistant to carbapenems, and *Staphylococcus aureus* resistant to methicillin and vancomycin [10]. Due to the increased antibiotic resistance found in these pathogens, the aim of the present study was to evaluate, in vitro, the antibacterial and hemolytic response of *Crotalus triseriatus* and *Crotalus ravus* venoms on bacteria of public health importance.

2. Material and Methods

2.1. Field Sampling

Two field outings were carried out per month during each of the months of August, September, October and November 2018 in the state of Hidalgo, Mexico; covering the municipalities of Acatlán, Almoloya, Cuautepec de Hinojosa, Mineral del Chico, Mineral del Monte, Santiago Tulantepec, Singuilucan, Tula de Allende and Zacualtipán.

A direct search was conducted according to the methodology described by McDiarmid et al. in 2012. The rattlesnakes were trapped in accordance with the official norms for wildlife protection (NOM-059-SEMARNAT-2010) established by the government of Mexico and with a scientific collecting permit issued by General Directorate of Wildlife of the Secretariat of Environment and Natural Resources of Mexico (Office N° SGPA/SGVS/003613/18) [11,12].

2.2. Obtaining Venom Samples

Four samples were collected in the field, two of them belonging to the species *Crotalus triseriatus* (CT1 and CT2) and the remaining from the species *Crotalus ravus* (CR3 and CR4). A record of each individual was noted.

Once the samples were obtained, they were subjected to a lyophilization process and kept at −70 °C until further evaluation.

2.3. Antibacterial Activity

The venom's antibacterial activity was determined through the Minimum Inhibitory Concentration (MIC) and the Minimum Bactericidal Concentration (MBC) procedures, in accordance with the CLSI guidelines and with the standards published by Olmedo-Juárez et al., in 2019 and by Morales-Ubaldo et al., in 2020 [13–15].

Escherichia coli ATCC^{35218}, *Pseudomonas aeruginosa* ATCC^{9027}, and *Staphylococcus aureus* ATCC^{6538} strains were used to perform the evaluation. These samples were the same which were reactivated from cryopreservation in Müller–Hinton agar (BD Bioxon, Heidelberg, Germany) through simple strain technique to obtain isolated colonies. A Gram staining was performed to corroborate their morphology.

Once the purity was confirmed, one colony of each strain was inoculated in nutritive broth (BD Bioxon), and incubated under constant agitation at 70 rpm for 24 h at 37 °C. The bacterial cell suspension was adjusted to a 0.5 McFarland (Remel, R20421, Kansas, U.S.A.) standard (approximately 1.5×10^6 Colony Forming Units (CFU) per mL).

2.3.1. Minimal Inhibitory Concentration (MIC)

Micro-dilution was used to determine the MIC, evaluating different venom concentrations (100, 50, 25, 12.5, 6.25, 3.12, 1.56, 0.78 µg/mL).

In a sterile 96- well plate, 100 µL of each venom concentrations were added along with 10 µL of bacterial cell suspension previously adjusted to a 0.5 McFarland standard. The plates were incubated at 37 °C for 24 h at 70 rpm. Kanamycin (AppliChem 4K10421, Darmstadt, Germany) was used as a positive control (128 to 1 µg/mL) and nutritive broth as the negative control. Treatments were evaluated by triplicate.

After incubation 20 µL of a 0.04% (*w/v*) p-iodonitrotetrazolium (Sigma-Aldrich I8377, Missouri, U.S.A.) solution was added into each well and incubated for 30 min. The MIC was determined by the concentration at which the solution turned to a pinkish color.

2.3.2. Minimal Bactericidal Concentration (MBC)

After incubation and previous addition of p-iodonitrotetrazolium, 5 µL from each well was inoculated in Müller–Hinton agar (BD Bioxon) and incubated at 37 °C for 24 h. The MBC was considered as the lowest concentration where no visible growth of the bacteria was observed on the plates.

2.4. Indirect Hemolytic Activity

In accordance with the protocols described by Pirela et al., in 2006 with modifications, the venom's indirect hemolytic activity was evaluated [16]. A donor donkey blood sample was collected. The blood sample was stored in 10 mL sodium citrate (3.2%) tubes (DB Vacutainer) and in 3 mL EDTA (10.8 mg) tubes (BD Vacutainer).

Blood agar was used (Merck©, Darmstadt, Germany). To obtain plates with 8% blood concentration, 250 mL of agar base was prepared, and 20 mL of blood was added.

One hundred micrograms (100 µg) of each treatment were weighed out (lyophilized venom) and reconstituted in 1 mL of nutritive broth (BD Bioxon). Dilutions were made (100, 50, 25, 12.5, 6.25, 3.12 µg/mL) from this concentrated solution for further evaluation.

Four wells were made (6 mm diameter) on the plate's surface. Twenty micrograms (20 µL) were added of each concentrate to be evaluated. Treatments were performed by triplicate. Tween 80 at 100% (Sigma-Aldrich) and nutritive broth (BD Bioxon) were used as positive and negative controls, respectively. Plates were incubated for 24 h at 37 °C. Once the incubation period elapsed, hemolysis halos were measured (mm).

2.5. Statistical Analysis.

Obtained data were analyzed using two-way variance analysis (ANOVA) and a means comparison by Tukey at a significance level of 0.05% through Minitab 18 statistical package [17].

3. Results and Discussion

3.1. Individuals Data

A record of each individual was made with the following information: length, weight, age, and gender (Table 1). The characteristics of the rattlesnakes in the study coincided with those reported by Campbell and Lamar in 2004 [1], as seen in Figure 1.

Table 1. Individual data of trapped rattlesnakes in Hidalgo State.

Species	Species Characteristics	Individual Identification	Gender	Age	Length (cm)	Weight (g)
Crotalus triseriatus	Triangular head 8–10 rattles	CT1	Male	Adult	37	210
	Postocular strip	CT2	Male	Adult	25	175
Crotalus ravus	Triangular head Thin rattle	CR3	Male	Adult	25	180
	Symmetric scales in head	CR4	Male	Adult	25	175

(**a**) *Crotalus triseriatus*

(**b**) *Crotalus ravus*

Figure 1. Captured species (**a**) *Crotalus triseriatus* and (**b**) *Crotalus ravus*.

3.2. Antibacterial Activity

A MIC of 50 µg/mL and an MBC of 100 µg /mL were determined as effective for treatments CT2 and CR3 over *P. aeruginosa* (Table 2, Figure 2). Nevertheless, antibacterial activity was not detected for *E. coli and S. aureus.*

Table 2. Minimal Inhibitory Concentration (MIC) and Minimal Bactericidal Concentration (MBC) of *Crotalus triseriatus and Crotalus ravus* venoms.

Evaluated Bacteria	Evaluated Treatments µg/mL (MIC/MBC)				Controls (MIC/MBC)	
	CT1	CT2	CR3	CR4	Nutritive Broth	Kanamycin (µg/mL)
E. coli	-	-	-	-	-	2/4
P. aeruginosa	-	50 [a]/100 [A]	50 [a]/100 [A]	-	-	16 [b]/64 [B]
S. aureus	-	-	-	-	-	1/4

CT1 *Crotalus triseriatus* 1, CT2 *Crotalus triseriatus* 2, CR3 *Crotalus ravus* 3, CR4 *Crotalus ravus* 4 [a,b] Different small letters indicate significant statistical differences between MIC ($p < 0.05$) [A,B] Different capital letters indicate significant statistical differences between MBC ($p < 0.05$)

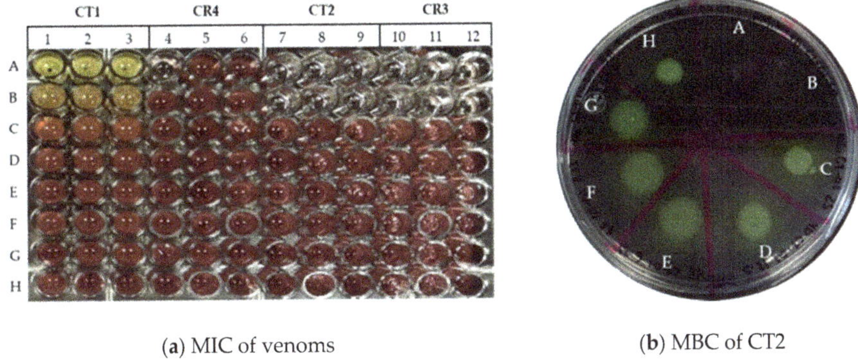

(a) MIC of venoms   (b) MBC of CT2

Figure 2. Antibacterial activity of rattlesnake's venoms against *P. aeruginosa*: (**a**) columns 1–3, CT1 from 100 at 0.78 µg/mL, columns 4–6, CR4 from 100 at 0.78 µg/mL, columns 7–9, CT2 from 100 at 0.78 µg/mL, columns 10–12, CR3 from 100 at 0.78 µg/mL. The MIC value is read at the minimal concentration in which the color changes to pink; (**b**) Plate with *P. aeruginosa* + CT2 in Müller–Hinton agar; **A** CT2 to 100 µg/mL, **B** CT2 to 50 µg/mL, **C** CT2 to 25 µg/mL, **D** CT2 to 12.5 µg/mL, **E** CT2 to 6.25 µg/mL, **F** CT2 to 12.5 µg/mL, **G** CT2 to 6.25 µg/mL, **H** CT2 to 0.78 µg/mL. The MBC is read to the lowest concentration where no visible growth of the bacteria.

It was determined that the antibacterial response seen in treatments CT2 and CR3 were bactericidal, since the relation between MIC and MBC is less than 4, in accordance with González-Alamilla et al., in 2019 [18].

Boda et al., in 2019 evaluated the antibacterial activity of eleven crude venoms from different snake species including *Crotalus atrox* and *Crotalus polystictus* against *Staphylococcus aureus*, *Escherichia coli* and *Pseudomonas aeruginosa* among others, at varied concentrations of 500 to 1.95 µg/mL, determining a MIC and MBC of 125 and 500 µg/mL against *S. aureus* for *Crotalus atrox* and *Crotalus polystictus*, respectively. In the present study, antibacterial activity was not found for *S. aureus* and *E. coli* but was determined for *Pseudomonas aeruginosa*, obtaining a MIC of 50 and a MBC of 100 µg/mL for *C. triseriatus* and *C. ravus* (CT2 and CR3). According to Boda et al., 2019, the antibacterial activity of venoms from viperid species is probably due to their content of proteins with proteolytic activity [19].

Samy et al., in 2014, evaluated CaTx-II a toxin isolated from *Crotalus adamanteus* venom, determining a MIC of 7.8 µg/mL for *S. aureus* and 62.5 to 125 µg/mL for *P. aeruginosa*. Oguiura et al., in 2011, evaluated crotamine, a myotoxin from *Crotalus durissus* venom against different bacteria strains which included *E. coli*, *S. aureus*, and *P. aeruginosa*. They report a MIC of 100 µg/mL for *E. coli* and >200 µg/mL for the other two [20,21], a contrast with the results obtained in our present study since the antibacterial activity was not determined for *E. coli* or *S. aureus*. Since it was determined that a MIC of 50 µg/mL from *C. triseriatus* and *C. ravus* (CT2 and CR3) occurred in crude venom, the activity could be attributed to the presence of these bioactive compounds in the venom of the individuals used for this evaluation since both compounds were isolated from snake venom of the same genus (*Crotalus*).

Although the aim of the study did not include identifying the venom's active mechanism, it has been reported that phospholipase A_2 (CaTx-II) interacts with lipopolysaccharide (LPS), particularly with lipid A, a Gram-negative bacteria component, causing membrane permeabilization. Crotamine also has effects over some bacteria through membrane permeabilization, so it could be suggested that CT2 and CR3 treatments antibacterial activity is related to this mechanism [21,22].

In this respect, the efficiency of these compounds, specially phospholipase A_2 against antibiotic-resistant bacteria, holds promise for biotechnological applications, in this case, new medical treatment alternatives, however, it should be understood there are different antibacterial activity mechanisms from venom-based drugs [23,24].

In accordance with WHO, *P. aeruginosa* actually is in the critical priority group of the list of antibiotic-resistant pathogens. WHO has been expressing its interest by promoting the research and development of new antibiotics for this bacterium [10]. These results obtained herein show that CT2 and CR3 treatments demonstrated bactericidal activity against this pathogen showing its importance, since rattlesnake venoms or compounds thereof could be used to develop effective therapeutic agents to treat infections caused by *P. aeruginosa*.

3.3. Hemolytic Activity

With respect to the hemolysis produced, the generated halos showed significant statistical differences between them ($p < 0.05$) (Table 3). It was observed that CT1, CT2, and CR3 showed the highest hemolytic potential and there were no statistically significant differences between them at 100 µg/mL concentration compared with the other treatments (Figure 3).

Table 3. Hemolysis halos generated by *C. triseriatus* and *C. ravus* venoms.

Concentration (µg/mL)	Evaluated Treatments					
	CT1	CT2	CR3	CR4	Nutritive Broth	Tween 80
100	18.67 ± 1.53 [a,A,*]	17.00 ± 1.00 [a,A]	18.67 ± 1.15 [a,A,*]	0.0 [b]		
50	15.00 ± 0.0 [b,B]	12.33 ± 0.58 [c,B]	16.67 ± 0.58 [a,B]	0.0 [d]		
25	13.67 ± 0.58 [a,B,C]	13.67 ± 1.15 [a,B]	13.33 ± 0.58 [a,C]	0.0 [b]	0.00	20.33 ± 0.58 *
12.5	12.00 ± 1.00 [a,C]	10.00 ± 0.00 [a,b,C]	10.33 ± 0.58 [b,D]	0.0 [c]		
6.25	8.67 ± 0.58 [a,D]	8.33 ± 0.58 [a,C]	7.33 ± 0.58 [a,E]	0.0 [b]		
3.12	0.00 ± 0.00 [a,E]	0.00 ±0.00 [a,D]	0.00 ± 0.00 [a,F]	0.0 [a]		

[a,b,c] Different letters indicate significant statistical differences between treatments ($p < 0.05$). [A,B,C] Different letters indicate significant statistical differences between concentrations ($p < 0.05$). * No statistical differences between treatments ($p > 0.05$). CT1 *Crotalus triseriatus* 1, CT2 *Crotalus triseriatus* 2, CR3 *Crotalus ravus* 3, and CR4 *Crotalus ravus* 4.

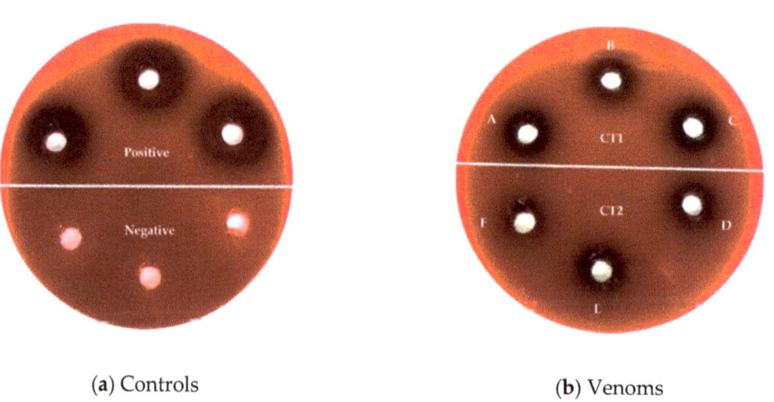

(a) Controls (b) Venoms

Figure 3. Indirect hemolytic activity of the rattlesnake's venoms: (a) hemolytic activity of controls, positive Tween 80, negative nutritive broth; (b) hemolytic activity of venoms **A, B, C** 100 µg/mL of CT1; **D, E, F** 100 µg/mL of CT2.

Macías-Rodríguez et al., in 2014 [25] evaluated the hemolytic activity of *Crotalus molossus* venom (*C. molossus molossus* and *C. molossus nigrescens*) at a 50 µg/mL concentration. In the present study, hemolytic halos were measured over different periods of time, (1, 2, 3, and 14 h). The results obtained showed that at 14 h halos generated measured 19.2 ± 1.5 and 17.00 ± 1.2 mm for *C. molossus molossus* and *C. m. molossus nigrescens*, respectively, whereas at the same concentration over a longer period of time (24 h) with the venoms of *C. triseriatus* and *C. ravus* generated smaller halos 15.00 ± 0.0 (CT1), 12.33 ± 0.58 (CT2), 16.67 ± 0.58 (CR3), and 0.00 ± 0.00 (CR4), showing the hemolytic potential of these species is lower.

On the other hand, Pirela et al., in 2006, determined that the indirect hemolytic dose of *Crotalus durissus cumanensis* venom to produce a 20 mm hemolytic halo was 379.51 ± 67.67 µg of venom [16]. In a similar study, Dos Santos et al. in 1993 obtained a dose of approximately 310 µg for the white venom and 350 µg for the yellow venom of *Crotalus durissus ruruima* to produce hemolytic halos of 20 mm [26]. With respect to *Crotalus triseriatus* and *Crotalus ravus* venoms, an average of 18.67 ± 1.53 mm was obtained at 100 µg/mL concentration of venom. Although there are no equivalent values in the measurements of hemolytic halos, the venom of *C.triseriatus* and *C. ravus* have close values in the measure of their halos in comparison with the other studies and in a lower venom concentration.

In accordance with Macías-Rodríguez et al., in 2014 [27], during the fall months, there exists a high proteomic concentration in rattlesnake venom. *C. ravus* and *C. triseriatus* were sampled in September and November, respectively, months which correspond to the autumn, while the individuals sampled by Pirela et al. in 2006 [16] were sampled in May, June, and July, months that have been shown to have decreased protein concentration. On the other hand, in 2010 Chippaux et al., [6] reported that the species *C. durissus durissus* and *C. durissus terrificus* have myotoxic and neurotoxic venoms compared to other species of the genus *Crotalus*, which mostly have hemotoxic and histologic venoms [28]. Therefore, due to this, in *C. triseriatus* and *C. ravus*, the highest concentration evaluated in this study (100 µg/mL) was enough to produce halos with measurements similar to those of the aforementioned study.

Treatment CR4, characterized by its transparent color, did not show hemolytic activity. This variation in color has been observed in other viperids. In the study carried out by Macías-Rodríguez et al., in 2014 [25], *C. molossus* presented a yellowish venom which turned out to be more hemolytic than *Crotalus tigris* venom, which was transparent in appearance, similar to *C. ravus* (CR4). Galán et al., in 2004 [29], reported that yellowish venoms have greater toxicity compared to white venoms. Lourenço et al., in 2013 [30], reported that the yellow coloration of the venom is due to the presence of crotamine, a myotoxin from rattlesnakes.

Snake venom complexity produces a source of bioactive molecules with different activities. The results obtained in this study confirm rattlesnake's crude venom contains compounds that could be used as therapeutic models, in this case, molecules with antibacterial activity. Although the venom cannot be used directly due to its high toxicity, some of its compounds will serve as prototypes for the development of new drugs.

4. Conclusions

Until today, there are no studies reporting on the antibacterial and hemolytic activity of the venoms of *C. triseriatus* and *C. ravus*. The results of the present study indicate that both rattlesnakes produce venoms rich in bioactive compounds with a bactericidal effect against *Pseudomonas aeruginosa*. These compounds could also serve as new antimicrobial drugs for the treatment of diseases caused by this bacterium; however, the isolation, identification, and evaluation of these molecules is necessary since it could present hemolytic activity.

Author Contributions: Conceptualization and Methodology, N.R.-P. and A.Z.-B.; Validation and Formal analysis, A.Z.-B., A.L.M.-U. and L.M.A.-C.; Investigation and Resources, N.R.-P., A.Z.-B. and L.R.-L.; Data curation, S.C.F.-A., C.E.R.-M. and A.O.-J.; Writing—original draft preparation, S.C.F.-A., A.L.M.-U and N.R.-P.; Writing—review and editing, A.Z.-B., B.V.-C., A.O.-J. and L.R.-L.; Supervision, N.R.-P., C.E.R.-M. and A.Z.-B.; Project administration, N.R.-P. and A.Z.-B.; Funding acquisition, N.R.-P., A.Z.-B. and L.M.A.-C. All authors have read and agreed to the published version of the manuscript.

Funding: This research did not receive any specific grant from funding agencies in the public, commercial, or not-for-profit sectors.

Acknowledgments: The authors would like to acknowledge to the Universidad Autónoma del Estado de Hidalgo (UAEH) by for the support provided for carrying out the study in its facilities.

Conflicts of Interest: The authors declare no conflicts of interest.

References

1. Campbell, J.A.; Lamar, W.W.; Brodie, E.D. *The Venomous Reptiles of the Western Hemisphere*; Cornell University Press: Ithaca, NY, USA, 2004; p. 528.
2. Mata-Silva, V.; Johnson, J.D.; Wilson, L.D.; García-Padilla, E. The herpetofauna of Oaxaca, Mexico: Composition, physiographic distribution, and conservation status. *Mesoam. Herpetol.* **2015**, *2*, 6–62.
3. Palacios-Aguilar, R.; Flores-Villela, O. An updated checklist of the herpetofauna from Guerrero, Mexico. *Zootaxa* **2018**, *4422*, 1–24. [CrossRef] [PubMed]
4. Chippaux, J. Snakebite in Africa: Current situation and urgent needs. In *Handbook of Venoms and Toxins of Reptiles*; CRC Press Inc.: London, UK, 2009; pp. 453–473.
5. Vetter, I.; Davis, J.L.; Rash, L.D.; Anangi, R.; Mobli, M.; Alewood, P.F.; Lewis, R.J.; King, G.F. Venomics: A new paradigm for natural products-based drug discovery. *Amino Acids* **2011**, *40*, 15–28. [CrossRef] [PubMed]
6. Prashanth, J.R.; Brust, A.; Jin, A.H.; Alewood, P.F.; Dutertre, S.; Lewis, R.J. Cone snail venomics: From novel biology to novel therapeutics. *Future Med. Chem.* **2014**, *6*, 1659–1675. [CrossRef]
7. Fox, J.W.; Serrano, S.M. Approaching the golden age of natural product pharmaceuticals from venom libraries: An overview of toxins and toxin-derivatives currently involved in therapeutic or diagnostic applications. *Curr. Pharm. Des.* **2007**, *13*, 2927–2934. [CrossRef] [PubMed]
8. Beeton, C. Targets and Therapeutic Properties. In *Handbook of Biologically Active Peptides*, 2nd ed.; Kastin, A.J., Ed.; Academic Press: Cambridge, MA, USA, 2013; pp. 473–482.
9. Vargas-Muñoz, L.J.; Estrada-Gomez, S.; Vásquez-Escobar, J. Toxinas de venenos de serpientes y escorpiones, una fuente natural de moléculas con actividad antimicrobiana. *Curare* **2015**, *2*. [CrossRef]
10. World Health Organization. WHO Publishes List of Bacteria for Which New Antibiotics are Urgently Needed. 2017. Available online: https://www.who.int/news-room/detail/27-02-2017-who-publishes-list-of-bacteria-for-which-new-antibiotics-are-urgently-needed (accessed on 15 December 2019).
11. McDiarmid, R.; Foster, M.; Guyer, C.; Gibbons, W.; Chernoff, N. *Reptile Biodiversity: Standard Methods for Inventory and Monitoring*; University of California Press: Berkely, CA, USA, 2012; p. 412.
12. SEMARNAT: Secretaría del Medio Ambiente y Recursos Naturales. México: Gobierno de México. 2016. Licencia de Colecta Científica con Propósitos de Enseñanza en Materia de vida Silvestre. Available online: https://www.gob.mx/tramites/ficha/licencia-de-colecta-cientifica-con-propositos-de-ensenanza-en-materia-de-vidasilvestre/SEMARNAT442 (accessed on 15 December 2019).
13. Clinical and Laboratory Standards Institute. *Methods for Dilution Antimicrobial Susceptibility Tests for Bacteria that Grow Aerobically (Approved Standards)*; CLSI document M7-A5; CLSI: Wayne, PA, USA, 2012.
14. Olmedo-Juárez, A.; Briones-Robles, T.I.; Zaragoza-Bastida, A.; Zamilpa, A.; Ojeda-Ramírez, D.; Mendoza-de Gives, P.; Olivares-Pérez, J.; Rivero-Perez, N. Antibacterial activity of compounds isolated from *Caesalpinia coriaria* (Jacq) Willd against important bacteria in public health. *Microb. Pathog.* **2019**, *136*, 103660. [CrossRef] [PubMed]
15. Morales-Ubaldo, A.; Hernández-Alvarado, J.; Valladares-Carranza, B.; Velázquez-Ordoñez, V.; Delgadillo-Ruiz, L.; Rosenfeld-Miranada, C.; Rivero-Pérez, N.; Zaragoza-Bastida, A. Actividad antibacteriana del extracto hidroalcohólico *Croton draco* sobre bacterias de importancia sanitaria. *Abanico Vet.* **2019**, *10*, 10. [CrossRef]
16. Pirela De las Salas, R.d.C.; López-Jonsthon, J.C.; Hernández Rangel, J.L. Caracterización toxinológica del veneno total de la serpiente de cascabel *Crotalus durissus cumanensis* (viperidae), presente en la localidad de Porshoure, Guajira Venezolana. *Rev. Cient.* **2006**, *16*, 232–238.
17. *Minitab 18 Statistical Software 2010*; Minitab, Inc.: State College, PA, USA, 2010; Available online: www.minitab.com (accessed on 15 November 2019).
18. González-Alamilla, E.N.; Gonzalez-Cortazar, M.; Valladares-Carranza, B.; Rivas-Jacobo, M.A.; Herrera-Corredor, C.A.; Ojeda-Ramírez, D.; Zaragoza-Bastida, A.; Rivero-Perez, N. Chemical Constituents of *Salix babylonica* L. and Their Antibacterial Activity Against Gram-Positive and Gram-Negative Animal Bacteria. *Molecules* **2019**, *24*, 2992. [CrossRef]
19. Boda, F.A.; Mare, A.; Szabó, Z.I.; Berta, L.; Curticapean, A.; Dogaru, M.; Man, A. Antibacterial activity of selected snake venoms on pathogenic bacterial strains. *Rev. Rom. Med. Lab.* **2019**, *27*, 305. [CrossRef]

20. Samy, R.P.; Kandasamy, M.; Gopalakrishnakone, P.; Stiles, B.G.; Rowan, E.G.; Becker, D.; Shanmugam, M.K.; Sethi, G.; Chow, V.T. Wound healing activity and mechanisms of action of an antibacterial protein from the venom of the eastern diamondback rattlesnake (Crotalus adamanteus). *PLoS ONE* **2014**, *9*, e80199-e. [CrossRef] [PubMed]

21. Oguiura, N.; Boni-Mitake, M.; Affonso, R.; Zhang, G. In vitro antibacterial and hemolytic activities of crotamine, a small basic myotoxin from rattlesnake Crotalus durissus. *J. Antibiot.* **2011**, *64*, 327–331. [CrossRef] [PubMed]

22. Santamaría, C.; Larios, S.; Angulo, Y.; Pizarro-Cerda, J.; Gorvel, J.P.; Moreno, E.; Lomonte, B. Antimicrobial activity of myotoxic phospholipases A2 from crotalid snake venoms and synthetic peptide variants derived from their C-terminal region. *Toxicon* **2005**, *45*, 807–815. [CrossRef] [PubMed]

23. Canhas, I.; Heneine, L.G.; Fraga, T.; Assis, D.; Borges, M.; Chartone-Souza, E.; Amaral Nascimento, A.M. Antibacterial activity of different types of snake venom from the Viperidae family against *Staphylococcus aureus*. *Acta Sci. Biol. Sci.* **2017**, *39*, 309. [CrossRef]

24. Bocian, A.; Hus, K. Antibacterial properties of snake venom components. *Chem. Pap.* **2019**. [CrossRef]

25. Macias-Rodríguez, E.; Martínez-Martínez, A.; Gatica-Colima, A.; Bojórquez-Rangel, G.; Plenge-Tellechea, L. Análisis comparativo de la actividad hemolítica entre las subespecies *Crotalus molossus* y *Crotalus molossus nigrescens*. *Revista Bio Ciencias* **2014**, *2*, 302–312.

26. Dos Santos, M.C.; Ferreira, L.C.L.; Da Silva, W.D.; Furtado, M.d.F.D. Caracterizacion de las actividades biologicas de los venenos 'amarillo' y 'blanco' de *Crotalus durissus ruruima* comparados con el veneno de *Crotalus durissus terrificus*. Poder neutralizante de los antivenenos frente a los venenos de *Crotalus durissus ruruim*. *Toxicon* **1993**, *31*, 1459–1469.

27. Macias-Rodríguez, E.; Díaz-Cárdenas, C.O.; Gatica-Colima, A.; Plenge, L. Variación estacional del contenido proteico y actividad de la PLA2 del veneno de *Crotalus molossus molossus* entre especímenes de origen silvestre y en cautiverio. *Acta Univ.* **2014**, *24*, 38–47. Available online: http://www.redalyc.org/articulo.oa?id=41630112002 (accessed on 15 December 2019).

28. De Roodt, A.R.; Estévez-Ramírez, J.; Paniagua-Solís, J.F.; Litwin, S.; Carvajal-Saucedo, A.; Dolab, J.A.; Robles–Ortiz, L.E.; Alagón, A. Toxicidad de venenos de serpientes de importancia médica en México. *Gac. Méd. Méx.* **2005**, *141*, 13–21.

29. Galán, J.A.; Sánchez, E.E.; Rodríguez-Acosta, A.; Pérez, J.C. Neutralization of venoms from two Southern Pacific Rattlesnakes (*Crotalus helleri*) with commercial antivenoms and endothermic animal sera. *Toxicon* **2004**, *43*, 791–799. [CrossRef] [PubMed]

30. Lourenço, A.; Zorzella-Creste, C.F.; Curtolo de Barros, L.; Delazari-dos Santos, L.; Pimenta, D.C.; Barraviera, B.; Ferreira, R.S., Jr. Individual venom profiling of *Crotalus durissus terrificus* specimens from a geographically limited region: Crotamine assessment and captivity evaluation on the biological activities. *Toxicon* **2013**, *69*, 75–81. [CrossRef] [PubMed]

MDPI

St. Alban-Anlage 66

4052 Basel

Switzerland

Tel. +41 61 683 77 34

Fax +41 61 302 89 18

www.mdpi.com

Animals Editorial Office

E-mail: animals@mdpi.com

www.mdpi.com/journal/animals